Lectures on Euclidean Geometry - Volume 2

Paris Pamfilos

Lectures on Euclidean Geometry - Volume 2

Circle measurement, Transformations, Space Geometry, Conics

Paris Pamfilos
Department of Mathematics
& Applied Mathematics
University of Crete
Heraklion, Greece

ISBN 978-3-031-48912-9   ISBN 978-3-031-48910-5 (eBook)
https://doi.org/10.1007/978-3-031-48910-5

Mathematics Subject Classification (2020): 51-01, 51M04, 51M15, 51M25

© The Editor(s) (if applicable) and The Author(s), under exclusive license to Springer Nature Switzerland AG 2024

This work is subject to copyright. All rights are solely and exclusively licensed by the Publisher, whether the whole or part of the material is concerned, specifically the rights of reprinting, reuse of illustrations, recitation, broadcasting, reproduction on microfilms or in any other physical way, and transmission or information storage and retrieval, electronic adaptation, computer software, or by similar or dissimilar methodology now known or hereafter developed.

The use of general descriptive names, registered names, trademarks, service marks, etc. in this publication does not imply, even in the absence of a specific statement, that such names are exempt from the relevant protective laws and regulations and therefore free for general use.

The publisher, the authors, and the editors are safe to assume that the advice and information in this book are believed to be true and accurate at the date of publication. Neither the publisher nor the authors or the editors give a warranty, expressed or implied, with respect to the material contained herein or for any errors or omissions that may have been made. The publisher remains neutral with regard to jurisdictional claims in published maps and institutional affiliations.

This Springer imprint is published by the registered company Springer Nature Switzerland AG
The registered company address is: Gewerbestrasse 11, 6330 Cham, Switzerland

Paper in this product is recyclable.

To Kostas, to Michalis, to Iasonas, to Odysseas, to Danae

Preface

> If I had to live my life again, I would have made a rule to read some poetry and listen to some music at least once every week; for perhaps the parts of my brain now atrophied would thus have been kept active through use. The loss of these tastes is a loss of happiness, and may possibly be injurious to the intellect, and more probably to the moral character, by enfeebling the emotional part of our nature.
>
> C. Darwin

> All things are mutually intertwined, and the tie is sacred, and scarcely anything is alien the one to the other. For all things have been ranged side by side, and together help to order one ordered Universe.
>
> Marcus Aurelius, To himself VII, 9

    This book is the result of processed notes from courses in *Geometry, Euclidean Geometry, Geometry at School* and *Geometry and Computers*, which I repeatedly taught during the last twenty five years at the University of Crete in Herakleion Greece. Although the book is intended for school teachers and courses, the material it contains is much more extended than what can be naturally taught in school classes. The material however is developed gradually from simple and easy stuff to more complex and difficult subjects. This way, in the beginning chapters, I even avoid using negative numbers and the notion of transformation, so that the book can be used in all school courses.

    In its content and organization the whole work complies with the philosophy of having one book of reference for every school course on a specific subject: the book of Geometry, the book of Physics, the book of Chemistry, etc. If not for the student, at least for the teacher. The book's intention is to offer a solid and complete foundation to both student and teacher, so that both can consult it for studying, understanding and examining various problems and extensions of elementary geometry.

The book's intention is not to develop a structural exposition involving mathematical structures such as vectors, groups, etc., but to proceed in the traditional, synthetic method and familiarize the reader with the basic notions and the problems related to them. Mathematics has definitely passed from the art of calculations to the discovery and investigation of structures. Although this is a long term accepted development, I do not necessarily consider this path appropriate for a beginning and an introduction to Geometry, its related notions, the elementary simple shapes and the problems they lead to. I reckon that the pupil must first have some elementary experience and incoming impressions of the simplest possible kind, without the interjection of notions of abstract structures, which in my opinion at a beginning stage would make the student's approach more difficult.

Thus, I spend a minimal time with axioms so that the reader has a reference point, and proceed quickly towards their logical consequences, so as to speed up the reader's contact with more complex and more interesting shapes. The proposed axioms represent very basic properties, some of which could conceivably be proved from others, even simpler. Such a practice however would cause a slightly more involved discussion on trivial consequences and conclusions, which may appear boring and repel the pupil from the course. In my opinion, a discussion of the foundations of geometry is the work of a late wisdom and must be done after one starts to love the material. In first place one has to examine what exactly is this stuff, practice and, slowly, depending on one's interest and capabilities, proceed to theory. This book therefore has an elementary character, avoiding the use of complex mathematical structures, yet exposing the reader to a wide range of problems.

I think that a teacher can use this book as a reference and road-map of Geometry's material in several geometry courses. Individual courses for specific classes, must and should be supported with companion aids of instructional character (practical exercises, additional exercises, drawing exercises for consolidation of ideas, spreadsheets, drawing software, etc.). The book contains many (over 1400) exercises, most of them with solutions or hints.

In writing this book, my deeper desire is to see Geometry return to its ancient respectable place in the school. This, because, Geometry, with all its beautiful shapes, offers a strong motivation and help for inductive thinking and many tangible pedagogical benefits. I will mention four main ones.

The first is the realization that there are things in front of you, which you do not see. Simple things, simple relations exposed to public view, initially invisible, which start to reveal themselves after lots of work and effort. Attention therefore increases, as does capability for correct observation. To anyone who asks "did I miss something" the answer is everywhere and always, "many (things)". Not asking this question, avoiding to answer it or answering carelessly, is completely incompatible with Geometry's pedagogy.

The second is the all encompassing power of detail, in other words, the accuracy of thought. Real, creative work, means involving yourself with details. With Geometry, this is accomplished with the exercises. Good intentions, visions and abstractions are void of content, when they are not emerging from the sea of details. Speaking on generalities is characteristic of rhetorical speech, the art of words. It is not a coincidence that Geometry has been pushed aside while flourishing the rhetorical and political speech.

The third and most important is the very essence of thought, as well as the human's character, if there is one, consistency. Mathematics and Geometry especially, with the assistance of its figures, is the great teacher of consistency. You begin from certain notions and basic properties (axioms) and start building, essentially, ad infinitum, without ever diverging from the initial principles and the simple rules of logic. This way the work done is always additive towards a building of absolute validity, which has nothing to do with lame improvisations and products, which are the result of continuously changing rules. Man, often, to please his desire, changes arbitrarily the rules of the game. He creates thus a certain culture in which, he is either an abuser, when he himself changes the rules, or a victim, when he suffers from external arbitrary rule changes. This way the work done is sometimes additive and other times negative, canceling the previous work done. This mode is the widespread culture of non-thought, since clear thought is virtually synonymous to mathematical thought and mathematical prototyping, the discovery and the respect of accepted rules.

The fourth and very crucial benefit is the acquaintance with the general problem of knowledge, and the balance between the quantity and quality. In the process of learning Geometry we realize with a particular intention the infinity of the directions towards it extends, but also the unity and the intimate relations of its parts. There results a question of approach, a question of psychology, a question of philosophy. How you can approach this whole, an immense body of knowledge? The question is crucial and posed at an early age. A correct or wrong attitude sets the foundation for a corresponding evolution of the whole life of the student. A big part of the failure of the learning is due to the misconceptions about its nature. If we look at the established practice, we realize that the mainstream behavior is that of the hunter. You target the language, history, chemistry, etc. you find the crucial paths and passages and you shoot. Unfortunately though, knowledge consists not of woodcocks. Any subject we may consider consists not of isolated entities. It is not a large herd of birds containing some rare and some exotic individuals. It is rather a connected continuous and consistent body, accepting only one approach, through the feeling. You cannot learn something if you do not approach it with positive feeling. Besides, every subject of knowledge is similar to a musical instrument. And as you cannot learn 10 instruments at the same time, so you cannot learn 10 subjects at the same time. The 10 instruments you can touch, put out of their cases, taste their sounds. You have to choose though and indulge into one. This is the characteristic property of the thinking man. He can indulge into his subject withdrawing from everything else. Like the virtuoso of the musical instrument absorbed to its music brings to unity the instrument, the music and his existence.

If we were asked to set the targets of education, one of the principals would be the ability to indulge into a subject. This, though it is developed by a mental process, it has an emotional basis. The teacher at the school, the college, the university, meets invariably the same stereotyped problem. The failure of the student is not due to missing mental forces. It is due to underdeveloped or totally missing emotional basis for its subject. The failure of the teacher is not due to what he teaches or lefts aside. It results from not recognizing the role of the emotional basis, not highlighting it, not cultivating it.

Motivated by these ideas I proceed with my proposition, proposing this foundation for the organization of lessons as simple as possible or, at any rate, as simple as the subject allows. I wish that the teacher and the studious pupil will read the book with the same and even greater pleasure than I had while studying for a long time the exquisite material. A material perfect and timeless, the only responsible for imperfections and mistakes being myself.

At this point I wish to thank the colleagues Georgia Athanasaki, Ioanna Gazani for the correction of many mistakes and also many suggestions for improvements. I thank also Dimitris Kontokostas for his numerous interventions on the first chapters of the book and the supervisors of the two greek editions, John Kotsopoulos and John Papadogonas. I am also very indebted to my collegue Manolis Katsoprinakis for many corrections of formulas as well as of figures. I thank also my colleagues Stylianos Negrepontis and George Stamou for their encouragement to continue my work and the colleague Antonis Tsolomitis for his help on "latex". I also express my gratitude to my colleague Giannis Galidakis for his assistance in the translation from the Greek. The suggestion to publish the English text with Springer came from my colleague Athanase Papadopoulos from Strasbourg and the heavy work of organization and management of the editorial work has been done by Springer's senior editor Elena Griniari. To them both I would like to express my deepest gratitude.

Regarding my sources, I have included all bibliographical references, which existed in my notes and were used to backtrack and complete the material. I believe they will be useful to those who wish to dig deeper and compare with other sources. It is especially interesting to search and enter in discussion with great minds, who came before and created material in the subject.

Looking back at the times of school, when I was initiated to Geometry by my excellent teachers, like the late Papadimitriou, Kanellos and Mageiras, I wish to note that the book binds, hopefully worthily, to a tradition we had on the area, which was cultivated, at that time, by the strong presence of Geometry in high schools and gymnasiums. Many of the problems herein are the ones I encountered in books and notes of the teachers I mentioned, as well as these found in Papanikolaou, Ioannidis, Panakis, Tsaousis. Several other problems were collected from classical Geometry texts, such as Catalan [32], Lalesco [44], Legendre [84], Lachlan [43], Coxeter [34], [35], Hadamard [42], F.G.M. [67] and many others, which are too numerous to list.

Concluding the preface, I mention, for those interested, certain books which contain historical themes on the subject [63], [38], [13], [41], [82], [16], [80], [22], [33].

In the second edition of the book there are substantial changes having to do with the material's articulation and expansion, the addition of exercises, figures, and the return of epigrams, which were omitted in the first edition, because of some mistaken reservations concerning the total volume of the book. Going through the first edition, I felt somewhat guilty and had a strong feeling that these small connections with the other non-mathematical directions of thinking and their creators are necessary and very valuable to be omitted for the benefit of a minimal space reduction. In my older notes I used to put them also at the end or even the middle of lengthy proofs. I had thus the feeling to be a member of the international and timeless university and the ability to ask and discuss with these great teachers of all possible subjects. Perhaps some of the readers will eventually keep in mind some of the many figures of the book and some of the epigrams, which are a sort of figures in this "invisible geomery", as the great poet says.

Inatos, July 2003   *Paris Pamfilos*

Symbol index

$(AB\Gamma)$	Circle going through the points A, B, Γ
$\kappa = (AB\Gamma\Delta)$	Circle going through the points A, B, Γ, Δ
$X = (AB, \Gamma\Delta)$	Intersection point X of the lines $AB, \Gamma\Delta$
$(AB; \Gamma\Delta)$	Cross ratio $\frac{\Gamma A}{\Gamma B} : \frac{\Delta A}{\Delta B}$
$(AB; \Gamma\Delta) = -1$	Harmonic quadruple of four collinear points
$(A, B) \sim (\Gamma, \Delta) \Leftrightarrow (AB; \Gamma\Delta) = -1$	Harmonic pairs
$\Delta = \Gamma(A, B) \leftrightarrow (AB; \Gamma\Delta) = -1$	Δ harmonic conjugate to Γ w.r.t. (A, B)
$|AB|$	Length of AB
$\kappa(O, \rho)$	Circle κ, center O, radius ρ
$\kappa(O)$	Circle κ with center O
$\kappa(\rho)$	Circle κ of radius ρ
$O(\rho)$	Circle: center O, radius ρ
$p(X)$ or $p(X, \kappa)$ or $p_\kappa(X)$	Power of point X relative to circle κ
$|\widehat{AB\Gamma}|$	Angle measure $\widehat{AB\Gamma}$
$O(A, B, \Gamma, \Delta)$	Pencil of four lines $OA, OB, O\Gamma, O\Delta$ through O
$\varepsilon(AB\Gamma)$	Area of triangle $AB\Gamma$
$\varepsilon(AB\Gamma...)$	Area of polygon $AB\Gamma...$
$o(AB\Gamma...)$	Volume of polyhedron $AB\Gamma...$
τ	Half perimeter of triangle
α, β, γ	Measures of angles of triangle $AB\Gamma$
R	Radius(circumradius) of $AB\Gamma$
v_A, v_B, v_Γ	Altitudes of triangle $AB\Gamma$
μ_A, μ_B, μ_Γ	Medians of triangle $AB\Gamma$
$\delta_A, \delta_B, \delta_\Gamma$	Inner bisectors of triangle $AB\Gamma$
a, b, c	Lengths of sides of triangle $AB\Gamma$
r, r_A, r_B, r_Γ	Radius of inscribed/escribed circles of $AB\Gamma$
$\frac{AB}{\Gamma\Delta}$	Signed ratio of segments of the same line
$\phi = \frac{\sqrt{5}+1}{2} =\sim 1.61803398874989484820...$	Golden section ratio
$\mathbb{N}$	Set $\{1, 2, 3, ...\}$ of natural numbers
$\mathbb{Z}$	Set of integers (positive, negative, 0)
$\mathbb{Q}$	Set (field) of rational numbers
$\mathbb{R}$	Set (field) of real numbers

Contents

Part II Circle measurement, Transformations, Space Geometry, Conic sections

1 Circle measurement 3
   1.1 The difficulties, the limit 3
   1.2 Definition of the perimeter of the circle 8
   1.3 The number π 12
   1.4 Arc length of a circle, radians 14
   1.5 Definition of the area of the circle 17
   1.6 The area of a circular sector 22
   1.7 The isoperimetric inequality 27
   1.8 Anthyphairesis 32
   1.9 Comments and exercises for the chapter 37
   References 46

2 Transformations of the plane 47
   2.1 Transformations, isometries 47
   2.2 Reflections and point symmetries 50
   2.3 Translations 55
   2.4 Rotations 62
   2.5 Congruency or isometry or equality 69
   2.6 Homotheties 73
   2.7 Similarities 77
   2.8 Inversions 89
   2.9 The hyperbolic plane 92
   2.10 Archimedean tilings 106
   2.11 Comments and exercises for the chapter 115
   References 125

3 Lines and planes in space 127
- 3.1 Axioms for space 127
- 3.2 Parallel planes 133
- 3.3 Angles in space 136
- 3.4 Skew lines 139
- 3.5 Line orthogonal to plane 141
- 3.6 Angle between line and plane 150
- 3.7 Theorem of Thales in space 152
- 3.8 Comments and exercises for the chapter 156

4 Solids 161
- 4.1 Dihedral angles 161
- 4.2 Trihedral angles 166
- 4.3 Pyramids, polyhedral angles 177
- 4.4 Tetrahedra 180
- 4.5 Regular pyramids 184
- 4.6 Polyhedra, Platonic solids 188
- 4.7 Prisms 192
- 4.8 Cylinder 198
- 4.9 Cone, conical surface 200
- 4.10 Truncated cone, cone unfolding 206
- 4.11 Sphere 209
- 4.12 Spherical and circumscribed polyhedra 219
- 4.13 Spherical lune, angle of great circles 221
- 4.14 Spherical triangles 223
- 4.15 The supplementary trihedral 228
- 4.16 Axonometric projection, affinities 234
- 4.17 Perspective projection 240
- 4.18 Comments and exercises for the chapter 244
- References 260

5 Areas in space, volumes 261
- 5.1 Areas in space 261
- 5.2 Area of the sphere 269
- 5.3 Area of spherical polygons 272
- 5.4 Euler Characteristic 276
- 5.5 Volumes 279
- 5.6 Volume of prisms 282
- 5.7 Volume of pyramids 287
- 5.8 Volume of cylinders 294
- 5.9 Volume of cones 296
- 5.10 Volume of spheres 298
- 5.11 Comments and exercises for the chapter 306
- References 316

6 Conic sections ... 317
- 6.1 Conic sections ... 317
- 6.2 Dandelin's spheres ... 320
- 6.3 Directrices ... 325
- 6.4 General characteristics of conics ... 334
- 6.5 The parabola ... 344
- 6.6 The ellipse ... 354
- 6.7 The hyperbola ... 366
- 6.8 Comments and exercises for the chapter ... 379
- References ... 398

7 Transformations in space ... 399
- 7.1 Isometries in space ... 399
- 7.2 Reflections in space ... 401
- 7.3 Translations in space ... 405
- 7.4 Rotations in space ... 407
- 7.5 Congruence or isometry in space ... 411
- 7.6 Homotheties in space ... 420
- 7.7 Similarities in space ... 421
- 7.8 Archimedean solids ... 424
- 7.9 Epilogue ... 432
- References ... 433

Index ... 435

Part II
Circle measurement, Transformations, Space Geometry, Conic sections

Chapter 1
Circle measurement

1.1 The difficulties, the limit

> Upon the whole, I am inclined to think that the far greater part, if not all, of those difficulties which have hitherto amused philosophers, and blocked up the way to knowledge, are entirely owing to our selves. That we have first raised a dust and then complain we cannot see.
>
> G. Berkeley, *Concerning the Principles of Human Knowledge*

The problem with the circle is, that before measuring it, one has to prove that it has length, called **perimeter**). This appears difficult to understand by the novice, it is however a problem which results from the fact that we don't have, up to now, a definition for the length of a curve. Currently, we know how to measure length only for line segments and its derivatives which are broken lines and polygons. The circle however is something different. We have to clarify things, to define what exactly we mean by *length of a circle* or *perimeter of a circle* and to prove that our definition is meaningful and doesn't contradict our axioms. After all this is accomplished, we can subsequently proceed in the determination of the length of the circle. In the subject of the existence of length we need to resort to an axiom of the real numbers, called **axiom of completeness**, which relates to *sequences* and *limits*. This axiom says:

Axiom 1.1 *Every increasing sequence of real numbers $\alpha_1, \alpha_2, \alpha_3, \ldots$ which is also bounded has limit a real number A.*

The word **sequence** (of numbers) means a set of numbers, each one of which is characterized by an integer number (index):

$$\alpha_1, \alpha_2, \alpha_2, \ldots \alpha_\nu, \ldots.$$

Each one of these numbers is called **member** of the sequence. The word **increasing** means that these numbers are increasing with each step:

$$\alpha_1 \leq \alpha_2 \leq \alpha_3 \leq \alpha_4....$$

That this sequence is **bounded** means that there exists a specific number M, such that all the sequence numbers are less than M. We could write briefly:

$$\alpha_1 \leq \alpha_2 \leq \alpha_3 \leq \alpha_4... \leq M.$$

M is called an **upper bound** of the sequence. A number $M' > M$ is also such a bound, and usually the exact value of M is immaterial. It suffices to find some M, which satisfies the preceding inequalities. For example, the numbers

$$0 < \frac{1}{2} < \frac{2}{3} < \frac{3}{4} < \frac{4}{5} < \frac{5}{6} < ...$$

define an increasing sequence, which is also bounded, since all numbers are less than $M = 1$. Notice that the terms of this sequence are given by the formula

$$\alpha_\nu = \frac{\nu - 1}{\nu},$$

by substituting in it the values 1, 2, 3, 4, 5, ... etc.

The last word which requires explanation, in this short formulation of the axiom, is the **limit** of a sequence. I'll not give here the definition for general sequences but only the particular one for the limit of an increasing sequence. The limit of an increasing sequence, then, is a number A for which are satisfied two inequalities:

$$A - \varepsilon < \alpha_\nu \leq A.$$

The right one for all α_ν. The left, for any $\varepsilon > 0$ we choose, is satisfied for all α_ν, with the exception of some, which are nevertheless finite in number (number depending on the size of ε). In the preceding example say, we can easily see that the limit is $A = 1$. Indeed the preceding inequality becomes

$$1 - \varepsilon < \frac{\nu - 1}{\nu} \leq 1.$$

The right inequality holds true for all $\nu = 1, 2, 3, 4, ...$ The left inequality can be written

$$1 - \varepsilon < 1 - \frac{1}{\nu},$$

and is equivalent to

$$\frac{1}{\nu} < \varepsilon \quad \text{which in turn is equivalent to} \quad \frac{1}{\varepsilon} < \nu.$$

1.1. THE DIFFICULTIES, THE LIMIT

The last however is satisfied exactly as required by the definition of the limit, that is for all ν with the exception of some, whose number is finite (and depends on the magnitude of ε). Thus, for $\varepsilon = 0.001$ equivalently $\frac{1}{\varepsilon} = 1000$, the inequality is satisfied for all $\nu > 1000$ and is not satisfied for $\nu < 1000$, which are however finite in number. In practice, to show that A is the limit of the increasing sequence we proceed as follows:

1. We find the (candidate) limit A,
2. We show that $\alpha_\nu \leq A$ holds for all ν,
3. We consider $\varepsilon > 0$ and we "solve" the inequality $A - \varepsilon < \alpha_\nu$ for ν,
4. We verify that the α_ν which are not solutions to the preceding inequality are finite in number.

The first difficult step is to find or guess A. The second step usually follows from the first. On the third step we must, like we do with equations, manage to bring to the one side of the inequality the ε and to the other the ν and to show that the inequality is equivalent to one of the form

$$N(\varepsilon) < \nu.$$

$N(\varepsilon)$ is usually a number which depends on ε, and transforming inequality (3.) in this form, means automatically: (i) that (3.) is satisfied for all ν which are greater than $N(\varepsilon)$ and (ii) the inequality is not satisfied for all ν which are less than $N(\varepsilon)$, which are nevertheless finite in number.

Remark 1.1. We need the notion of limit, because through it are defined the length of the circle and its area. As we'll see below, beginning with the square, we construct regular polygons $\Pi_1, \Pi_2, \Pi_3, \ldots$ each of which has double the number of sides than the preceding one and all are inscribed in the same circle κ. The perimeter of the circle is then defined as the limit of the perimeters of these polygons. Similarly the area also is defined as the limit of the areas of these polygons.

Often, in the lessons of Geometry, the discussion on the difficulties, which are involved with the circle, when we try to measure its perimeter and its area, are omitted. I think however that this is not correct. At this point we are faced with a difficulty, whose resolution is one of the major cultural achievements. In the investigation of circles appears the need for approximations, the need for limits, and through them are laid the foundations of calculus by *Archimedes*, later developed by *Newton* and *Leibnitz*. It is a pity for the student to not have a small idea about these concepts. Even if one has problems in understanding fully the notions, it is useful for one's education to feel, some more some less, that a difficulty leads to its transcendence, which opens new fields and new horizons. The connection between the three notions "difficulty", "limit" and "transcendence" is noteworthy and ever present in mathematics and, more general, in life.

Exercise 1.1. Show, that if the increasing sequence $\alpha_1 \leq \alpha_2 \leq \alpha_3...$ has as limit the number A, then if $A' < A$, then there exists an α_i with

$$A' < \alpha_i < A.$$

Hint: Write $A' = A - \varepsilon$ with $\varepsilon = A - A' > 0$. Next apply the definition of the limit of an increasing sequence, according to which, there are infinitely many α_i which satisfy $A - \varepsilon < \alpha_i < A$. Choose one of them.

Exercise 1.2. Show, that if the increasing sequence $\alpha_1 \leq \alpha_2 \leq \alpha_3...$ has as limit the number A, then the sequence $\rho\alpha_1 \leq \rho\alpha_2 \leq \rho\alpha_3...$, where $\rho > 0$, has as limit the number ρA.

Hint: The right inequality of the definition: $\rho\alpha_i \leq \rho A$ is a consequence of the corresponding $\alpha_i \leq A$ which holds by assumption. The left inequality of the definition of limit: $\rho A - \varepsilon < \rho\alpha_i$ is equivalent to $A - \frac{\varepsilon}{\rho} < \alpha_i$. However, by assumption the last one holds for all α_i except for finitely many, consequently the same will happen with the equivalent to the preceding one. Finally then, for every $\varepsilon > 0$ the inequalities $\rho A - \varepsilon < \rho\alpha_i \leq \rho A$ will hold for all the α_i with the exception of some, something which shows the truth of the claim.

Exercise 1.3. Show that the sequence

$$\alpha_\nu = \frac{3\nu + 2}{5\nu + 10}$$

is increasing. Show further that the limit of the sequence is $A = \frac{3}{5}$

Hint: The fact that it is increasing is equivalent to

$$\frac{3\nu + 2}{5\nu + 10} < \frac{3(\nu+1) + 2}{5(\nu+1) + 10},$$

which is proved easily. For the fact that the limit is the specific number we show first that for each $\nu = 1, 2, 3, ...$ holds

$$\frac{3\nu + 2}{5\nu + 10} < \frac{3}{5}.$$

Next we "solve" the inequality isolating ν :

$$A - \varepsilon < \alpha_\nu \Leftrightarrow A - \alpha_\nu < \varepsilon \Leftrightarrow$$
$$\frac{3}{5} - \frac{3\nu + 2}{5\nu + 10} < \varepsilon \Leftrightarrow \frac{4}{5} \cdot \frac{1}{\nu + 2} < \varepsilon \Leftrightarrow$$
$$\frac{4}{5} \cdot \frac{1}{\varepsilon} < \nu + 2 \Leftrightarrow \frac{4}{5} \cdot \frac{1}{\varepsilon} - 2 < \nu$$

The last holds for all ν except finitely many, the number of which depends on the magnitude of ε.

1.1. THE DIFFICULTIES, THE LIMIT

Exercise 1.4. Consider the sequence whose general term is given by the formula
$$\alpha_\nu = \frac{\alpha \nu + \beta}{\gamma \nu + \delta},$$
where α, β, γ and δ are constants. Show that this sequence is increasing, if and only if $\alpha \delta - \beta \gamma > 0$. Also show that when it is increasing, then the sequence has limit $\frac{\alpha}{\gamma}$.

Besides sequences of numbers, we'll also consider next *sequences of polygons*. As with numbers so with polygons, the word **sequence** means a set of polygons each one of which is characterized by an integer:

$$\Pi_1, \Pi_2, \Pi_2, ... \Pi_\nu, ...$$

Each one of the polygons is a **term** of the sequence. It is fairly obvious that the notion of the sequence and its terms can be transferred to sets of similar objects. This way we could define in some way sequences of points, sequences of squares, sequences of circles etc. Figure (See Figure 1.1-I), for ex-

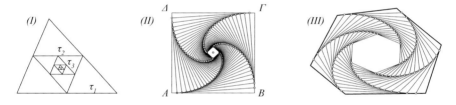

Fig. 1.1: Triangles seq.   Squares seq.   Hexagons seq.

ample, shows a few terms of a sequence of triangles $\tau_1, \tau_2, \tau_3, ...$ each defined by the middles of the sides of the preceding one, beginning with a specific triangle τ_1. Figure (See Figure 1.1-III), shows a sequence of polygons. The sequence is constructed using always the same orientation and taking on each side a point dividing the side into the ratio κ. Joining the points on the various sides results in a new polygon, which is inscribed in the preceding one. Repeating this process we create a sequence. The preceding sequence of triangles, is a special case of that more general sequence.

Exercise 1.5. In the square $\tau_1 = AB\Gamma\Delta$ we inscribe the square $\tau_2 = A'B'\Gamma'\Delta'$, the vertices of which divide the sides $\{AB, B\Gamma, \Gamma\Delta, \Delta A\}$ in the ratio $k < 0$: $k = \frac{A'A}{A'B} = \frac{B'B}{B'\Gamma} = \frac{\Gamma'\Gamma}{\Gamma'\Delta} = \frac{\Delta'\Delta}{\Delta'A}$. The same procedure is repeated with τ_2 and continuing this way we define the sequence $\{\tau_n\}$ (See Figure 1.1-II). Compute the area of the square τ_n, as well as the distance ε_n of its vertices from the center of τ_1. Show that ε_n, for appropriate n, becomes as small as we wish.

The archetype of a sequence is certainly the one of natural numbers $\mathbb{N} = \{1, 2, 3, ...\}$. One of the fundamental properties of this sequence is formulated by the so called **principle of mathematical induction** ([81, p.9]), according to which, if we have a sequence of propositions $\{A_1, A_2, \ldots\}$, each depending on the associated index, then to prove the validity of all these propositions it suffices:

1. To prove the validity of A_1.
2. With the assumption that A_k is true, for an arbitrary k, to prove that A_{k+1} is true.

A typical example of such a sequence of propositions is the so called *Bernoulli's inequality*:

$$A_n : (1+x)^n \geq 1 + nx, \quad \text{valid for } x > -1 \quad \text{and all integers} \quad n \geq 1. \quad (*)$$

The proof results by applying the principle of induction:
(1) For $n = 1, A_1 : (1+x)^1 = (1+x)$, is true.
(2) For $n = k$, assuming $A_k : (1+x)^k \geq (1+kx)$ is true, multiply it by $(1+x) > 0$, resulting in $(1+x)^{k+1} \geq (1+kx)(1+x) = 1 + (k+1)x + kx^2 > 1 + (k+1)x$, hence A_{k+1} is also true.

A typical application of this inequality is in the study of the so called **geometric progression**, which is the sequence of numbers $\{s_n = 1 + x + \cdots + x^n\}$, depending on x. Using induction we can prove that

$$s_n = 1 + x + \cdots + x^n = \frac{1 - x^{n+1}}{1 - x} \quad \Rightarrow \quad \left| s_n - \frac{1}{1-x} \right| = \frac{|x|^{n+1}}{|1-x|}. \quad (**)$$

Numbers $0 < x < 1$ can be written $x = \frac{1}{1+\theta}$ with $\theta > 0$. Then, $x^n = \frac{1}{(1+\theta)^n} \leq \frac{1}{1+n\theta}$, because of $(*)$. This, in view of $(**)$ implies easily, that, for $0 < x < 1$, the geometric progression converges to the limit $1/(1-x)$.

Exercise 1.6. Supply the details of the proof, that the geometric progression for $0 < x < 1$, converges to $1/(1-x)$.

1.2 Definition of the perimeter of the circle

> The second, to divide each of the difficulties I will examine into as many parts as possible and needed to solve it better.
>
> *Descartes, Discourse on the Method, II*

In this section we consider a circle of radius ρ and some regular polygons $\Pi_1, \Pi_2, \Pi_3, \ldots$ inscribed in it, each of which has as vertices those of the predecessor plus the middles of the arcs defined from the successive vertices of the predecessor. We begin then with the square Π_1 (See Figure 1.2-I). If $AB\Gamma\Delta$ labels the square, then the middles of the arcs AB, $B\Gamma$, $\Gamma\Delta$, ΔA define

1.2. DEFINITION OF THE PERIMETER OF THE CIRCLE

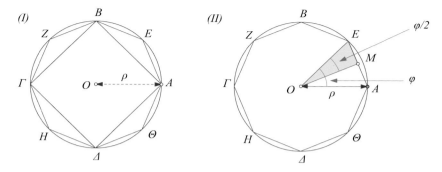

Fig. 1.2: Square and octagon $|AE|: \frac{1}{\mu}$ of the perimeter

four additional points E, Z, H, Θ respectively and $AEBZ\Gamma H\Delta\Theta$ is a regular octagon: Π_2. Π_3 will result similarly and will have 16 sides, Π_4 will have 32 and, in general, Π_ν will have $2^{\nu+1}$ sides.

Lemma 1.1. *The regular polygon with μ sides, inscribed in the circle of radius ρ has perimeter*

$$2\mu\rho \sin\left(\frac{180°}{\mu}\right).$$

Proof. Because all the sides of a regular polygon are equal, its perimeter will be $\mu \cdot |AE|$, where AE is one of its sides. If O is the center of the circle, then triangle AOE is isosceles with angle at O equal to (See Figure 1.2-II)

$$\phi = \frac{360°}{\mu} \Rightarrow \frac{\phi}{2} = \frac{180°}{\mu}$$

since the sum of μ such equal angles will give a full turn about point O. Also the median AM will be orthogonal at the middle M of AE (Corollary I-1.3) and the triangle OME will be right, therefore according to Theorem I-3.11

$$\frac{|AE|}{2} = |ME| = \rho \cdot \sin\left(\frac{\phi}{2}\right).$$

Combining the preceding relations gives the requested

Lemma 1.2. *The sequence of perimeters $p_1, p_2, p_3, \ldots$ of the polygons $\Pi_1, \Pi_2, \Pi_3, \ldots$ is increasing.*

Proof. Because each such polygon has sides double those of its predecessor, it suffices to show that the ratio of the perimeters $\frac{p}{p'}$ of two regular polygons Π and Π' from which the first has μ sides and the second has 2μ sides is less than 1. However according to the preceding lemma, taking into account the formula for the sine of the double angle (Exercise I-3.54), we have:

$$\frac{p}{p'} = \frac{2\mu\rho \sin(\frac{180°}{\mu})}{2(2\mu)\rho \sin(\frac{180°}{2\mu})} = \frac{\sin(\frac{180°}{\mu})}{2\sin(\frac{180°}{2\mu})} = \frac{2\sin(\frac{180°}{2\mu}) \cdot \cos(\frac{180°}{2\mu})}{2\sin(\frac{180°}{2\mu})} = \cos\left(\frac{180°}{2\mu}\right) < 1.$$

Lemma 1.3. *Every regular polygon with μ sides inscribed in a circle of radius ρ has perimeter less than that of the circumscribed square on the same circle.*

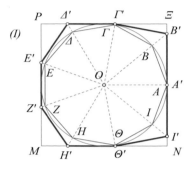
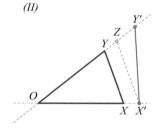

Fig. 1.3: Projection on the square · Triangle and its projection

Proof. The proof results by projecting the regular polygon from its center O onto a circumscribed square. To this, for each of the vertices A, B, Γ, ... of the inscribed polyhon we draw the corresponding half line OA, OB, $O\Gamma$, ... and we define its intersection point with the circumscribed square as A', B', Γ',... (See Figure 1.3-I). This defines a polygon $A'B'\Gamma'$... which has its vertices on the square and each of its sides is greater than the corresponding side of the initial polygon. This results from the comparison of the corresponding triangles $(AOB, A'OB')$, $(BO\Gamma, B'O\Gamma')$, ... For each such pair of triangles, let me call one $(XOY, X'OY')$ (See Figure 1.3-II), point X' is on the extension of OX and Y' is on the extension of OY, consequently $X'Y'$ is greater than XY.

For the proof of the last claim, suppose that from the distances XX' and YY', XX' is the less one and draw the parallel $X'Z$ to XY from X'. It obviously holds
$$|XY| \leq |X'Z| < |X'Y'|.$$
The first inequality holds because of similarity, $\frac{|XY|}{|X'Z|} = \frac{|OX|}{|OX'|} \leq 1$. The second inequality holds because the angle $X'ZY'$ is obtuse (Corollary I-1.10), therefore opposite to it lies the greater side of the triangle $X'ZY'$. It follows then that the perimeter of $A'B'\Gamma'$... is greater than that of $AB\Gamma$... but also less than the perimeter of the square. The latter because the sides of $A'B'\Gamma'$... are either parts of the sides of the square (like $A'B'$ in the figure) or they have their parts in different sides of the square (like $B'\Gamma'$ in the figure) and they are hypotenuses of right triangles, which are less than the sum of the two orthogonal sides. Concluding then, the perimeters p, p' and p'' of the regular polygon, of its projection to the square and of the square, satisfy:

1.2. DEFINITION OF THE PERIMETER OF THE CIRCLE

$$p < p' < p''.$$

Lemmata 1.2 and 1.3 guarantee, relying on Axiom 1.1, that the sequence $\alpha_1, \alpha_2, \ldots$ of the perimeters of the polygons $\Pi_1, \Pi_2, \ldots$ has a limit A. This A we define as length or **perimeter** of the circle.

Remark 1.2. The numbers α_i, which are mentioned above as perimeters of the Π_i, result from Lemma 1.1 by setting to it

$$\mu = 2^{i+1}.$$

This way the perimeter of Π_i is

$$\alpha_i = 2(2^{i+1}) \cdot \rho \cdot \sin\left(\frac{180°}{2^{i+1}}\right).$$

According to the definition of the limit, for all $\varepsilon > 0$ it follows, that for the perimeter of the circle A and the above numbers α_i hold the inequalities

$$A - \varepsilon < \alpha_i < A,$$

the right one for all α_i and the left for all the α_i except finitely many. How many α_i don't satisfy the left inequality depends on the magnitude of ε. Their number however is always finite.

Remark 1.3. The definition of the perimeter of the circle we gave is connected to the definition of length of a broken line and it can be proved that these two notions of length (broken and circle) can result as special cases of a more unified definition of *curve length*. Unfortunately or fortunately this definition belongs to the area of calculus and goes beyond the purposes of this lesson. An idea about the details which are involved is given in the next exercises.

Exercise 1.7. Show that every regular polygon, inscribed in a given circle κ, has perimeter less than that of the circle κ.

Hint: Here the problem is, that we consider a general regular polygon and not the special regular polygons Π_i with 2^{i+1} sides, for which we showed that their perimeters α_i make an increasing sequence with limit the perimeter of the circle A and $\alpha_i < A$. The proof in the general case follows from this special case and Exercise I-2.149, according to which the perimeter of the regular polygon with $\nu + 1$ sides is greater than this of the regular polygon with ν sides. This way, for given ν we find a power of 2 for which $\nu \leq 2^{i+1}$, something which is always possible. Then however the perimeter p of the regular polygon with ν sides will be less than the corresponding perimeter α_i of Π_i, which in turn is less than A.

Exercise 1.8. Show that the sequence of perimeters $\beta_1, \beta_2, \beta_3, \ldots$ of the regular polygons with $\nu = 3, 4, 5, \ldots$ sides and inscribed in the circle κ is increasing and bounded consequently has limit B. Show also that $B = A$, where A is the limit of the perimeters α_1, α_2, α_3, ... of the regular polygons Π_1, Π_2, Π_3, ... inscribed in κ.

Hint: The fact, that the sequence β_1, β_2, β_3, ... of the perimeters of *all* the polygons is increasing, is proved in Exercise I-2.149. The fact, that this sequence is bounded, is proved exactly as in (Lemma 1.3). The existence of the limit B is then a consequence of Axiom 1.1. The fact, that $B = A$, is proved by excluding the cases $B < A$ and $A < B$. Indeed, if we suppose that $B < A$, then, according to the definition of the limit, there will exist α_i, such that $B < \alpha_i < A$. Then, however, and for each polygon with number of sides $\nu > 2^{i+1}$ we'll have corresponding perimeter $\alpha_i < \beta_\nu$, which implies that $B < \alpha_i < \beta_\nu$. This though is contradictory, because for all the β_ν holds $\beta_\nu < B$. Similarly we prove that $A < B$ leads to a contradiction.

1.3 The number π

> My method to overcome a difficulty is to go round it.
>
> George Polya, *How to solve it*

Lemma 1.4. *For every regular polygon with μ sides the ratio of its perimeter to the diameter of its circumcircle is independent of its size, it depends only on μ and is given by the formula:*

$$\mu \cdot \sin\left(\frac{180°}{\mu}\right).$$

Proof. Follows immediately from Lemma 1.1 by dividing with the diameter 2ρ.

Theorem 1.1. *For every circle, the ratio of its perimeter to its diameter, is a constant π, independent of the radius of the circle.*

Proof. Suppose that the ratio of the perimeter p to the diameter $2r$ of a specific circle κ is π and another circle κ' has corresponding ratio of perimeter to diameter

$$\pi' = \frac{p'}{2r'} < \pi.$$

We show $\pi' = \pi$. From the definition of the perimeter of the circle as a limit of perimeters of inscribed polygons in the preceding section (also see Exercise 1.1), there is a regular polygon with μ sides inscribed in the circle κ with perimeter p_μ which satisfies:

$$2r \cdot \pi' < p_\mu < p = 2r \cdot \pi,$$

1.3. THE NUMBER π

which is equivalent with

$$\pi' < \frac{p_\mu}{2r} < \pi.$$

The corresponding regular polygon with μ sides and inscribed in the circle κ' will have, according to the preceding lemma, the same ratio of perimeter to diameter:

$$\frac{p'_\mu}{2r'} = \frac{p_\mu}{2r},$$

and the preceding inequality implies

$$\pi' = \frac{p'}{2r'} < \frac{p'_\mu}{2r'} \Leftrightarrow p' < p'_\mu.$$

The last inequality however leads to a contradiction, because according to our definitions, the perimeter p' is the limit of the increasing sequence of p'_μ and is always

$$p'_\mu < p'.$$

By interchanging the roles of κ and κ', we prove in exactly the same way that also the inequality $\pi < \pi'$ leads to a contradiction. Therefore it must be $\pi' = \pi$.

Corollary 1.1. *The perimeter of a circle of radius ρ is equal to $2\pi\rho$.*

Exercise 1.9. Show that the sequence of numbers

$$\mu \cdot \sin\left(\frac{180°}{\mu}\right), \quad \mu = 3, 4, 5, \ldots$$

is increasing and bounded and has limit the number π.

Hint: This sequence results from the perimeters $\beta_\mu = 2\rho \cdot \mu \cdot \sin(\frac{180°}{\mu})$ of the regular polygons which are inscribed in a circle of radius ρ (Lemma 1.1). This sequence, according to Exercise 1.8, has limit the number $A = 2\pi\rho$. The conclusion follows from Exercise 1.2, applying it to the sequence $\frac{\beta_\mu}{2\rho} = \mu \cdot \sin(\frac{180°}{\mu})$ which tends to $\frac{A}{2\rho} = \pi$.

Remark 1.4. It can be proved (1882 Lindemann (1852-1939)), that the number π is, not only irrational, but also **transcendental**, in other words it is not a root of a polynomial ([66, p.128]) with integer coefficients. This implies, that the, so called **squaring of the circle**, that is, the construction of π using ruler and compass, is *impossible*. Archimedes, using two polygons of 96 sides, one inscribed and another circumscribed to the circle, proved that the value of π is between:

$$3\frac{10}{71} \; (\sim 3.14084\ldots) < \pi < 3\frac{10}{70} \; (\sim 3.14285\ldots).$$

Today, with the use of software, we can approximate π, practically, with any accuracy we desire (see for example [1]). Next approximation gives its first

50 decimal places:

$\pi \sim 3.14159265358979323846264338327950288419716939937510....$

Archimedes' approximation method can be done with the usage of two sequences $\alpha_0, \alpha_1, ...$ and $\beta_0, \beta_1, ...$ which give the half perimeter of the regular polygons with $6 \cdot 2^v$ sides which are circumscribed (α_v), respectively, inscribed (β_v) in a circle of radius 1. The first terms are

$$\alpha_0 = 2\sqrt{3}, \quad \beta_0 = 3,$$

and for the rest it is proved easily ([8, p.2]) that they are given through the formulas

$$\alpha_{v+1} = \frac{2\alpha_v \beta_v}{\alpha_v + \beta_v}, \quad \beta_{v+1} = \sqrt{\alpha_{v+1}\beta_v}.$$

The number π is a limit of the sequence α_v but also simultaneously of β_v. The noteworthy with these formulas is that they lead into approximations of π and finally to the limit, without the mediation of measurements of the real half perimeters. This way, to find the corresponding half perimeters of the 12-gon we replace α_0, β_0 into

$$\alpha_1 = \frac{2\alpha_0 \beta_0}{\alpha_0 + \beta_0}, \quad \beta_1 = \sqrt{\alpha_1 \beta_0},$$

which give the approximations $\alpha_1 = 3.2153$, $\beta_1 = 3.1058$. The half perimeters of the 24-gon are calculated similarly as $\alpha_2 = 3.1597$, $\beta_2 = 3.1326$, of the 48-gon as $\alpha_3 = 3.1461$, $\beta_3 = 3.1394$, of the 96-gon as $\alpha_4 = 3.1427$, $\beta_4 = 3.1410$ etc.

1.4 Arc length of a circle, radians

> (While) evil is done without effort, *naturally*, fatally, the good is always the product of skill.
>
> Baudelaire, Eloge de Toilette

The **length** of an arc $\widehat{AB}$ of a circle κ of radius ρ is defined, analogously to the circle, as the limit of lengths of broken lines inscribed in the circle and with endpoints the points A and B. Using a procedure similar to that of the determination of the length of the circle, which I won't repeat here, we find that the length of the arc is expressed by the formula (See Figure 1.4)

$$|\widehat{AB}| = \theta \cdot \tfrac{\pi}{180°} \cdot \rho, \quad (\theta \text{ the measure of the angle in degrees}).$$

1.4. ARC LENGTH OF A CIRCLE, RADIANS

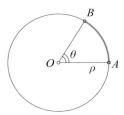

Fig. 1.4: Arc length: $\theta \frac{\pi}{180°} \rho$

In particular, for the angle of 360 degrees (full turn), the formula gives the length of the circle as $2\pi\rho$. I summarize the basic properties of the length of arc, which result from this formula and the properties of arcs:

1. Equal arcs on the same circle or on congruent circles have the same length and conversely, arcs of the same length on the same circle or in congruent circles are equal.
2. The length of the sum of two arcs is equal to the sum of the length of the two arcs.
3. The perimeter of the circle is $2\pi\rho$.

These properties transfer the properties which we met in I-§ 2.12 from the arcs to their corresponding lengths. The number

$$\phi = \theta \cdot \frac{\pi}{180°}, \quad (\theta \text{ the measure of the angle in degrees})$$

is called **measure of the angle in radians**. Obviously it is an expression of the measure of the angle using a different unit of measurement, which measures the percentage of 2π which corresponds to the central angle θ, in the sense that $\frac{\theta}{360°}$ is the percentage of the angle relative to the 360 degrees of the full turn. And with this percentage we multiply the number 2π. Using then the number ϕ as the measure of the angle, the formula for the length of arc is written as

$$|\widehat{AB}| = \phi \cdot \rho, \quad (\phi \text{ the measure of the angle in radians}).$$

In particular, for $\rho = 1$, i.e. a circle with radius the unit of length, the preceding formula becomes

$$|\widehat{AB}| = \phi, \quad (\rho = 1 \text{ and measure in radians}).$$

This formula, reading it from the right to the left, reveals that the measure of angle in radians is nothing more than the length of the arc of the unit circle (circle with radius the measurement unit) which has this angle as corresponding central (see I-§ 2.12). This way, an angle of one radian ($\phi = 1$) is the central angle, which results from arc length equal to one unit of measurement and taken on the unit circle. It is as if we bend the unit of measurement (without changing its length, as if it were made of wire or something similar) and attach it on the unit circle. The central angle of the arc which results is the angle with measure 1 radian. Next table shows a sample of the correspondence between degrees and radians for some special angles we meet often in applications.

degrees	0	30	45	60	90	120	135	180	270	360
radians	0	$\frac{\pi}{6}$	$\frac{\pi}{4}$	$\frac{\pi}{3}$	$\frac{\pi}{2}$	$\frac{2\pi}{3}$	$\frac{3\pi}{4}$	π	$\frac{3\pi}{2}$	2π

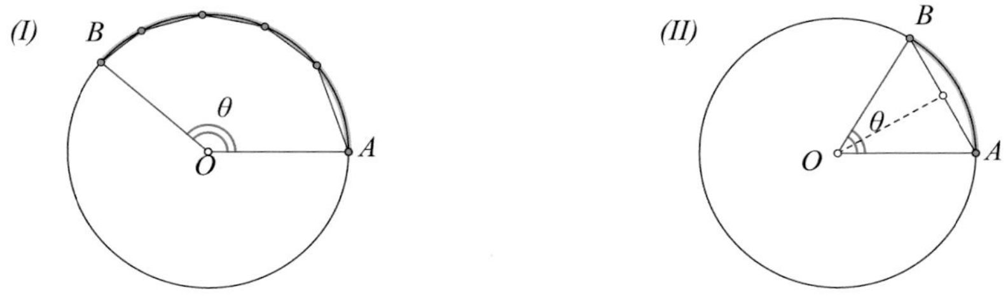

Fig. 1.5: Arc length > length of broken line $\sin(\theta) < \theta$

Remark 1.5. Remark 1.3 is valid here as well. The arc length is also included in a more general definition, which includes the lengths of broken lines, of the circle and of the arcs of the circle as special cases. A direct consequence of this definition is that the polygonal lines, which are inscribed in the arc with start point A and endpoint B, have length less than the length of the arc (See Figure 1.5-I). Specifically the length of the chord AB is less than the length of the arc. This way measuring angles in radians we have (See Figure 1.5-II):

$$|AB| = 2\rho \sin\left(\frac{\theta}{2}\right) < \rho\theta \Leftrightarrow \sin\left(\frac{\theta}{2}\right) < \frac{\theta}{2}.$$

Exercise 1.10. Construct an equilateral polygon circumscribing a circle κ, which is not regular. Show that an equilateral polygon circumscribing a circle κ, which has an odd number of sides, is regular.

Hint: Show that the succession of angles which are formed using the lines from the center, is exactly as in the preceding figure (See Figure 1.6-I) and see what happens when the polygon has an odd number of sides.

Exercise 1.11. Construct a polygon with equal angles inscribed in a circle which is not regular. Show that an inscribed in a circle polygon with equal angles and odd number of sides is regular.

Exercise 1.12. Consider a half circle with diameter the line segment AB. Divide the segment into ν segments and construct half circles $\{\alpha, \beta, \ldots\}$ with

1.5. DEFINITION OF THE AREA OF THE CIRCLE

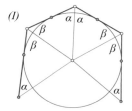
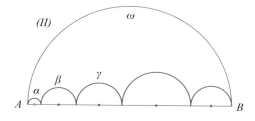

Fig. 1.6: Succession of angles   Circular polygonal line

diameters these segments (See Figure 1.6-II). Find the length of the line which consists of these half circles and show that it is independent of v and the length of the segments.

Exercise 1.13. Show that the sequence

$$\frac{\mu}{\pi} \cdot \sin\left(\frac{\pi}{\mu}\right), \quad \mu = 3, 4, 5, \ldots$$

is increasing and bounded and has as limit the number 1.

Hint: The exercise is a repeat of Exercise 1.9, replacing the 180° with π, that is counting the angles in radians. Caution is needed in the way we interpret the equation

$$\sin\left(\frac{180°}{\mu}\right) = \sin\left(\frac{\pi}{\mu}\right).$$

On the sine to the left we measure the angle in degrees. On the sine to the right we measure the same angle in radians. According to the aforementioned exercise the limit of $\mu \sin(\frac{\pi}{\mu})$ is the number π. Therefore the limit of $\frac{\mu}{\pi} \cdot \sin(\frac{\pi}{\mu})$ will be, according to Exercise 1.2, $\frac{\pi}{\pi} = 1$.

1.5 Definition of the area of the circle

> Therefore our main concentration will not be on how clever we are to have found it all out, but on how clever nature is to pay attention to it.
>
> R. Feynman, *The character of physical low*

The area of the circle, like its length or perimeter, requires a special approach. The basic properties of the areas (I-3.1-I-3.4) allow only polygon measurements. The circle is a new kind of shape and we must *define* what exactly we mean by area of this shape. For the definition or the area we need the same limit procedure of inscribed regular polygons $\{\Pi_1, \Pi_2, \Pi_3, ...\}$ with which we approximated the circle. Thus, in what follows, we'll do almost exactly the same we did in I-§ 1.2. That's why I use the same figure and the same symbols, changing only the way of angle measurement and *adopting from now on as the angle unit of measurement the radian*. Thus, instead of the angle measurement $\phi = \frac{360°}{\mu}$ used there, here we'll use the expression $\omega = \frac{2\pi}{\mu}$ (which is of course equal to $\frac{\pi}{180°}\phi$).

Lemma 1.5. *The area of a regular polygon with μ sides inscribed in a circle of radius ρ is*
$$\mu \cdot \frac{\rho^2}{2} \cdot \sin\left(\frac{2\pi}{\mu}\right).$$

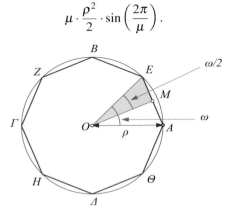

Fig. 1.7: $\varepsilon(AOE)$: the $\frac{1}{\mu}$-th of the area

Proof. Because the regular polygon is split into μ isosceli triangles, like AOE (See Figure 1.7), which is defined from the center and one of its sides, the area of the polygon will be (Property I-3.2) the sum of the areas of these, pairwise congruent triangles. The triangle AOE is isosceles with angle at O equal to
$$\omega = \frac{2\pi}{\mu} \Rightarrow \frac{\omega}{2} = \frac{\pi}{\mu}$$
since the sum of μ such equal agles will give a full turn about O. Also the median AM is orthogonal at the middle of AE (Corollary I-1.3) and the triangle OME is right, therefore according to Theorem I-3.11
$$|AE| = 2|ME| = 2\rho \cdot \sin\left(\frac{\omega}{2}\right)$$
$$|OM| = \rho \cdot \cos\left(\frac{\omega}{2}\right) \Rightarrow$$
$$\varepsilon(AOE) = \frac{1}{2}|AE||OM| = \rho^2 \cdot \sin\left(\frac{\omega}{2}\right)\cos\left(\frac{\omega}{2}\right)$$
$$= \frac{\rho^2}{2}\sin(\omega) = \frac{\rho^2}{2}\sin\left(\frac{2\pi}{\mu}\right).$$

1.5. DEFINITION OF THE AREA OF THE CIRCLE

The one before the last equality holds because of Exercise I-3.54. The result follows from the fact, that the area of the polygon is $\mu \cdot \varepsilon(AOE)$.

Lemma 1.6. *The sequence of areas* $\{\varepsilon_1, \varepsilon_2, \varepsilon_3, \ldots\}$ *of the polygons* $\{\Pi_1, \Pi_2, \Pi_3, \ldots\}$ *is increasing.*

Proof. Because each such polygon has a number of sides double those of its preceding, it suffices to show that the ratio of areas $\frac{\varepsilon}{\varepsilon'}$ of two regular polygons Π and Π', of which the first has μ sides and the second has 2μ sides is less than 1. However according to the preceding results, taking into account the formula for the sine of the double angle (Exercise I-3.54), we have:

$$\frac{\varepsilon}{\varepsilon'} = \frac{\mu(\frac{\rho^2}{2}\sin(\frac{2\pi}{\mu}))}{2\mu(\frac{\rho^2}{2}\sin(\frac{2\pi}{2\mu}))} = \frac{\sin(\frac{2\pi}{\mu})}{2\sin(\frac{\pi}{\mu})} = \frac{2\sin(\frac{\pi}{\mu})\cdot\cos(\frac{\pi}{\mu})}{2\sin(\frac{\pi}{\mu})} = \cos\left(\frac{\pi}{\mu}\right) < 1.$$

Lemma 1.7. *The sequence of areas* $\{\varepsilon_1, \varepsilon_2, \varepsilon_3, \ldots\}$ *of the polygons* $\{\Pi_1, \Pi_2, \Pi_3, \ldots\}$ *is bounded.*

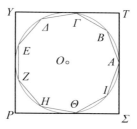

Fig. 1.8: $\varepsilon(\Pi_i) < \varepsilon(P\Sigma TY)$

Proof. According to Property I-3.3 the area of each of these polygons is less than the area of the square which is circumscribed on the circle with radius ρ (See Figure 1.8).

Lemmata 1.6 and 1.7 guarantee, relying on Axiom 1.1, that the sequence $\{\varepsilon_1, \varepsilon_2, \varepsilon_3, \ldots\}$ of the areas of the polygons $\{\Pi_1, \Pi_2, \Pi_3, \ldots\}$ has a limit E. This E we define as the area of the circle.

Theorem 1.2. *The area of a circle of radius ρ is $\pi\rho^2$.*

Proof. According to the definition, the area of the circle is the limit of the areas of the inscribed polygons $\{\Pi_1, \Pi_2, \Pi_3, \ldots\}$, ... which according to Lemma 1.5 is

$$\varepsilon(\Pi_i) = \mu\frac{\rho^2}{2}\sin(\tfrac{2\pi}{\mu}), \quad \text{where } \mu = 2^{i+1}.$$

The proof follows by relating this limit to the limit of the perimeter of the same polygons, which according to Lemma 1.1 is

$$\alpha_i = 2\mu\rho \sin(\tfrac{\pi}{\mu}), \quad \text{where } \mu = 2^{i+1},$$

and, according to the definition, their limit is the perimeter of the circle

$$A = 2\pi\rho.$$

The relation is very simple:

$$\begin{aligned}
\varepsilon(\Pi_i) &= \mu \frac{\rho^2}{2} \sin\left(\frac{2\pi}{\mu}\right) \\
&= \rho\left(\frac{\mu}{2}\rho \sin\left(\frac{\pi}{\frac{\mu}{2}}\right)\right) \\
&= \left(\frac{\rho}{2}\right)\left(2\frac{\mu}{2}\rho \sin\left(\frac{\pi}{\frac{\mu}{2}}\right)\right) \\
&= \left(\frac{\rho}{2}\right) \alpha_{i-1}.
\end{aligned}$$

The last follows from the fact, that we have set $\mu = 2^{i+1}$. Consequently $\frac{\mu}{2} = 2^i$. Because the limit of the α_i is the number $2\pi\rho$ the last formula, combined with Exercise 1.2, shows that the limit of the $\varepsilon(\Pi_i)$ is

$$\frac{\rho}{2} \cdot (2\pi\rho) = \pi\rho^2.$$

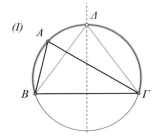
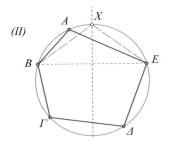

Fig. 1.9: Maximization of triangle area   Maximization of polygon area

Exercise 1.14. Show that from all triangles $AB\Gamma$ inscribed in the same circle κ and having a common base $B\Gamma$, the one having the maximum area is the isosceles.

Hint: Because all the triangles have the same base $B\Gamma$, the one with maximal area will have its apical vertex at the maximum possible distance from the base $B\Gamma$ (See Figure 1.9-I). This position, however, coincides with one of the two points (Δ in the figure) at which the medial line of $B\Gamma$ intersects the circle κ.

1.5. DEFINITION OF THE AREA OF THE CIRCLE

Exercise 1.15. Show that from all inscribed polygons $AB\Gamma\Delta...$ with ν sides in the same circle κ the regular polygon with ν sides has the maximum area.

Hint: Suppose that the polygon $\Pi = AB\Gamma\Delta...$ with ν sides has maximal perimeter p and is not regular (See Figure 1.9-II). Then there exist two successive sides of it, which are unequal for example BA and AE. According to the preceding exercise the triangle ABE will have area less than the corresponding inscribed isosceles XBE. By substituting then ABE with XBE, we find another polygon with ν sides and area $p' > p$, which is contradictory, since we supposed that p has the maximum possible area ([67, p.198]).

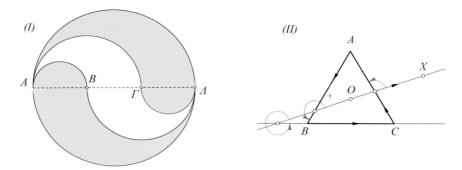

Fig. 1.10: Circle division   Oriented angles

Exercise 1.16. We divide a diameter $A\Delta$ of the circle into three equal parts, and drawing semicircles we divide the circle into three parts as in figure 1.10-I. Show that these three parts of the circle have equal areas.

Exercise 1.17. Divide a given circle κ into ν parts of equal area using circles concentric with κ.

Exercise 1.18 (Galileo's Theorem). The area $\varepsilon(\kappa)$ of the circle κ is the mean proportional of the area of one circumscribed to κ polygon Π and a similar one to it Π', whose perimeter is equal to the perimeter of the circle: $\varepsilon(\kappa) = \sqrt{\varepsilon(\Pi)\varepsilon(\Pi')}$.

Exercise 1.19. Given is a positively oriented regular polygon $A_1A_2...A_\nu$ and a line OX through its center (See Figure 1.10-II). If $\alpha_1, ..., \alpha_\nu$ are the positively oriented angles which are formed by the oriented line OX with (the oriented lines defined by) its sides $A_1A_2,...,A_\nu A_1$, show that it holds:

$$\cos(\alpha_1) + \cos(\alpha_2) + ... + \cos(\alpha_\nu) = 0.$$

Exercise 1.20. Show that for every point X at distance δ from the center of a regular ν-gon $A_1A_2...A_\nu$ inscribed in a circle of radius ρ, it holds:

$$|XA_1|^2 + |XA_2|^2 + ... + |XA_\nu|^2 = \nu(\rho^2 + \delta^2).$$

1.6 The area of a circular sector

> But in the case of a scholarly and cultivated man, on an occasion which requires a later dinner than usual, a mathematical problem on hand, or some pamphlet or musical instrument, will not permit him to be harried by his belly; on the contrary, he will steadily turn away and transfer his thoughts from the table to these other things, and scare away his appetites, like Harpies, by means of the Muses.
>
> *Plutarch, Advice about keeping well*

Circular sector is called the part of the circle, which is enclosed by an arc of the circle and the two radii at their ends (See Figure 1.11). The area of

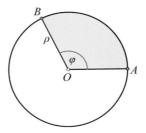

Fig. 1.11: Circular sector

a circular sector is defined, by analogy to the circle, as a limit of areas of broken lines which have the radii *OA*, *OB* as two successive sides and the rest forming a polygonal line inscribed in the circle, having as endpoints the points *A* and *B*. Using a process similar to that of the determination of the area of the circle, which I'll not repeat here, we find that the area of the sector can be expressed through the corresponding central angle of its arc through the formula

$$\varepsilon(AOB) = \tfrac{1}{2}\phi\rho^2, \quad (\phi \text{ the measure of the angle } \widehat{AOB} \text{ in radians}).$$

In paraticular, for the angle of 360 degrees (full turn, $\phi = 2\pi$) the formula gives the area of the circle $\pi\rho^2$. I summarize the basic properties for the area of the circular sector which follow from this definition and the properties of arcs (I-§ 2.12):

1. Equal arcs on the same circle or on congruent circles define circular sectors with the same area.
2. The area of an arc sector, whose arc is the sum of two arcs is equal to the sum of the areas of the sectors of the two arcs.
3. The area of the circle is $\pi\rho^2$.

These properties are completely similar to the properties of circular arcs we met in I-§ 2.12 and have corresponding consequences for areas. The remarks

1.6. THE AREA OF A CIRCULAR SECTOR

made in section 1.4 for the length of an arc hold also for the areas of sectors. It is proved, that there exists a more general notion of area of shapes, which includes that of polygons, circles and circular sectors as special cases (Jordan Measure [2, p.490]). This definition, however, belongs to the area of Mathematical Analysis and goes beyond the purposes of the lesson. The unified definition allows for comparisons between polygonal areas and areas between curves, like that of the sector. The properties of the areas, we described in I-§ 3.1, are transferred verbatim to the general definition, which includes curvilinear figures and as such result inequalities like, for example, the one suggested by the figure 1.12-I.

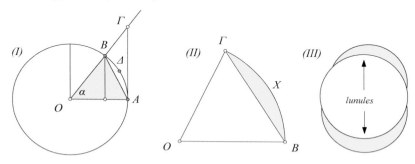

Fig. 1.12: Area comparison   Circular section   Lunules

$$\varepsilon(AOB) < \varepsilon(A\Delta BO) < \varepsilon(A\Gamma O).$$

The second area is that of the sector $A\Delta BO$ and the inequality holds for every angle α with $0 < \alpha < \frac{\pi}{2}$ (its measure in radians). Replacing with the formulas which express the area, we get

$$\frac{1}{2}\rho^2 \sin(\alpha) < \frac{1}{2}\rho^2 \alpha < \frac{1}{2}\rho^2 \tan(\alpha),$$

which after simplification leads to

$$1 < \frac{\alpha}{\sin(\alpha)} < \frac{1}{\cos(\alpha)}.$$

Next exercises use the notions **circular section** and **Lunule**. The first is defined as the part of the circle $BX\Gamma$, which is contained between an arc $\widehat{BX\Gamma}$ and the chord which joins its endpoints (See Figure 1.12-II). The second is defined as the non common part of a circle with another one which intersects it (See Figure 1.12-III) ([67, p.788]).

Exercise 1.21. Show that the area of a circular section $BX\Gamma$ is equal to $\varepsilon(BX\Gamma)$ $= \frac{r^2}{2}(\omega - \sin(\omega))$, where $\{r, \omega\}$ are respectively the radius and central angle of the arc $\widehat{BX\Gamma}$ defining it (See Figure 1.12-II).

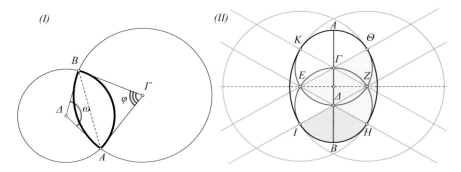

Fig. 1.13: Area of common part   Area of the ovoid

Exercise 1.22. Calculate the area of the common part of two circles, like those of figure (See Figure 1.13-I), in terms of their radii and the two central angles $\{\omega, \phi\}$, by which their common chord is seen from the respective centers.

Exercise 1.23. The line segment AB of length 3δ is divided into three equal parts with the points Γ and Δ. With center Γ and radius δ, the arc $\widehat{\Theta AK}$ of angle $\pi/3$ is constructed. With center Δ and radius δ, the arc $\widehat{HBI}$ of angle $\pi/3$ is constructed. With centers the points of intersection E, Z of the circles $\Gamma(\delta)$ and $\Delta(\delta)$ and radius 2δ, arcs $\widehat{\Theta H}$ and $\widehat{IK}$ of angle $\pi/6$ are constructed. Find the area of the resulting figure (See Figure 1.13-II).

Hint: The area ε of the figure (which is called an **ovoid**) is the sum of four circular sections which are defined by the arcs of the formulation minus the area of the rhombus $\Gamma E \Delta Z$, which is counted twice.

$$\varepsilon = 2\varepsilon(\Delta IBH) + 2\varepsilon(EH\Theta) - \varepsilon(\Gamma E \Delta Z)$$
$$= 2\left(\frac{1}{2}\left(\delta^2 \cdot \frac{\pi}{3}\right)\right) + 2\left(\frac{1}{2}\left((2\delta)^2 \frac{\pi}{6}\right)\right) - \delta \cdot \left(\frac{\sqrt{3}}{2}\delta\right)$$
$$= \delta^2 \left(\pi - \frac{\sqrt{3}}{2}\right).$$

Exercise 1.24. Calculate the area of a 3-foil, which is created in the interior of an equilateral triangle of side δ from circles congruent to its circumcircle, which have as centers the symmetric points of the centroid H of the equilateral relative to its sides(See Figure 1.14-I).

Exercise 1.25. Calculate the area of the 4-foil, which is created in the interior of a square of side δ from circles with diameter the sides of the square (See Figure 1.14-II).

1.6. THE AREA OF A CIRCULAR SECTOR

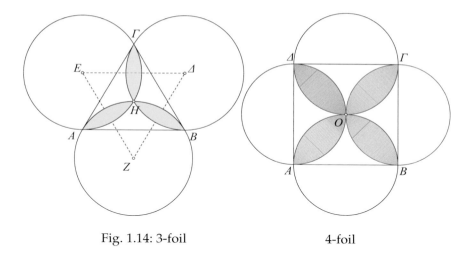

Fig. 1.14: 3-foil   4-foil

Exercise 1.26. In a given circle construct a curvilinear triangle (See Figure 1.15-I)/pentagon (See Figure 1.15-II), whose sides are three/five equal arcs, which are pairwise tangent at their endpoints. Calculate the area of one such triangle/pentagon.

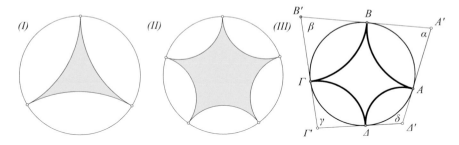

Fig. 1.15: Curvilinear triangle, pentagon and quadrilateral

Exercise 1.27. Show that every circumscriptible to a cirlce quadrilateral $q' = A'B'\Gamma'\Delta'$ defines a corresponding circular quadrilateral $q = AB\Gamma\Delta$, whose sides are mutually tangent circle arcs, and its vertices are the contact points of the sides of q' (See Figure 1.15-III). Calculate the area of such a quadrilateral q in terms of the inradius r and the angles of q'.

Exercise 1.28. In the **salinon** (from the name of an ancient Greek shield), which appears in figure (See Figure 1.16-I), the shadowed region is bounded by four half circles. The area of the shadowed region (shield) is equal to the area of the circle with diameter $B\Gamma$.

Exercise 1.29. Calculate the area of Greek-Trojan type shield in dependence of $\{a,b\}$, whose construction depends on a rectangle with sides $\{a,b\}$ and four semi-circles on its sides (See Figure 1.16-II).

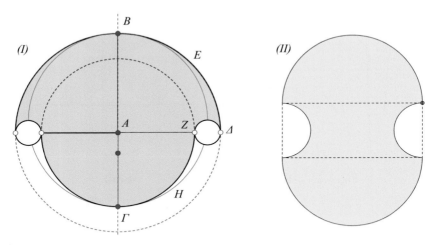

Fig. 1.16:   Greek and Greek-Trojan "Salinon" type shield

Exercise 1.30. Calculate the area of a lunule, considering it as the difference of two circular sections (See Figure 1.17-I).

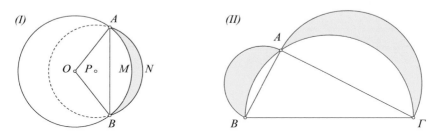

Fig. 1.17: Lunule calculation   Lunule of Hippocrates

Exercise 1.31 (Lunule of Hippocrates). (apprx. 470-410 B.C.) Consider a right triangle $AB\Gamma$ and the circles with diameters the sides of the triangle. Show that the sum of the areas of the lunules on its orthogonal sides is equal to the area of the triangle (See Figure 1.17-II).

Hint: The area in question ε is equal to the difference of two areas. The first is the area of the two half circles with diameters the orthogonal sides of $AB\Gamma$

$$\varepsilon_1 = \frac{1}{8}\pi(|AB|^2 + |A\Gamma|^2).$$

The second is the area of the two circular sections of the circumcircle, which are defined by the orthogonal sides. Let us call this area ε_2. Adding to ε_2 the area of the triangle, we get the whole area of the half circle with diameter the hypotenuse $B\Gamma$. We therefore have

$$\begin{aligned}\varepsilon &= \varepsilon_1 - \varepsilon_2 \\ &= (\varepsilon_1 + \frac{1}{2}|AB||A\Gamma|) - (\varepsilon_2 + \frac{1}{2}|AB||A\Gamma|) \\ &= (\frac{1}{8}\pi(|AB|^2 + |A\Gamma|^2) + \frac{1}{2}|AB||A\Gamma|) - \frac{1}{8}\pi|B\Gamma|^2 \\ &= \frac{1}{8}\pi(|AB|^2 + |A\Gamma|^2 - |B\Gamma|^2) + \frac{1}{2}|AB||A\Gamma| \\ &= \frac{1}{2}|AB||A\Gamma|.\end{aligned}$$

1.7 The isoperimetric inequality

> Of the forms which life is able to bestow on her creations, that of the birds egg is one of the simplest and loveliest. Nowhere do we find the beauty of the circle and the ellipse, the geometrical bases of organic bodies, combined with greater precision.
>
> J.H. Fabre, *The Life of the Scorpion*

The **isoperimetric inequality** connects the length L of a closed curve (polygon) κ with the area E, which this curve (polygon) encloses. From the formulas we proved in the preceding sections, for the circle holds

$$4\pi = \frac{L^2}{E}.$$

It is proved, that this is the only case where equality holds, while for every

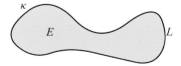

Fig. 1.18: Isoperimetric inequality $4\pi \leq \frac{L^2}{E}$

polygon but also for every closed curve different from a circle, of length L, which encloses area E (See Figure 1.18) holds the following strict inequality

$$4\pi < \frac{L^2}{E},$$

which is called *isoperimetric inequality* ([4, p.43]). Besides polygons and circles, the exact definition of the closed curves, for which the *length* L is defined and which bound a set of the plane, whose *area* E, can be measured, is beyond the purposes of this book. In this section we restrict ourselves to polygons of special categories and we examine the relation of their perimeter L to the area E they enclose. Restricting ourselves to polygons of a specific category, for example *convex* polygons with ν sides, we have an analogue of the isoperimetric inequality

$$k \leq \frac{L^2}{E},$$

where k is a constant dependent on the category of the polygons.

The isoperimetric inequality is invariant relative to similarity. In other words, if the polygon (curve) κ, has a ratio $\frac{L^2}{E} = \lambda$, then its similar κ' will also have ratio

$$\lambda' = \frac{L'^2}{E'} = \frac{(s \cdot L)^2}{s^2 \cdot E} = \frac{L^2}{E} = \lambda,$$

where s is the similarity ratio. This property helps to detect the value of the constant which is involved in the (special, depending on category) isoperimetric inequality through the process of the next lemma.

Lemma 1.8. *Suppose that, from all polygons with ν sides and fixed area E_0, there exists one, p_0 say, with least perimeter L_0. Then for every other polygon p with ν sides, of area E and perimeter L, will hold*

$$\frac{L_0^2}{E_0} \leq \frac{L^2}{E}.$$

Proof. Indeed, if the polygon p has area $E > E_0$, then a homothety f with ratio $s = \sqrt{\frac{E_0}{E}}$ defines polygon $p' = f(p)$ similar to p, with area E_0 and ratio $\frac{L^2}{E} = \frac{L'^2}{E_0}$, for which, according to the hypothesis, will hold

$$\frac{L_0^2}{E_0} \leq \frac{L'^2}{E_0} = \frac{L^2}{E}.$$

The same argument holds also in the case where $E < E_0$.

Theorem 1.3. *For every convex polygon with ν sides, perimeter L and area E holds*

$$4\nu \cdot \tan\left(\frac{\pi}{\nu}\right) \leq \frac{L^2}{E}.$$

The equality is valid, if and only if the polygon is regular.

1.7. THE ISOPERIMETRIC INEQUALITY

Proof. For the proof we use lemma 1.8, choosing an arbitrary area E_0 and considering all the v-gons whose area is E_0. We search between these polygons the one which has the least perimeter L_0. The left side of the inequality is the ratio $\frac{L^2}{E}$ for a regular polygon with v sides. It suffices, then, to show that the polygon with the least perimeter is the regular, in other words, the polygon, which has all its sides equal and all its angles equal. That the one with the least perimeter L_0 must have all sides equal, can be seen easily. Indeed, if the polygon p_0, with area E_0 and least perimeter L_0, had sides which were not all equal, then it would also have two *consecutive* sides AB, $B\Gamma$ unequal (See Figure 1.19-I). Then, by drawing the diagonal $A\Gamma$, we would

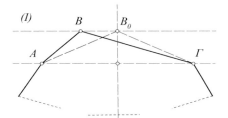 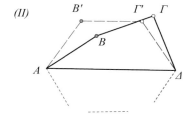

Fig. 1.19: Replace $AB\Gamma$ with $AB_0\Gamma$ Replace $AB\Gamma\Delta$ with $AB'\Gamma'\Delta$

be able to excise the triangle $AB\Gamma$ and to replace it with an isosceles triangle $AB_0\Gamma$ of the same area and smaller perimeter. Consequently, this would result in a polygon p_1 of area E_0 but of perimeter $L_1 < L_0$, contrary to the assumption that p_0 has the least possible perimeter. The fact that all the angles of p_0 would have to be equal also follows easily.

Indeed, let us suppose that p_0, for which we know now that it has all its sides equal, does not have all its angles equal. Then there will exist two successive angles $\beta = \widehat{AB\Gamma}$, $\gamma = \widehat{B\Gamma\Delta}$ which would be unequal (See Figure 1.19-II). Drawing then the diagonal $A\Delta$, we excise the quadrilateral $\tau = AB\Gamma\Delta$ and we replace it with the inscriptible in circle $\tau' = AB'\Gamma'\Delta$, which has the same side lengths, is a trapezium and, according to Theorem I-3.26, has area $E'(\tau') > E(\tau)$. The new polygon p_1 which results has the same perimeter L_0 with p_0 but total area $E_1 > E_0$. For its ratio we'll have $s = \sqrt{\frac{E_0}{E_1}} < 1$ and therefore a homothety f with ratio s will map polygon p_1 into another $p_2 = f(p_1)$ having area $E_2 = s^2 E_1 = E_0$ and perimeter $L_2 = sL_1 = sL_0 < L_0$. The latter, however, is contradictory, since we supposed that L_0 is the least possible perimeter between all polygons having area E_0.

Exercise 1.32. Show that, if from all the polygons with v sides and fixed perimeter L_0, there exists one with maximal area E_0, then for every other polygon with v sides, of area E and perimeter L, will hold

$$\frac{L_0^2}{E_0} \leq \frac{L^2}{E}.$$

Hint: Proof similar to that of lemma 1.8.

Theorem 1.4. *The regular polygon with ν sides, perimeter L_0 and area E_0, is characterized by the following three properties.*

1. *Every other convex polygon of ν sides and area E_0, has perimeter $L > L_0$.*
2. *Every other convex polygon of ν sides and perimeter L_0, has area $E < E_0$.*
3. *Every other convex polygon of ν sides, has ratio $\frac{L^2}{E} > \frac{L_0^2}{E_0}$.*

Proof. The proof is a combination of the preceding lemma, theorem and exercise.

Remark 1.6. In the context of the isoperimetric inequality, the restriction to convexity is not essential and could be omitted in the formulation of the preceding propositions. For a non-convex polygon p with perimeter L and area E, for example, it is proved, fairly easily, that there exists another convex polygon p' with corresponding perimeter $L' < L$ and area $E' > E$ which contains p. Such a p', which can be constructed through a precisely defined procedure from p, is the so called **convex hull** of p. This is defined as the set of common points of all the half planes which contain p. As it is also seen

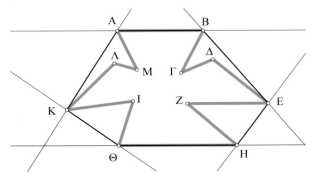

Fig. 1.20: The convex hull $p' = ABEH\Theta K$ of $p = AB\Gamma\Delta EZH\Theta IK\Lambda M$

in figure 1.20, the convex hull of the polygon p is a polygon p' which has vertices some of the vertices of p ([3, p.2]) and, because of the preceding inequalities, the ratios which are involved in the constant of the isoperimetric inequality satisfy

$$\frac{L'^2}{E'} < \frac{L^2}{E}.$$

This has as a consequence, the fact that in the search for the minimum value of these ratios, we can confine ourselves to convex polygons. This way, for example theorem 1.3, since it is valid for all the convex polygons with ν sides, will also hold for the non-convex case.

Remark 1.7. The isoperimetric inequality for triangles follows from Theorem 1.3 by replacing the constant $\nu = 3$ and $\tan\left(\frac{\pi}{\nu}\right) = \tan(60°) = \sqrt{3}$. Squaring

1.7. THE ISOPERIMETRIC INEQUALITY

both sides of the inequality and doing the calculations, results in the equivalent inequality

$$(b+c-a)(c+a-b)(a+b-c) \leq \left(\frac{a+b+c}{3}\right)^3,$$

where a, b, c are the lengths of the sides of the triangle. From the theorem also follows that equality holds exactly when the triangle is equilateral.

Such inequalities, like the preceding one, as well as the basic triangle inequality (Theorem I-1.6), which involve the lengths a, b, c of the sides of a triangle are called **triangle inequalities** and are used often in competitions and mathematical Olympiads ([19]).

I list below a few triangle inequalities, which could be useful in some applications. For their proofs I refer to the books [19], [5], [6]. In these $2\tau = a+b+c$, E denotes the area of the triangle and the equality (where inequality is shown and not a strict inequality) holds, if and only if the triangle is equilateral.

$$a^2 + b^2 + c^2 \geq 4E\sqrt{3}, \tag{1.1}$$

$$a(b+c-a) < 2bc, \tag{1.2}$$

$$(a^2 + b^2 + c^2)(a+b+c) \geq 9abc, \tag{1.3}$$

$$(a+b+c)\left(\frac{1}{a} + \frac{1}{b} + \frac{1}{c}\right) \geq 9, \tag{1.4}$$

$$3 \leq \frac{(a+b+c)^2}{bc+ca+ab} < 4, \tag{1.5}$$

$$8(\tau - a)(\tau - b)(\tau - c) \leq abc, \tag{1.6}$$

$$a^2 b(a-b) + b^2 c(b-c) + c^2 a(c-a) \geq 0, \tag{1.7}$$

$$(a+b)(b+c)(c+a) \geq 8abc, \tag{1.8}$$

$$3(a+b)(b+c)(c+a) \leq 8(a^3 + b^3 + c^3), \tag{1.9}$$

$$\frac{a+b+c}{abc} \leq \frac{1}{a^2} + \frac{1}{b^2} + \frac{1}{c^2}, \tag{1.10}$$

$$\frac{3}{2} \leq \frac{a}{b+c} + \frac{b}{c+a} + \frac{c}{a+b} < 2, \tag{1.11}$$

$$\frac{1}{3} \leq \frac{a^2 + b^2 + c^2}{(a+b+c)^2} < \frac{1}{2}, \tag{1.12}$$

$$\sqrt{b+c-a} + \sqrt{c+a-b} + \sqrt{a+b-c} \leq \sqrt{a} + \sqrt{b} + \sqrt{c}. \tag{1.13}$$

Exercise 1.33. Show that the next inequality follows as a special case of the theorem 1.3 for quadrilaterals with side lengths a, b, c, d

$$(b+c+d-a)(c+d+a-b)(d+a+b-c)(a+b+c-d) \leq \left(\frac{a+b+c+d}{2}\right)^4.$$

Exercise 1.34. The convex pentagon $AB\Gamma\Delta E$ is inscribed in a circle of unit radius and AE is a diameter of this circle. Show the inequalities for its sides $\{a = |AB|, b = |B\Gamma|, c = |\Gamma\Delta|, d = |\Delta E|\}$ ([12, p.34]):

$$a^2 + b^2 + c^2 + d^2 + abc + bcd < 4,$$
$$a^2 + b^2 + c^2 + d^2 + bcd + cda + dab + abc > 4.$$

1.8 Anthyphairesis

> But here another difficulty arises. While experiment convinces me of the correctness of this conclusion, my mind is not entirely satisfied as to the cause to which this effect is to be attributed.
>
> *Galilei, Dialogues concerning two new sciences*

Anthyphairesis is a procedure pertaining more to arithmetic than to geometry. In the history though, it is one of the first examples showing how intimatelly are related these two branches of mathematics. The procedure was used by the greeks in order to prove the asymetry of the roots of some integers, roots which appear as lengths of diagonals of polygons with integer side-lengths. A second example of use of algebra, which is the evolution of arithmetic, for geometric problems, is the so called *theory of fields*. Using this, one can clarify, which segments of a given non-rational length, can be constructed using ruler and compass only ([81, p.117], [10, p.293]).

Anthyphairesis begins with the euclidean algorithm for the determination of the greatest common divisor of two integers. According to this, if two integers $\alpha(56), \beta(33)$ are given, we find the integral quotient and remainder and we repeat the process with the divisor and resulting quotient, until, after successive divisions, we arrive at remainder 0.

$$56 = \boxed{1} \cdot 33 + 23$$
$$33 = \boxed{1} \cdot 23 + 10$$
$$23 = \boxed{2} \cdot 10 + 3$$
$$10 = \boxed{3} \cdot 3 + 1$$
$$3 = \boxed{3} \cdot 1 + 0.$$

The quotients $(1,1,2,3,3)$ of the successive divisions can be used for the representation of the number $\frac{\alpha}{\beta} = \frac{56}{33}$ as a compound fraction.

1.8. ANTHYPHAIRESIS

$$\frac{56}{33} = 1 + \frac{23}{33} = 1 + \frac{1}{\frac{33}{23}} = 1 + \cfrac{1}{1 + \frac{10}{23}} = 1 + \cfrac{1}{1 + \cfrac{1}{\frac{23}{10}}} = 1 + \cfrac{1}{1 + \cfrac{1}{2 + \frac{3}{10}}}$$

$$= 1 + \cfrac{1}{1 + \cfrac{1}{2 + \cfrac{1}{\frac{10}{3}}}} = \boxed{1} + \cfrac{1}{\boxed{1} + \cfrac{1}{\boxed{2} + \cfrac{1}{\boxed{3} + \cfrac{1}{\boxed{3}}}}}$$

A similar process in the next example gives the corresponding representation of a fraction:

$$\left. \begin{array}{rl} 115 &= \boxed{3} \cdot 34 + 13 \\ 34 &= \boxed{2} \cdot 13 + 8 \\ 13 &= \boxed{1} \cdot 8 + 5 \\ 8 &= \boxed{1} \cdot 5 + 3 \\ 5 &= \boxed{1} \cdot 3 + 2 \\ 3 &= \boxed{1} \cdot 2 + 1 \\ 2 &= \boxed{2} \cdot 1 + 0 \end{array} \right\} \Leftrightarrow \quad \frac{115}{34} = \boxed{3} + \cfrac{1}{\boxed{2} + \cfrac{1}{\boxed{1} + \cfrac{1}{\boxed{1} + \cfrac{1}{\boxed{1} + \cfrac{1}{\boxed{1} + \cfrac{1}{\boxed{2}}}}}}}$$

This process of repeated divisions is called **anthyphairesis** and the expression for the number obtained above, is called a **continued fraction** ([9], [21], [15], [14]). For a continued fraction like the preceding one, we often use the symbol $[3; 2, 1, 1, 1, 1, 2]$. One more example:

$$\left. \begin{array}{rl} 376 &= 5 \cdot 67 + 41 \\ 67 &= 1 \cdot 41 + 26 \\ 41 &= 1 \cdot 26 + 15 \\ 26 &= 1 \cdot 15 + 11 \\ 15 &= 1 \cdot 11 + 4 \\ 11 &= 2 \cdot 4 + 3 \\ 4 &= 1 \cdot 3 + 1 \\ 3 &= 3 \cdot 1 + 0 \end{array} \right\} \Leftrightarrow \quad \frac{376}{67} = 5 + \cfrac{1}{1 + \cfrac{1}{1 + \cfrac{1}{1 + \cfrac{1}{1 + \cfrac{1}{2 + \cfrac{1}{1 + \frac{1}{3}}}}}}}.$$

It is easily proved ([11, p.11], [15, p.16]), that every rational number can be written this way and, obviously, if a number can be written this way, then it is rational. Anthyphairesis has a geometric interest, because with its help one can prove the relative asymmetry of two line segments, like, for example, this of the segment of length 1 relative to a segment of length $\sqrt{2}$. This is equivalent to the fact that $\sqrt{2}$ is not rational. The method of the proof relies on the generalization of anthyphairesis from integers to real numbers, which may not be integers. In the case of non integral anthyphairesis the process may not terminate and this shows that the two segments which are subject to anthyphairesis are relatively asymmetrical.

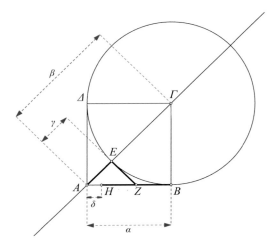

Fig. 1.21: Asymmetry of $\sqrt{2}$ to 1 through anthyphairesis

Theorem 1.5. *The diagonal of a square is asymmetrical relative to its side (the number $\sqrt{2}$ is irrational).*

Proof. ([7, p.5], [20, p.84]) Referring to figure 1.21, we use the square with $\alpha = 1$, $\beta = \sqrt{2}$, $\gamma = \beta - \alpha$, $\delta = \alpha - 2 \cdot \gamma$. The key in the proof is the equality, which results from the similarity of the shapes:

$$\frac{\gamma}{\alpha} = \frac{\delta}{\gamma},$$

which implies the equation for $\frac{\gamma}{\alpha}$:

$$\boxed{\frac{\gamma}{\alpha}} = \frac{1}{\frac{\alpha}{\gamma}} = \frac{1}{\frac{2\cdot\gamma+\delta}{\gamma}} = \frac{1}{2+\frac{\delta}{\gamma}} = \frac{1}{2+\boxed{\frac{\gamma}{\alpha}}} \qquad (*)$$

Subjecting then β and α to anthyphairesis we have:

$$\frac{\beta}{\alpha} = \frac{\alpha+\gamma}{\alpha} = 1+\frac{\gamma}{\alpha} = 1+\cfrac{1}{2+\cfrac{\gamma}{\alpha}} = 1+\cfrac{1}{2+\cfrac{1}{2+\cfrac{\gamma}{\alpha}}} = 1+\cfrac{1}{2+\cfrac{1}{2+\cfrac{1}{2+\cfrac{\gamma}{\alpha}}}}$$

$$= 1+\cfrac{1}{2+\cfrac{1}{2+\cfrac{1}{2+\cfrac{1}{2+\cfrac{\gamma}{\alpha}}}}} = \ldots$$

It is obvious that the process does not terminate, but continues to infinity in exactly the same way. It follows that the ratio $\frac{\beta}{\alpha} = \sqrt{2}$ is not rational, for, if it were, the process of anthyphairesis would terminate after a finite number of steps.

Exercise 1.35. Show, that the number $\sqrt{2}$ is irrational, using reduction to a contradiction.

Hint: This is the method we find in Euclid's Elements ([13, III, p.2]). We suppose, that $\sqrt{2}$ can be written as a quotient of *relatively prime* integers (that is integers with no common divisors except ± 1): $\sqrt{2} = \frac{\alpha}{\beta} \Leftrightarrow 2\beta^2 = \alpha^2$. In the last equation 2 divides the left side, therefore it will divide the right side as well. This means that α will be even, $\alpha = 2\gamma$, therefore the equation becomes $2\beta^2 = 4\gamma^2 \Leftrightarrow \beta^2 = 2\gamma^2$. Applying again the preceding argument, we see that β will also be even, which is contradictory, because then α and β would have 2 as a common divisor, contrary to the hypothesis.

Remark 1.8. In some cases of other roots of integers a similar relation to (∗) can result easily algebraically, through calculations, without utilizing a figure. For example, in the anthyphairesis of $\sqrt{3}$ to 1 the following easily proved relation can be used

$$\frac{1}{\sqrt{3}-1} = 1 + \cfrac{1}{2 + \cfrac{1}{\cfrac{1}{\sqrt{3}-1}}}. \tag{**}$$

According to the corresponding anthyphairesis, we'll therefore obtain

$$\sqrt{3} = 1 + \sqrt{3} - 1 = 1 + \cfrac{1}{\cfrac{1}{\sqrt{3}-1}} \stackrel{(**)}{=} 1 + \cfrac{1}{1 + \cfrac{1}{2 + \cfrac{1}{\cfrac{1}{\sqrt{3}-1}}}}$$

$$\stackrel{(**)}{=} 1 + \cfrac{1}{1 + \cfrac{1}{2 + \cfrac{1}{1 + \cfrac{1}{2 + \cfrac{1}{\cfrac{1}{\sqrt{3}-1}}}}}} = \ldots$$

It is obvious that again the process does not terminate and consequently the number $\sqrt{3}$ is irrational.

Exercise 1.36. What is the relation between the continued fractions which represent the inverse rationals $\frac{\alpha}{\beta}$ and $\frac{\beta}{\alpha}$?

Hint: See example $\frac{56}{33}$ above, with the corresponding anthyphairesis for the inverse fraction $\frac{33}{56}$:

$$33 = 0 \cdot 56 + 33$$
$$56 = 1 \cdot 33 + 23$$
$$33 = 1 \cdot 23 + 10$$
$$23 = 2 \cdot 10 + 3$$
$$10 = 3 \cdot 3 + 1$$
$$3 = 3 \cdot 1 + 0.$$

Exercise 1.37. Show that the continued fraction with infinitely many terms $[1; n, n, \ldots]$, where n is a positive integer, represents the positive number x which satisfies the equation

$$x^2 + (n-2)x - n = 0.$$

Conclude that for $n = 1$ the continued fraction $[1; 1, 1, \ldots]$ represents the ratio of the golden section $\phi = \frac{\sqrt{5}+1}{2}$ and for $n = 2$ the fraction $[1; 2, 2, \ldots]$ represents the number $\sqrt{2}$.

Hint: Use the fact that $x = 1 + \frac{1}{n-1+x}$.

1.9 Comments and exercises for the chapter

> Now, this means that, like all bodies, they have depth, and anything with depth is necessarily surrounded by surfaces, and any rectilinear surface consists of triangles.
>
> *Plato, Timaeus 53c*

Exercise 1.38. Calculate the total area of circles $\kappa_1, \ldots, \kappa_n$ inscribed and successively tangent to the arbelos, as a function of the arbelos radii r_1, r_2, r_3 (Theorem I-5.20).

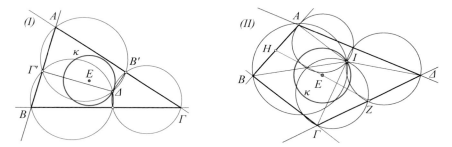

Fig. 1.22: Circles with constant sum of areas

Exercise 1.39. Point Δ varies on a circle κ with center the centroid E of triangle $AB\Gamma$ (See Figure 1.22-I). Show that the sum of the areas of the three circles with diameters respectively ΔA, ΔB, $\Delta \Gamma$ is fixed.

Exercise 1.40. Point I varies on a circle κ with center the centroid E of the quadrilateral $AB\Gamma\Delta$, which is the middle of segment HZ, joining the middles of two opposite sides (See Figure 1.22-II). Show that the sum of areas of the four circles with diameters respectively $\{IA, IB, I\Gamma, I\Delta\}$ is fixed.

Exercise 1.41. For a given convex quadrilateral $AB\Gamma\Delta$, consider the circumscribed circles (EAB), $(EB\Gamma)$, $(E\Gamma\Delta)$, $(E\Delta A)$, of the triangles formed from its sides and the intersection point E of its diagonals (See Figure 1.23-I). Show that the centers of these circles are vertices of a parallelogram, whose area ε' to the area ε of the quadrilateral has ratio

$$\frac{\varepsilon}{\varepsilon'} \leq 2,$$

where the equality holds, if and only if $AB\Gamma\Delta$ is orthodiagonal, i.e. when its diagonals intersect orthogonally.

Exercise 1.42. Given the convex quadrilateral $AB\Gamma\Delta$, show that the sum of areas $\varepsilon' = (EAB) + (EB\Gamma) + (E\Gamma\Delta) + (E\Delta A)$, of the circumscribed circles of the triangles formed from the intersection point of the diagonals E and its sides (See Figure 1.23-I) has ratio to the area ε of the quadrilateral

$$\frac{\varepsilon'}{\varepsilon} \geq \pi$$

and the equality holds, if and only if the quadrilateral is a square.

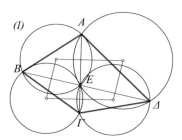
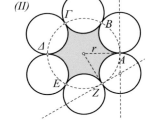

Fig. 1.23: Area comparison   Hexagon with circular arcs

Exercise 1.43. Calculate the area and the perimeter of the curvilinear polygon $AB\Gamma\Delta EZ$, which results from a regular hexagon inscribed in a circle of radius r (See Figure 1.23-II).

Exercise 1.44. Calculate the area and the perimeter of the curvilinear polygon $AB\Gamma...$, which results from a regular n-gon inscribed in circle of radius r (generalization of preceding exercise).

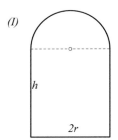
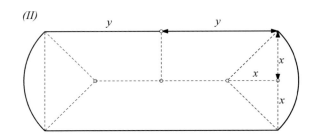

Fig. 1.24: Area maximization

Exercise 1.45. Find the relation between $\{r, h\}$ for which the area of the window in figure (See Figure 1.24-I) becomes maximum, when the length of the perimeter is fixed ([18, p.31]).

Hint: The area E and the perimeter p are given respectively by the formulas

$$E = \frac{r}{2}(4h + \pi r), \qquad p = 2r + 2h + \pi r.$$

Apply the method of exercise I-3.98 by putting $\{x = \frac{r}{2}, y = 4h + \pi r\}$ and maximizing

$$x \cdot y \quad \text{under condition} \quad ax + by = 2r + 2h + \pi r = p \text{ (fixed)}.$$

It follows that $\{a = 4 + \pi, b = \frac{1}{2}\}$ and the maximum is gotten when $ax = by \Leftrightarrow r = h$.

Exercise 1.46. Find the relation (ratio) of $\{x, y\}$ for which the area of figure (See Figure 1.24-II) becomes maximum, when the length of the perimeter is fixed. The circular arcs on the sides are $90°$.

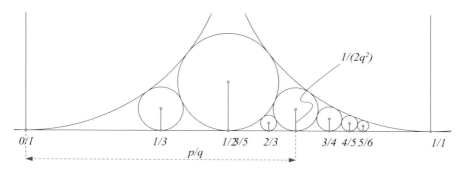

Fig. 1.25: Ford circles

Exercise 1.47. The **Ford circles** (1886-1967) are tangent to a line ε on the same side. They are constructed by selecting a system of coordinates on ε and considering the points with coordinates the irreducible fractions $\frac{p}{q}$ of the interval $[0, 1]$. To every such point a circle $\kappa_{p,q}$ is tangent (we often say "the circle $\frac{p}{q}$") of radius $\frac{1}{2q^2}$ (See Figure 1.25). Show that

1. Two Ford circles are tangent, if and only if $d^2 = 4r \cdot r'$, where d is the distance of the points of contact and $\{r, r'\}$ are the radii of the circles.
2. The preceding equality for the circles $\{\kappa_{p,q}, \kappa_{p',q'}\}$ is equivalent to $p \cdot q' - q \cdot p' = -1$.
3. Two Ford circles are either disjoint or they are tangent.
4. If the circles which correspond to the fractions $\{\frac{p}{q}, \frac{p'}{q'}\}$ are tangent, then the circle which corresponds to $\frac{p''}{q''}$, is tangent to two others, if and only if $\frac{p''}{q''} = \frac{p+p'}{q+q'}$.

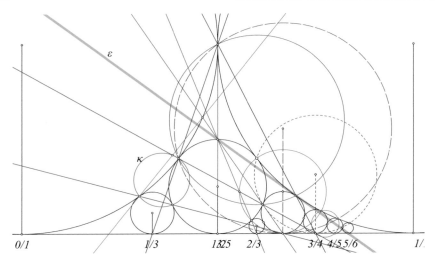

Fig. 1.26: Collinear, concyclic points on Ford circles

Exercise 1.48. How many and which relations exactly may one notice in figure 1.26? For example, that the circle κ is orthogonal to the circles which correspond to the fractions $\{\frac{0}{1}, \frac{1}{3}, \frac{1}{2}\}$. That the common tangent ε of the circles $\{\frac{2}{3}, \frac{1}{1}\}$ passes through the contact point $\frac{5}{6}$ and its diametrically opposite $\frac{1}{2}$ relative to the circle $\frac{1}{2}$.

The sequence of coordinates of the points of contact of the Ford circles with the line is called **sequence of Haros - Farey**. Its terms are ordered in a natural way in the so called *Haros-Farey series of order N*, which are subsets $\{F_1, F_2, ...\}$ of the sequence and are constructed, each one from the preceding through the operation, which in two fractions assigns the so called **middle** term (which is different from the geometric middle of the interval $[\frac{p}{q}, \frac{p'}{q'}]$).

$$\frac{p}{q} \oplus \frac{p'}{q'} = \frac{p+p'}{q+q'}.$$

The sets F_N consist of the irreducible fractions $0 \leq \frac{p}{q} \leq 1$ with $q \leq N$.

$$F_1 = \left\{\frac{0}{1}, \frac{1}{1}\right\},$$

$$F_2 = \left\{\frac{0}{1}, \frac{1}{2}, \frac{1}{1}\right\},$$

$$F_3 = \left\{\frac{0}{1}, \frac{1}{3}, \frac{1}{2}, \frac{2}{3}, \frac{1}{1}\right\},$$

1.9. COMMENTS AND EXERCISES FOR THE CHAPTER

$$F_4 = \left\{ \frac{0}{1}, \frac{1}{4}, \frac{1}{3}, \frac{1}{2}, \frac{2}{3}, \frac{3}{4}, \frac{1}{1} \right\}, \ldots$$

Next exercise ([17, p.114]), which presents the basic properties of the Haros - Farey sequence, is for those who are attracted to the theory of numbers.

Exercise 1.49. 1. For two elements $\frac{p}{q}, \frac{p'}{q'}$ of F_N, $\frac{p}{q} \oplus \frac{p'}{q'} = \frac{p+p'}{q+q'}$ is irreducible.
2. If $\left\{\frac{p}{q}, \frac{p'}{q'}, \frac{p''}{q''}\right\}$, are consecutive terms of F_N, then $\frac{p'}{q'} = \frac{p}{q} \oplus \frac{p''}{q''}$.
3. Each F_N contains the preceding F_{N-1}, and results from F_{N-1} by taking some of its middles.
4. Each term of the Haros - Farey sequence is contained in some F_N.
5. For every N the number of elements $|F_N|$ of F_N satisfies the equation $|F_N| = |F_{N-1}| + \phi(N)$, where $\phi(N)$ represents the number of integers $1 \leq x \leq N$, which share no common factor with N ([17, p.133]).

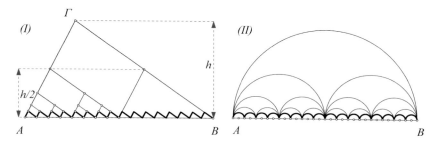

Fig. 1.27: Broken line of constant length   Curves of constant length

Exercise 1.50. Show that for every line segment AB and every $\varepsilon > 0$, there is a broken line from A to B, the length of whose is as big as we please and whose all of its points X satisfy $|X - X'| < \varepsilon$, where X' the projection of X on AB.

Hint: Start with a triangle $AB\Gamma$ and construct two similars to it on the two halves of its base AB (See Figure 1.27-I). Repeat this procedure with the new triangles etc. At each stage, joining the sides of the resulting triangles, except the bases, results to a broken line of length $|A\Gamma| + |\Gamma B|$. After ν steps the distance of the points X of this broken line from AB is less than $h/2^{\nu-1}$, where h is the altitude of the initial triangle.

Exercise 1.51. Show that for every line segment AB and every $\varepsilon > 0$, there is a curve with endpoints $\{A, B\}$, consisting of successive semicircles, having constant length $\pi|AB|$ and such that its points X satisfy $|X - X'| < \varepsilon$, where X' the projection of X on AB.

Hint: The construction is similar to the one of the preceding exercise (See Figure 1.27-II). See also the exercise 1.12.

Exercise 1.52. Let point Δ be outside the side-lines of triangle $AB\Gamma$ and $\Delta_1\Delta_2\Delta_2$ be the pedal triangle of Δ, whose vertices are the projections of Δ on the sides of the triangle (See Figure 1.28-I). Compute the area of the pedal triangle $\Delta_1\Delta_2\Delta_3$, the area of its circumcircle κ and show that the orthogonals to the sides, at their second intersection point with κ, pass through a point E, which is the isogonal conjugate of Δ.

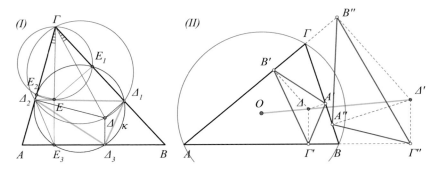

Fig. 1.28: Pedal triangle, isogonal points, Pedal of inverse point

Exercise 1.53. Let Δ be a point not lying on the side-lines of the triangle $AB\Gamma$. Show that the pedal triangle $t = A'B'\Gamma'$ of Δ and the pedal triangle $t' = A''B''\Gamma''$ of the inverse Δ' of Δ relative to the circumcircle of $AB\Gamma$ are similar triangles (See Figure 1.28-II). Compute the ratio of the areas of the triangles $\{t,t'\}$, as well as the ratio of areas of their circumcircles.

One of the best known sequences of integer numbers is the **Fibonacci sequence** (1175-1250)

$$\{1, 1, 2, 3, 5, 8, 13, 21, 34, 55, 89, 144, 233, 377, 610, 987, 1597, 2584...\}.$$

Its terms $\{a_i\}$ are determined by the relations:

$$a_1 = a_2 = 1,$$
$$a_n = a_{n-1} + a_{n-2}, \quad \text{for } n > 2.$$

This sequence appears in various problems, practical and theoretical ([24], [25]). We encounter it also in the study of the golden section $\phi = \frac{\sqrt{5}+1}{2}$, when we try to express the powers of ϕ in terms of ϕ using the basic equation of its definition: $\phi^2 = \phi + 1$ (I-§ 4.2):

1.9. COMMENTS AND EXERCISES FOR THE CHAPTER

$$\phi^2 = \phi + 1$$
$$\phi^3 = \phi^2 + \phi = 2 \cdot \phi + 1$$
$$\phi^4 = 2\phi^2 + \phi = 3 \cdot \phi + 2$$

$$\phi^5 = 3\phi^2 + 2\phi = 5 \cdot \phi + 3$$
$$\phi^6 = 5\phi^2 + 3\phi = 8 \cdot \phi + 5$$
$$\phi^7 = 8\phi^2 + 5\phi = 13 \cdot \phi + 8$$
$$\ldots = \ldots$$
$$\phi^n = a_n \cdot \phi + a_{n-1}.$$

A similar relation holds true for the inverse of ϕ, $x = \frac{1}{\phi} = \frac{\sqrt{5}-1}{2}$, which satisfies the equation $x^2 + x - 1 = 0$, from which it results analogously that

$$(-x)^n = a_n \cdot (-x) + a_{n-1}.$$

Combining the two relations, we obtain the Binet formula (1786-1856):

$$\phi^n - (-x)^n = a_n \cdot (\phi + x) = a_n \cdot \sqrt{5} \quad \Rightarrow$$
$$a_n = \frac{1}{\sqrt{5}}(\phi^n - (-x)^n) = \frac{1}{\sqrt{5}}\left(\left(\frac{1+\sqrt{5}}{2}\right)^n - \left(\frac{1-\sqrt{5}}{2}\right)^n\right).$$

The figure 1.29 below shows a sequence of rectangles, which start with

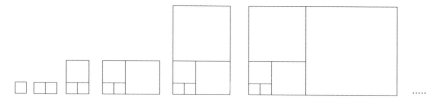

Fig. 1.29: Approximation of the golden rectangle

a square of side-length 1, and whose side-lengths are expressed with two consecutive Fibonacci numbers. Selecting the rectangles of odd order, i.e. those having their short side horizontal, we obtain the sequence of the ratios of their side-lengths:

$$b_n = \frac{a_{2n}}{a_{2n+1}}, \quad n = 1, 2, \ldots.$$

It is proved (Exercise 1.56), that this sequence is bounded and increasing, and its limit is the number $x = \frac{1}{\phi}$. Consequently, the ratios of these side-lengths approximate the ratio of the sides of the golden rectangle. The corresponding rectangles, while becomming bigger and bigger, simultaneously tend to be similar to the golden rectangle.

Exercise 1.54. Show that for every $n > 1$ holds true the formula $a_{n+1}a_{n-1} = a_n^2 + (-1)^n$.

Hint: Write $a_n = k(\phi^n - (-x)^n)$, where $k = 1/\sqrt{5}$ and observe that $\phi^m \cdot (-x)^m = (-1)^m$, for every $m \geq 1$. Then,

$$a_{n+1}a_{n-1} - a_n^2 = k^2((\phi^{n+1} - (-x)^{n+1})(\phi^{n-1} - (-x)^{n-1}) - (\phi^n - (-x)^n)^2)$$
$$= -k^2 \phi^{n-1}(-x)^{n-1}(\phi - x)^2 = (-1)^n.$$

Exercise 1.55. Show that the sequence $\{b_n\}$ is bounded and increasing, hence converges to a limit.

Hint: The sequence is bounded by 1. To see that it is increasing, observe that

$$b_{n+1} - b_n = \frac{a_{2n+2}}{a_{2n+3}} - \frac{a_{2n}}{a_{2n+1}} = \frac{a_{2n+2}a_{2n+1} - a_{2n+3}a_{2n}}{a_{2n+1}a_{2n+3}}$$
$$= \frac{a_{2n+2}a_{2n+1} - (a_{2n+2} + a_{2n+1})a_{2n}}{a_{2n+1}a_{2n+3}} = \frac{1}{a_{2n+1}a_{2n+3}} > 0.$$

Exercise 1.56. Show that the sequence $\{b_n\}$ converges to $x = 1/\phi$.

Hint: Since $x < 1 \Rightarrow x^n < 1$, we have:

$$(-x)^n = a_n(-x) + a_{n-1} \Leftrightarrow \frac{(-x)^n}{a_n} = \frac{a_{n-1}}{a_n} - x \Rightarrow \left|\frac{a_{n-1}}{a_n} - x\right| < \frac{1}{a_n}.$$

Exercise 1.57. Show that the numbers x which satisfy an equation of the form $x^2 - nx - 1 = 0$, where n a positive integer, can be expanded into an infinite continued fraction of the form $x = [n; n, n, n, \ldots]$.

Hint: $x = n + \frac{1}{x}, \ldots$

Exercise 1.58. Show that the numbers of the form $x = \sqrt{n^2 + 1}$, where n is a positive integer, can be expanded in continued fraction of the form $x = [n; 2n, 2n, 2n, \ldots]$.

Hint: $\sqrt{n^2 + 1} - n = \frac{(\sqrt{n^2+1}-n)(\sqrt{n^2+1}+n)}{\sqrt{n^2+1}+n} = \frac{1}{\sqrt{n^2+1}+n}$. This implies

$$x = n + \frac{1}{x+n} \quad \Rightarrow \quad x = n + \frac{1}{2n + \frac{1}{x+n}} \quad \text{etc.}$$

1.9. COMMENTS AND EXERCISES FOR THE CHAPTER

Exercise 1.59. Determine the irrational number x, whose expansion into continued fraction is the periodic one $x = [2; 5, 1, 3, 5, 1, 3, 5, 1, 3 \ldots] = [2; \overline{5, 1, 3}]$.

Hint: ([20, p.46]) Consider the periodic part of the fraction

$$y = 5 + \cfrac{1}{1 + \cfrac{1}{3 + \cfrac{1}{y}}} = 5 + \cfrac{1}{1 + \frac{y}{3y+1}} = 5 + \cfrac{3y+1}{4y+1} \Leftrightarrow 2y^2 - 11y - 3 = 0,$$

and use the relation $x = 2 + \frac{1}{y}$.

Exercise 1.60. Show that the positive number x, which satisfies $x = \frac{3x+2}{13x+9}$, has the expansion in continued fraction $x = [0; \overline{4, 2, 1}]$. Construct and study examples of expansions of numbers y, which satisfy an equation of the form $y = \frac{ay+b}{cy+d}$, where $\{a,b,c,d\}$ positive integers satisfying $ad - bc = 1$.

Exercise 1.61. Given the positive integers $\{a, b\}$. Determine the form of numbers x whose expansion in continued fraction has the form $x = [0; \overline{a, b}]$.

Knowing the decimal representation of an irrational number up to to a certain precision, we can find its expansion in a continued fraction up to a certain order. The number π for example, has an infinite and, not a kind of regularity showing, expansion in a continued fraction ([23, I, p.18]) whose first 40 terms are $[3; 7, 15, 1, 292, 1, 1, 1, 2, 1, 3, 1, 14, 2, 1, 1, 2, 2, 2, 2, 1, 84, 2, 1, 1, 15, 3, 13, 1, 4, 2, 6, 6, 99, 1, 2, 2, 6, 3, 5]$. Figure (See Figure 1.30) shows a geometric con-

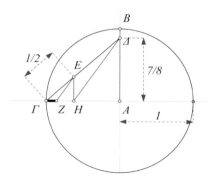

Fig. 1.30: Approximation of π

struction of the approximation

$$\pi \approx [3; 7, 15, 1] = \frac{355}{113} = 3 + \frac{4^2}{7^2 + 8^2} = 3.141592\ldots$$

due to Jacob de Gelder (1764-1848).

Exercise 1.62. Prove the construction of the segment $|\Gamma Z| = 4^2/(7^2+8^2)$ suggested by the figure, in which $|AB| = 1$, $|A\Delta| = 7/8$, $|\Gamma E| = 1/2$, E is projected on $A\Gamma$ to H and EZ is parallel to BH.

References

1. D. Bailey, 1988. The computation of π to 29,360,000 Decimal Digits Using Borweins' Quartically Convergent Algorithm, *Mathematics of Computation*, 50:283-296
2. P. Barker, R. Howe (2007) Continuous Symmetry, From Euclid to Klein. American Mathematical Society
3. M. Berg (2011) Computational Geometry. Springer, Heidelberg
4. W. Blaschke (1936) Kreis und Kugel. Walter de Gruyter, Berlin
5. O. Bottema, R. Djordjevic, R. Janic, D. Mitrinovic, P. Vasic (1969) Geometric Inequalities. Wolters-Noordhoff Publishers, Groningen
6. Dusan Djukic and all (2006) The IMO Compedium, International Mathematical Olympiads: 1959-2004. Springer, Heidelberg
7. S. Driankos (2011) Irrational Numbers and Continued Fractions. Diploma Dissertation, University of Athens
8. P. Eymard, J. Lafon (1999) Autour du nombre π. Hermann Editeurs, Paris
9. D. Fowler, 1979. Ratio in Early Greek Mathematics, *Bulletin of the American Mathematical Society*, 1:807-846
10. J. Fraleigh (2002) A first course in abstract algebra,7th edition. Pearson , London
11. O. Fuchs (2006) Mathematical Omnibus, Thirty lectures on Classic Mathematics. American mathematical society
12. S. Greitzer (1988) Arbelos, Special Geometry issue. Mathematical Association of America
13. T. Heath (1908) The thirteen books of Euclid's elements vol. I, II, III. Cambridge University Press, Cambridge
14. O. Karpenkov (2013) Geometry of Continued Fractions. Springer, Heidelberg
15. A. Khinchin (1961) Continued Fractions. The University of Chicago press, Chicago
16. W. Knorr (1993) The ancient tradition of geometric problems. Dover, New York
17. R. Knuth, D. Knuth, O. Patashnik (1995) Concrete Mathematics. Addison-Wesley, New York
18. A. Lohwater (1982) Introduction to inequalities. Cambridge university press, Cambridge
19. R. Manfrino, J. Ortega, R. Delgado (2009) Inequalities, A Mathematical Olympiad Approach. Birkhaeuser, Berlin
20. Ch. Moore (1964) An Introduction to Continued Fractions. The National Council of Teachers of Mathematics, Washington
21. S. Negrepontis, 2006. Plato's theory of ideas is the philosophic equivalent of the theory of continued fraction expansions of lines commensurable in power only, *Manuscript*, 1:1-70
22. A. Ostermann, G. Wanner (2012) Geometry by its history. Springer, Berlin
23. O. Perron (1957) Die Lehre der Kettenbrüchen, Band I, II. B.G Teubner, Stuttgart
24. A. Posamentier, I. Lehmann (2007) The (Fabulous) FIBONACCI Numbers. Prometheus Books, New York
25. H. Walser (2001) The Golden Section. The Mathematical Association of America

Chapter 2
Transformations of the plane

2.1 Transformations, isometries

> To properly know the truth is to be in the truth; it is to have the truth for one's life. This always costs a struggle. Any other kind of knowledge is a falsification. In short, the truth, if it is really there, is a being, a life.
>
> Kierkegaard, *Truth is the way*

Transformation of the plane is called a process f, which assigns to every point X of the plane, with a possible exception of some special points, another point Y of the plane which we denote by $f(X)$. We write $Y = f(X)$ and we call X a **preimage** of the transformation and Y the **image** of X through the transformation. We often say that the transformation f **maps** X to Y. For the process f we accept that it satisfies the requirement

$$X \neq X' \implies f(X) \neq f(X').$$

In other words, different points also have different images. Equivalently, this means that, if for two points X, X' holds $f(X) = f(X')$, then it will also hold $X = X'$. The set of points X, on which the transformation f is defined, is called **domain** of the transformation f, while the set consisting of all Y, such that $Y = f(X)$, when X varies in the domain of f, is called **range** of f. For transformation examples the reader may consult the beginnings of the subsequent sections. Here we limit ourselves in a description of common characteristics of these concepts.

For every shape of the plane Σ the set of images $f(X)$, where X runs through Σ, is called **image** of Σ and is denoted as $f(\Sigma)$.

Applying the processes one after the other, we create the concept of **composition of transformations**. For two given transformations f and g, we call **composition** of f and g, the transformation whose process results by the successive application of the processes of f and g. The composition of trans-

formations is denoted by
$$g \circ f.$$

By definition, the process of composition $g \circ f$ first corresponds $Y = f(X)$ to X and then $Z = g(Y) = g(f(X))$ to Y. Totally then, it corresponds $Z = g(f(X))$ to X. There are some details, which we must be careful with in compositions. These have to do with the domains and ranges of the transformations, which participate in the composition. For all of it to be meaningful, the range of the first transformation (f) must be contained in the domain of the second (g). Things are considerably simplified for transformations which have domain and range the entire plane.

Since the composition $g \circ f$ is a new transformation, we may consider its composition with a third transformation h:
$$h \circ (g \circ f),$$

and more generally we can define the composition of as many transformations $f_1, f_2, f_3, \ldots, f_k$ we want, which, for simplicity, let us consider that they are defined on the entire plane:
$$f_k \circ f_{k-1} \circ \ldots \circ f_1.$$

The meaning of such a composition of transformations is that we apply successively the processes of the transformations which participate in sequence from right to left. f_1 maps point X_1 to $X_2 = f_1(X_1)$, f_2 next maps X_2 to $X_3 = f_2(X_2)$, and so on and so forth. This process can be denoted pictorially by the diagram
$$X_1 \xmapsto{f_1} X_2 \xmapsto{f_2} X_3 \xmapsto{f_3} X_4 \ldots \xmapsto{f_k} X_{k+1}.$$

A very simple and insignificant, regarding its action, transformation is the so called **identity transformation**, which we denote with e and which, to every point X corresponds X itself. This one resembles the unit in the familiar multiplication, which leaves numbers unchanged. The same way, this transformation does nothing. It leaves every point fixed. Its structural meaning however is as important as that of the multiplication unit. With its help we can define immediately the **inverse transformation** of a transformation f which we denote with f^{-1}. This one performs exactly the opposite process to that of f and by definition holds
$$f^{-1} \circ f = e.$$

If we confine ourselves to transformations f, g, h, ..., defined for all points on the plane, then their totality together with composition, presents a noteworthy similarity with the set of positive numbers and multiplication. I list the similarities (and one difference) in two parallel columns:

Numbers
$z = x \cdot y$ (product)
$x \cdot y = y \cdot x$ (commutativity)
$x \cdot (y \cdot z) = (x \cdot y) \cdot z$ (associativity)
1 (unit)
$y = x^{-1} \Leftrightarrow y \cdot x = 1$ (inverse)

Transformations
$h = g \circ f$ (composition)
$g \circ f \neq f \circ g$ (in general)
$h \circ (g \circ f) = (h \circ g) \circ f$
e (identity transformation)
$g = f^{-1} \Leftrightarrow g \circ f = e$

We'll apply these rules in the next sections. I underline here the associative property $h \circ (g \circ f) = (h \circ g) \circ f$, which holds for transformations. This is due to their very nature as a correspondence process, which remains the same, any way we choose to group them (insert parentheses) in a particular composition of more than one transformations.

A special category of transformations, we'll deal with in the next sections, is that of **isometries** or **congruences** of the plane. With this naming we mean transformations, which are defined on the entire plane and additionally have the property of preserving distances [34, p.39], [53]). In other words, transformations $X' = f(X)$, such that, for every pair of points X, Y and their images X', Y' will hold
$$|X'Y'| = |XY|.$$
Theorem 2.1. *An isometry preserves angles.*

Proof. The short wording means that for three points X, Y, Z and their images X', Y', Z' through the transformation the angles $\widehat{YXZ}$ and $\widehat{Y'X'Z'}$ are equal. This however is a consequence of the property of the isometry of the transformation, on the basis of which $|X'Y'| = |XY|$, $|Y'Z'| = |YZ|$, $|Z'X'| = |ZX|$. In other words the triangles $X'Y'Z'$ and XYZ are congruent. From the congruence of triangles follows the equality of the angles as well.

Exercise 2.1. Show that the composition $g \circ f$ of two isometries f and g is again an isometry. Show also that the inverse transformation f^{-1} of an isometry is an isometry.

Theorem 2.2. *An isometry maps a line ε onto a line ε'. If the isometry fixes two points A and B of ε, then it also fixes all the points of ε.*

Proof. The position of a point X of the line AB is completely determined by the ratio $\frac{|XA|}{|XB|}$ and the fact that $||XA| \pm |XB|| = |AB|$. The latter gives the necessary and sufficient condition so that X is on the line. If therefore the isometry f fixes points A, B then for the images X', A', B' will hold:
$$\frac{|X'A'|}{|X'B'|} = \frac{|XA|}{|XB|}, \quad \text{and} \quad |X'A'| \pm |X'B'| = |XA| \pm |XB|.$$
Consequently if X is contained in line $\varepsilon = AB$, then also X' will be contained in line $\varepsilon' = A'B'$. The second part follows immediately from the preceding

equalities, if we take into account that $A' = A$, $B' = B$. Then from these follows that for every point X of ε point X' coincides with X.

Theorem 2.3. *An isometry, which fixes three non collinear points, coincides with the identity transformation.*

Proof. Suppose that the isometry f fixes the points A, B and Γ. Then, according to the preceding theorem, it also fixes the lines AB and $A\Gamma$. If X is a point not lying on these lines, we draw a line ε through X, which intersects AB and $A\Gamma$ respectively at Δ and E, which are fixed by f. By the preceding theorem f fixes all the points of ΔE therefore also X

Corollary 2.1. *Two isometries coincident at three points are coincident everywhere.*

Proof. Indeed, if f, g are the two isometries, then $g^{-1} \circ f$ will fix the three points, therefore it will coincide with the identity transformation $g^{-1} \circ f = e \Leftrightarrow f = g$.

Exercise 2.2. Show that an isometry f maps a circle κ onto a circle $\kappa' = f(\kappa)$ of equal radius.

Proposition 2.1. *If an isometry f satisfies the relation $f \circ f = e$, then it fixes at least one point.*

Proof. Obviously the identity transformation has the property of the psoposition. Let us suppose then that f is not coincident with the identity transformation and let us denote with X' the point $f(X)$, so that, according to the hypothesis $X'' = X$. We consider now an arbitrary point X such that $X' = f(X) \neq X$. Such a point exists, for otherwise f would be the identity. suppose M is the middle of XX'. We show that f fixes M. By hypothesis f exchanges X and X', therefore maps the line XX' to itself (Theorem 2.2). Also

$$|X'M'| = |XM| = |X'M| = |X''M'| = |XM'|.$$

The first equality holds because f is an isometry. The second because M is the middle of XX', the third because we have $|XM'| = |X'M'|$. This means that point M' lies on the medial line of XX', but also, as we noted, it is a point of the line XX', therefore it coincides with M

2.2 Reflections and point symmetries

> The thing that doesn't fit is the thing that's the most interesting, the part that doesn't go according to what you expected.
>
> R. Feynman, *The Pleasure of Finding Things Out*

A line of the plane ε defines a simple transformation called **reflection** or **mirroring** relative to the line ε, which is called **axis** or **mirror** of the reflec-

2.2. REFLECTIONS AND POINT SYMMETRIES

tion. The process for this transformation is described as follows: i) To every

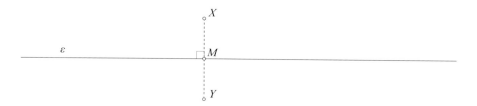

Fig. 2.1: Mirroring or reflection relative to ε

point X not contained in the line ε corresponds the point Y, such that ε is the medial line of XY. In other words, point X is projected orthogonally to ε at its point M and XM is extended to its double towards M, until Y.
ii) To every point X contained in the line ε the process corresponds the point X itself. In this case then, point X is, as we say, a **fixed point of the transformation** (See Figure 2.1).

Thus, the reflection is well defined for every point of the plane or, in other words, its domain is the entire plane. The same happens also with its range. It also coincides with the entire plane, since for every Y there exists one X such that $f(X) = Y$. The line ε, through which a reflection is defined, consists of all the fixed points of the reflection. Every point X not contained in ε is in correspondence with Y which is on the other side of ε than that where X is to be found. A reflection then interchanges the two sides of ε and leaves the points of ε fixed.

The reflection underlies the notion of *axial symmetry* which we examined in I-§ 1.16: The shape Σ is axially symmetric, if there exists a reflection f, such that $f(\Sigma) = \Sigma$. Next theorem is proved in exactly the same way as the corresponding Theorem I-1.14.

Theorem 2.4. *Every reflection is a plane isometry (See Figure 2.2).*

Proposition 2.2. *For every reflection f holds $f \circ f = e$, in other words the inverse of a reflection is the same transformation of the reflection.*

Proof. Indeed, if $Y = f(X)$, then point Y is the symmetric of X relative to the line ε, which defines the reflection. Then, however, point X is also the symmetric of Y relative to ε, consequently $X = f(Y)$ hence, for every X will hold $f(Y) = f(f(X)) = X$, which means that the composition $f \circ f$ coincides with the identity transformation e.

The important characteristic of reflections is that, as we say, they **generate** all the isometries of the plane. In other words, every isometry of the plane may be written as a composition of reflections. With a little more work (Exercise 2.22) we'll prove later the theorem ([34, p.46]):

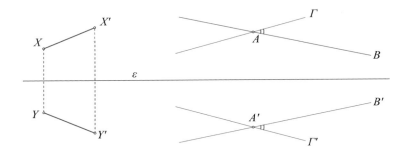

Fig. 2.2: Reflections are isometries: $|XX'| = |YY'|$ and $\widehat{BA\Gamma} = \widehat{B'A'\Gamma'}$

Theorem 2.5. *Every isometry of the plane is either a reflection or a composition of two or three reflections.*

Closely connected with reflections is also the other simple transformation we met, the point symmetry. A point O of the plane defines the **point symmetry** f relative to O (I-§ 1.16). This, to every $X \neq O$ corresponds point $X' = f(X)$ to X, which is the symmetric of X relative to O. In other words the point X' for which O is the middle of segment XX' (See Figure 2.3).

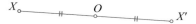

Fig. 2.3: Transformation of point symmetry relative to O

Theorem 2.6. *The composition $g \circ f$ of two reflections whose axes intersect orthogonally at O coincides with the point symmetry relative to O.*

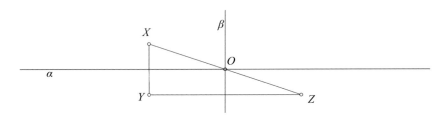

Fig. 2.4: Point symmetry as composition of two reflections

Proof. The proof, suggested by the figure 2.4, is the same as that of Theorem I-1.15. If Y is symmetric relative to line α and Z is symmetric of Y relative to

2.2. REFLECTIONS AND POINT SYMMETRIES

line β, which intersects line α orthogonally at O, then Z is also symmetric of X relative to point O.

Theorem 2.7. *If an isometry f fixes two points A and B, then it also fixes all the points of the line AB and coincides with either the identity transformation or the reflection relative to the line AB.*

Proof. The first part of the theorem follows from Theorem 2.2. For the second, suppose that $f \neq e$. It suffices to consider one point X off the line AB and see what is the point $X' = f(X)$. Triangles ABX and ABX' will be congruent

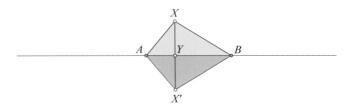

Fig. 2.5: Isometry with two fixed points

and we see easily that they either coincide or they will be symmetric relative to AB (See Figure 2.5). The first is excluded by assumption, therefore point X' will be the symmetric of X relative to AB.

Proposition 2.3. *If an isometry of the plane f, different from the identity, satisfies the relation $f \circ f = e$ and has exactly one fixed point O, then it is coincident with the point symmetry relative to O.*

Proof. The proof is contained in Proposition 2.1. We showed there that for every X with $X' = f(X) \neq X$ the middle M of the segment XX' is fixed. Then if there exists exactly one fixed point O, then all XX' will have the same middle O, which is the essense of point symmetry.

Corollary 2.2. *An isometry of the plane f, different from the identity, which satisfies the relation $f \circ f = e$ is coincident with a symmetry relative to a point O or with a reflection relative to line ε.*

Remark 2.1. This section repeats, in essence, the material related to the symmetry of shapes, which we studied in I-§ 1.16. It brings, however, in the foreground the abstract notion of transformation, with which we can describe more general shape symmetries. Indeed, for a given transformation f, the shape Σ of the plane is called **symmetric** relative to f, when $f(\Sigma) = \Sigma$. The shapes Σ which are symmetric relative to a point O are precisely those which satisfy $f(\Sigma) = \Sigma$, where f is the symmetry transformation relative to O. The shapes Σ which are symmetric relative to an axis ε are precisely those which satisfy $f(\Sigma) = \Sigma$, where f is the reflection relative to the line ε.

Remark 2.2. Transformations which coincide with their inverse ($f^{-1} = f \Leftrightarrow f \circ f = e$) are called **involutions** and play an important role in Euclidean, as well as in other Geometries ([56, I, p.102]).

Exercise 2.3. Show that, for two different lines of the plane ε and ε', there always exists a reflection f which transforms one to the other ($f(\varepsilon) = \varepsilon'$). In fact, depending on the position of the lines there exist exactly two or exactly one reflection which has this property. When exactly do these cases happen?

Exercise 2.4. Show that for two different line segments AB, $\Gamma\Delta$ of the same length, sometimes there exists a reflection which transforms one to the other $f(AB) = \Gamma\Delta$ and sometimes there doesn't. When exactly does either case happen?

Exercise 2.5. Given two different lines (equal segments, congruent circles) examine when there exists a point symmetry which transforms one to the other.

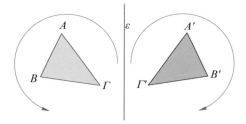

Fig. 2.6: Reflections reverse orientation

Reflections are closely connected to the reversal of *orientation* of triangles and more generally of polygons, which we noted in I-§ 1.7. Every reflection f maps a triangle $AB\Gamma$ onto a triangle $A'B'\Gamma'$, which has the opposite orientation (See Figure 2.6).

The transformations we consider in this book (are, as it is said, "continuous" and) have the property that, if they preserve the orientation of a triangle, then they preserve the orientation of every other triangle. Respectively, if they reverse the orientation of a triangle, they will reverse the orientation of every other triangle. It suffices then to examine what happens to the orientation of a single triangle, for us to conclude if the specific transformation preserves or reverses the orientation.

Isometries which reverse the orientation of triangles, we say that they *reverse the orientation of the plane* and we sometimes distinguish with the name **anti-isometries**, while those that preserve orientation, we say that they *preserve the orientation of the plane* and we call them **direct isometries**. An example of a direct isometry is the point symmetry (See Figure 2.7), which as we saw (Theorem 2.6) can be written as the composition of two reflections.

More generally than single reflections, every composition $f = f_k \circ ... \circ f_1$ of an odd number of reflections reverses the orientation of the plane, while every composition of an even number of reflections preserves the orientation.

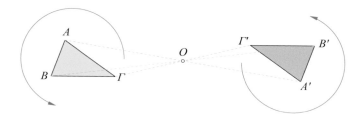

Fig. 2.7: Point symmetries of the plane preserve orientation

2.3 Translations

> Metaphor (the greek name for translation) is perhaps one of the more productive abilities of man. Its efficiency reaches the limits of magical art, and it looks like a tool of creation that God forgot in the interior of one of His creations, when He made it.
>
> *Ortega y Gasset, The antipopularity of new art*

Translation of the plane by AB is called the transformation f, which is defined by an oriented line segment AB. This transformation, to every point X

Fig. 2.8: Translation by AB : $X \mapsto Y$

of the plane, corresponds a point Y, such that the line segments XY and AB are parallel, equal and equally oriented (See Figure 2.8). Obviously the domain and range of this transformation is the entire plane. It is also obvious that parallel, equally oriented and equal line segments AB, $\Gamma\Delta$ define the same translation. From the definition follows immediately, that the inverse transformation f^{-1} is the translation by the inversely oriented line segment BA. Finally, the **null translation** is the identity transformation, considered as a translation by a segment whose endpoints coincide.

Theorem 2.8. *Every translation is an isometry.*

Proof. We must show that a translation f preserves distances. If X, Y are different points and $X' = f(X)$, $Y' = f(Y)$ then $|XY| = |X'Y'|$. This follows immediately from the fact that $XX'Y'Y$ is a parallelogram.

Segments XX' and YY' are by definition parallel, equal and equally oriented to AB, which defines the translation (See Figure 2.9).

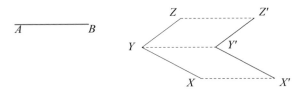

Fig. 2.9: Translations are isometries

Theorem 2.9. *The composition $g \circ f$ of two translations by the oriented segments AB and $\Gamma\Delta$ is a translation by the oriented segment EZ. The segment EZ is defined as the side of the triangle EHZ which results from an arbitrary point E and from $H = f(E)$ and $Z = g(H)$ (See Figure 2.10).*

Fig. 2.10: Composition of translations is a translation

Proof. Indeed, if X is another arbitrary point different from E and $Y = f(X)$, $\Omega = g(Y)$, then the triangles $XY\Omega$ and EHZ will have respective sides parallel equally oriented and equal: $|XY| = |EH|$, $|Y\Omega| = |HZ|$, therefore they will be congruent and will also have their third sides parallel, equally oriented and equal: $|X\Omega| = |EZ|$.

Corollary 2.3. *Consider the broken line with vertices $\{A_1, A_2, \ldots, A_k\}$. This defines $k-1$ translations $\{f_1, f_2, \ldots, f_{k-1}\}$ relative to its respective oriented sides $\{A_1A_2, A_2A_3, A_3A_4, \ldots, A_{k-1}A_k\}$. The composition of these translations is equal to the translation f which is defined by the oriented segment A_1A_k (See Figure 2.11):*

$$f_{k-1} \circ f_{k-2} \circ \ldots \circ f_2 \circ f_1 = f.$$

Proof. The proof results (by induction on k) by applying the preceding theorem and reducing gradually the number of the sides of the broken line. If for example, the broken line has four vertices $\{A_1, A_2, A_3, A_4\}$, then $f_2 \circ f_1 = g$ where g is the translation by A_1A_3 and $f_3 \circ g = h$, where h is the translation by A_1A_4. Totally then $f_3 \circ f_2 \circ f_1 = f_3 \circ g = h$.

Corollary 2.4. *The composition of translations $f = f_k \circ f_{k-1} \circ \ldots \circ f_2 \circ f_1$ parallel to the oriented sides $A_1A_2, A_2A_3, \ldots, A_{k-1}A_k, A_kA_1$ of the polygon $A_1A_2\ldots A_k$ is the identity transformation.*

2.3. TRANSLATIONS

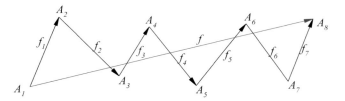

Fig. 2.11: Composition of several translations

Proof. In this corollary, which is a direct consequence of the preceding one, we consider that the identity transformation e is a translation by a line segment whose endpoints coincide (*null translation*).

Theorem 2.10. *The composition of two reflections relative to two parallel lines $\{\alpha, \beta\}$ lying at distance δ, is a translation by a line segment of length 2δ and direction orthogonal to that of the parallel axes, from α to β.*

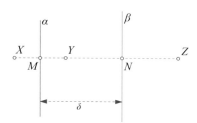

Fig. 2.12: Translation from reflections

Proof. Figure 2.12 suggests the proof. If the axes α, β of the reflections are parallel at distance δ, then for every point X and its image $Y = f(X), Z = g(Y)$ the distance will be $|XZ| = 2\delta$, since the middles M, N respectively of XY, YZ will be on α and β respectively.

Remark 2.3. In the preceding theorem the parallel axes of the two reflections can be positioned on any part of the plane. It suffices that they are orthogonal to the line segment AB, which defines the translation and their distance is equal to $\delta = \frac{|AB|}{2}$. The proof of the next proposition is not the simplest one, but demonstrates the use of composition of transformations and applies the preceding remark.

Theorem 2.11. *The composition $g \circ f$ of a point symmetry f relative to point O and a translation g by the segment AB is the symmetry relative to a point O', where O' is the translation of O by AM, where M is the middle of AB. Similarly, the composition $f \circ g$ is a symmetry relative to the point O'', which is the translation of O by MA.*

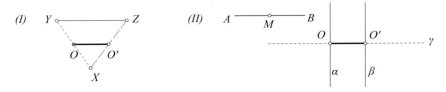

Fig. 2.13: Translation and point symmetry, simple ... and with compositions

Proof. Consider the composition order $g \circ f$ (See Figure 2.13-I) (the proof for the order $f \circ g$ is similar). According to Theorem 2.6, the symmetry f relative to a point O coincides with the composition of two reflections with axes intersecting orthogonally at O (See Figure 2.13-II). We therefore choose these axes γ and α, so that the first passes through O and is parallel to AB and the second passes through O and is orthogonal to AB. Then f is written as $f = f_\alpha \circ f_\gamma$, where f_α, f_γ are the reflections relative to the lines α and γ respectively. Also the translation is written as the composition $g = f_\beta \circ f_\alpha$, where f_β is the reflection relative to line β parallel of α and passing through point O', where OO' is parallel, equal and equally oriented to AM. Then, the composition that interests us is written:

$$g \circ f = (f_\beta \circ f_\alpha) \circ (f_\alpha \circ f_\gamma) = f_\beta \circ (f_\alpha \circ f_\alpha) \circ f_\gamma = f_\beta \circ e \circ f_\gamma = f_\beta \circ f_\gamma.$$

The equality between initial and final term gives the proof.

Remark 2.4. On the last formula we use the fact that placement of parentheses may be arbitrary. This, because the composition of transformations is, as we say, *associative* (§ 2.1). In other words for three transformations always holds

$$h \circ (g \circ f) = (h \circ g) \circ f.$$

This is a direct consequence of the definition of transformation, as a process of correspondence of points. The question of how these processes are grouped, that is, where the parentheses will be, is irrelevant, as long as we don't change the order of application of these processes. The thing changes if we change the order of application of these processes. As it is seen also from the preceding theorem, in general, for two transformations the order of application plays an important role, so

$$g \circ f \neq f \circ g.$$

In some cases, however, equality holds. When it holds $g \circ f = f \circ g$, we say that the transformations **commute**. One such case, for example, occurs in the case of a symmetry f_O relative to a point O. This, according to Theorem 2.6, is written as the composition of two reflections

$$f_O = f_\beta \circ f_\alpha,$$

2.3. TRANSLATIONS

relative to lines α and β respectively, which pass through O and are orthogonal. Besides the fact, that these lines may have an arbitrary orientation, provided they are orthogonal at O, so in this case it is easy to see that additionally holds

$$f_\beta \circ f_\alpha = f_\alpha \circ f_\beta.$$

The interesting characteristic of point symmetries and translations is that they are represented as compositions of reflections. The fact that this representation may be done in many ways is one additional characteristic, useful in many applications. Next theorem gives one such application.

Theorem 2.12. *The composition $f_P \circ f_O$ of two symmetries relative to two different points O and P, is a translation by the double of OP (See Figure 2.14-I).*

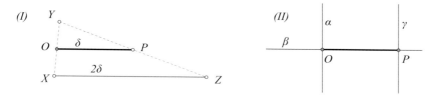

Fig. 2.14: Two point symmetries, simple ... and with compositions

Proof. The proof follows directly by writing $f_O = f_\beta \circ f_\alpha$ and $f_P = f_\gamma \circ f_\beta$, where f_α, f_γ are reflections relative to the lines orthogonal to OP and f_β is the reflection relative to the line $\beta = OP$ (See Figure 2.14-II). We have then

$$f_P \circ f_O = (f_\gamma \circ f_\beta) \circ (f_\beta \circ f_\alpha) = f_\gamma \circ (f_\beta \circ f_\beta) \circ f_\alpha = f_\gamma \circ e \circ f_\alpha = f_\gamma \circ f_\alpha.$$

The last composition, however, is exactly (Theorem 2.10) that one which defines the translation mentioned in the theorem.

Corollary 2.5. *The composition of ν symmetries relative to ν points $A_1, A_2, ..., A_\nu$ is, for even ν a translation and for odd ν a symmetry.*

Proof. Indeed, let us denote by $f_1, f_2, ..., f_\nu$ the respective point symmetries. Then we can group their compositions in pairs

$$f = f_\nu \circ ... \circ (f_4 \circ f_3) \circ (f_2 \circ f_1).$$

If ν is even, then we have exactly $\mu = \nu/2$ pairs, representing each a translation (Theorem 2.10). Then their composition will also be a translation (Corollary 2.3). If ν is odd, then in the aforementioned composition participate $\mu = \frac{\nu-1}{2}$ pairs, which represent translations, therefore their composition will also be a translation. This translation is then composed with a symmetry f_ν and gives finally a symmetry (Theorem 2.11).

Exercise 2.6. Show that the composition of a reflection f_ε relative to line ε and translation f_{AB} relative to a line segment orthogonal to ε is a reflection $f_{\varepsilon'}$ relative to line ε' parallel to ε and at distance $\frac{|AB|}{2}$ from it.

Exercise 2.7. For which pairs of lines ε, ε' does there exist a translation which maps one to the other?

Exercise 2.8. Show that for two circles of equal radius there exist both a reflection and a translation which maps one to the other.

Theorem 2.13. *Given ν different points A_1, A_2, ..., A_ν, there exists exactly one polygon which has these points as middles of successive sides if ν is odd. If ν is even, in general, there is no such polygon. If however there exists one, then there exist infinitely many and, in fact, every point of the plane may be considered to be a vertex of such a polygon.*

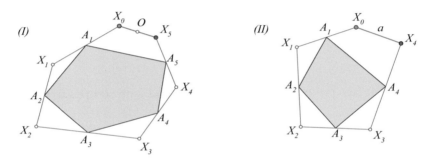

Fig. 2.15: Polygons with given middles of sides

Proof. The theorem generalizes Exercise I-2.74 and Exercise I-2.75. The proof results by tracing the orbit of an arbitrary point X_0, to which we apply successively the symmetry transformations $\{f_1,...,f_\nu\}$ relative to the vertices $\{A_1,...,A_\nu\}$:

$$X_1 = f_1(X_0),\ X_2 = f_2(X_1),...,\ X_\nu = f_\nu(X_{\nu-1}).$$

According to Corollary 2.5, if the polygon has an odd number of vertices, then the composition $f = f_\nu \circ ... \circ f_1$ of these symmetries will be a new symmetry relative to some point O of the plane (See Figure 2.15-I). Consequently the last point $X_\nu = f(X_0)$ will always be the symmetric of X_0 relative to O and we'll have coincidence $X_0 = X_\nu$ and therefore a closed polygon with the requested properties, exactly then, when X_0 coincides with O. This shows the first part of the theorem.

The second part is proved by a similar argument. In this case the aforementioned theorem guarantees that f is a translation by a fixed line segment a (See Figure 2.15-II). Consequently, no matter which X_0 we use to start, the final X_ν will always be an image of X_0 relative to the translation by a. If,

2.3. TRANSLATIONS

therefore, there exists one closed polygon ($X_0 = X_v$), then $a = X_0X_v$ will collapse to a point and the translation will coincide with the identity transformation e. Then, however, for every point X_0 the corresponding polygon will close and will satisfy the requirements of the theorem

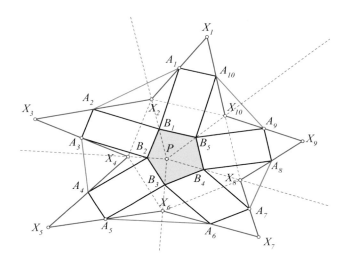

Fig. 2.16: Polygons with an even number of sides

Next proposition points out a category of polygons of an even number of sides, for which the special case of the preceding theorem applies: Every point X_0, produces a polygon with one of its vertex at X_0 and with predetermined middles of sides ([58, I, p.88]).

Proposition 2.4. *Let the polygon $a = A_1A_2...A_v$ for an even number $v = 2\mu$ be constructed from the polygon $b = B_1...B_\mu$, by attaching parallelogramms to the sides of b (See Figure 2.16). Then for every point X_1 of the plane there is a polygon $x = X_1...X_v$, having for middles the vertices of a.*

Proof. Figure 2.16 shows one of these special polygons. Polygon a is a decagon and the respective b a pentagon. a was constructed by attaching to b parallelograms. The endpoints of the opposite sides of these parallelograms define the vertices of the decagon a. Let us suppose then that we have one such polygon a and the respective b and let us consider an arbitrary point X_1 and the successively symmetric points relative to the vertices of a. We extend X_2B_1 towards B_1 by doubling it until point P. Because A_1 is also the middle of X_1X_2, X_1P will be parallel to and double of A_1B_1. Because A_2 is the middle of X_3X_2, X_3P will be parallel to and double of A_2B_1. Similarly X_5P will be parallel to and double of A_4B_2, and so on. This way we arrive at point X_{v-1} (point X_9 in the figure) and we prove that $X_{v-1}P$ is parallel to and double of $A_{v-1}B_\mu$ (A_9B_5 in the figure). This implies that points X_v, B_μ and P are collinear and point B_μ is the middle of the line segment X_vP. This, in turn, implies that the points X_v, A_v and X_1 are collinear and the polygon x has the desired properties

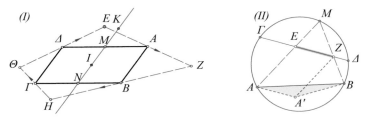

Fig. 2.17: Symmetries composition   Chord problem

Exercise 2.9. Let $AB\Gamma\Delta$ be a parallelogram and E a point. We define successively point Z symmetric of E relative to A, point H symmetric of Z relative to B, Θ symmetric of H relative to Γ (See Figure 2.17-I). Show that the symmetric Θ' relative to Δ always coincides with the initial point E. Also show that the created quadrilateral $EZH\Theta$ is a parallelogram, exactly in the case, in which the point E coincides with the symmetric K of the center I of the parallelogram, relative to the middle M of the side $A\Delta$.

Exercise 2.10. Given is a circle and two non-intersecting chords of it AB and $\Gamma\Delta$ (See Figure 2.17-II). Locate a point M on the circle, such that the chords $\{MA, MB\}$ intersect on $\Gamma\Delta$ a segment EZ of given length λ ([28]).

Exercise 2.11. Show that the inverse transformation of a translation by (the oriented) segment AB is the translation by the segment BA.

Exercise 2.12. Show that a translation preserves the orientation of the plane.

Exercise 2.13. Show that the convex quadrilateral $AB\Gamma\Delta$ with $\{M,N\}$ middles respectively of the sides $\{\Delta A, B\Gamma\}$ satisfying $|MN| = (|AB| + |\Gamma\Delta|)/2$ is a trapezium ([28] converse of exercise I-2.117).

2.4 Rotations

> No one will get very far or become a real mathematician without certain indispensable qualities. He must have hope, faith, and curiosity, and prime necessity is curiosity.
>
> L. J. Mordell, *Reflections of a Mathematician*

In order to define the rotation we need the notions of **oriented angle** and of its **signed measure**. The oriented angle $\widehat{XOY}$ is an angle in which we distinguish the order of its sides OX, OY (See Figure 2.18). If the transition from OX

2.4. ROTATIONS

to OY is opposite to the direction of the clock's hands movement, then we consider the angle as being **positively oriented**, or simply a *positive angle*. If the transition is in the same direction as the clock's, we consider the angle

Fig. 2.18: Positively (+) and negatively (-) oriented angle $\widehat{XOY}$

as being **negatively oriented**, or simply a *negative angle*. The signed measure of an oriented angle $\widehat{XOY}$, which we denote by (XOY), coincides with $\pm |\widehat{XOY}|$, where $|\widehat{XOY}|$ is its usual measure. The sign is taken to be positive for positively oriented angles and negative for negatively oriented ones.

From the definition follows immediately, that for successive oriented angles $\widehat{XOY}$ and $\widehat{YOZ}$ the following rule is valid

$$(XOZ) = (XOY) + (YOZ).$$

Rotation of the plane, relative to the center O and by the (oriented) angle ω, is called the transformation f, which is defined by the rules: i) the center O of the rotation remains fixed ($f(O) = O$), ii) to every other point X of the plane corresponds the point Y such that $|OY| = |OX|$ and the angle $\widehat{XOY}$ has signed measure ω.

Remark 2.5. As it is suggested by the figure 2.19, different rotations may produce the same result. This way for example, for the same center O, the rotation by $\theta = \frac{\pi}{2}$ and the rotation by an angle of opposite orientation $-(2\pi - \theta) = -\frac{3\pi}{2}$ produce the same result. If we denote these rotations with f and g respectively, then $f(X) = g(X)$ for every point of the plane. The same happens also for every other positively oriented angle θ. For negatively oriented angles θ the same result can be had also with the angle $\theta' = 2\pi + \theta$. Consequently, if we are interested in the result and not in the process used to reach it, we may suppose that a specific rotation f is done by an angle θ with $|\theta| \leq \pi$. The special case $|\theta| = \pi$ defines the so called **half turn**, which coincides with the symmetry relative to the center O of the rotation. This may be done either by rotating X by π or by rotating X by $-\pi$. In all other cases ($|\theta| \neq \pi$) we may suppose that the rotation takes place by the unique angle which satisfies the inequality $|\theta| < \pi$. A special case is also the identity transformation e. This may be considered as a rotation by a zero angle with center any point of the plane.

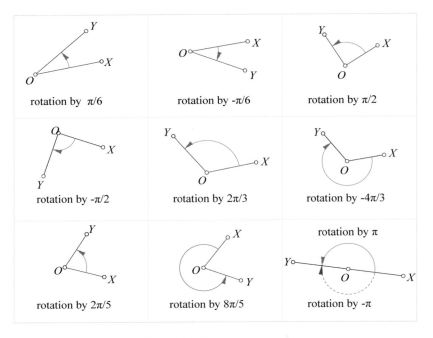

Fig. 2.19: Some rotations

Theorem 2.14. *Rotations are isometries of the plane.*

Proof. If f denotes the rotation by ω relative to the center O (See Figure 2.20), it suffices to show, that for two points X, Y and their images X', Y', holds $|XY| = |X'Y'|$. However this follows from the fact that the triangles XOY

Fig. 2.20: Rotations are isometries

and $X'OY'$ are congruent, because they have by definition of rotation $|OX| =$

2.4. ROTATIONS

$|OX'|$, $|OY| = |OY'|$ and the angle $\widehat{XOY}$ is equal to $\widehat{X'OY'}$. Indeed $(X'OY') = (XOY') - (XOX') = (XOY) + (YOY') - (XOX') = (XOY) + \omega - \omega = (XOY)$.

Exercise 2.14. Given two points X, Y, show that there exist infinitely many rotations f with the property $f(X) = Y$. Also show that the centers of these rotations lie on the medial line of the segment XY.

Proposition 2.5. *The composition of two rotations with the same center O and angles α and β is a rotation with center also O and rotation angle $\alpha + \beta$.*

Fig. 2.21: Composition of rotations with the same center

Proof. The proof follows immediately from the definitions (See Figure 2.21). If point X is rotated first by α to Y, then $(XOY) = \alpha$. If, next, point Y is rotated by β to Z, then $(YOZ) = \beta$ and because the angles are successive $(XOZ) = (XOY) + (YOZ) = \alpha + \beta$. The figure to the right underlines that the relation holds also for negatively oriented angles.

Exercise 2.15. Show that the inverse transformation of a rotation f with center O and angle ω is the rotation with the same center and angle $-\omega$.

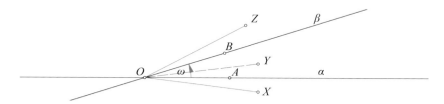

Fig. 2.22: Rotation as the composition of two reflections

Proposition 2.6. *The composition of two reflections $f = f_\beta \circ f_\alpha$, whose axes intersect at point O at an angle of signed measure ω with $|\omega| \leq \frac{\pi}{2}$, is a rotation with center point O and rotation angle 2ω (See Figure 2.22).*

Proof. The proof follows directly from the fact that, if $Y = f_\alpha(X)$, $Z = f_\beta(Y)$, then the angles $\widehat{XOY}$ and $\widehat{YOZ}$ are bisected by the axes of the reflections α and β respectively. Consequently $(XOZ) = 2\omega$.

Remark 2.6. In the last proposition the order of composition of the reflections is important. In the composition $f_\beta \circ f_\alpha$ we must rotate from α to β. In the composition $f_\alpha \circ f_\beta$ we must rotate from β to α. Also from the two angles of different measure, which are formed by the two lines we consider the one which has the smaller measure. In the last figure, where we consider the composition $f_\beta \circ f_\alpha$, the angle which rotates the line α onto β is $\omega = \widehat{AOB}$, where A, B are points on α and β respectively. Here too, we can choose the smaller in absolute value oriented angle, which rotates α onto β. Besides, the case, where the two lines intersect orthogonally, this restriction determines uniquely the signed angle which does the work.

Proposition 2.7. *The composition of two rotations $g \circ f$ with different centers O, O' and angles respectively ϕ and ψ is a rotation when $\phi + \psi \neq 2k\pi$ (k integer), with center a point P, which is determined from the given data, and rotation angle $\phi + \psi$. When $\phi + \psi = 2k\pi$, then the composition of the rotations is a translation.*

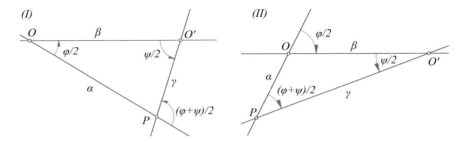

Fig. 2.23: Composition of rotations with different centers

Proof. Let us express each rotation as a composition of two reflections. The first rotation f as a composition of two reflections relative to the lines α and β (See Figure 2.23-I). We can choose these lines to have any orientation we want, provided they pass through O and form there the angle $\frac{\phi}{2}$ (Proposition 2.6). We choose them then so that β coincides with the line OO' which joins the centers of the two rotations. We consider the second rotation g as a composition of two reflections relative to two lines, the first of which coincides with β. Then the second required line for the expression of the rotation g will form with β at O' an angle equal to $\frac{\psi}{2}$. If f_α, f_β, f_γ denote the reflections relative to the corresponding lines, then we have:

$$g \circ f = (f_\gamma \circ f_\beta) \circ (f_\beta \circ f_\alpha) = f_\gamma \circ (f_\beta \circ f_\beta) \circ f_\alpha = f_\gamma \circ e \circ f_\alpha = f_\gamma \circ f_\alpha.$$

The last is a composition of two reflections, which defines a rotation, when the respective axes α and γ intersect (Proposition 2.6). The intersection condition of these lines will be exactly $\frac{\phi+\psi}{2} \neq k\pi \Leftrightarrow \phi + \psi \neq 2k\pi$. In the case

2.4. ROTATIONS

$\phi + \psi = 2k\pi$, the lines α and γ will be parallel and consequently the composition of the rotations $g \circ f = f_\gamma \circ f_\alpha$ will be a translation (Theorem 2.10). Figure 2.23-II underlines the case where one rotation is negatively oriented. The sum takes account of the signed measures of the angles. The proof also gives the procedure by which the center P of the rotation $g \circ f$ (if it exists) can be constructed.

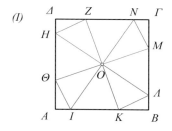
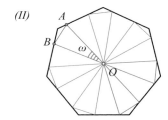

Fig. 2.24: Square construction   Regular polygon construction

Exercise 2.16. Construct a square $AB\Gamma\Delta$ whose given is its center O and two points Z, H on the sides respectively $\Gamma\Delta$ and ΔA, with $|ZO| \neq |HO|$ (See Figure 2.24-I).

Exercise 2.17. Construct a regular polygon with n sides, whose given is the center O and two points $\{A, B\}$ lying on two successive sides and such that the angle $\omega = \widehat{AOB} < \frac{2\pi}{n}$ (See Figure 2.24-II).

Exercise 2.18. Show that a rotation preserves the orientation of the plane.

Theorem 2.15. *Given two equal and non parallel line segments AB and $A'B'$ there exists exactly one rotation which maps A onto A' and B onto B'.*

Proof. Since the rotation will map A onto A' its center will be on the medial line of AA' (See Figure 2.25-I). Similarly, its center will also be on the medial line of BB', therefore the center of the rotation will coincide with the intersection point O of these lines. The medial lines cannot be parallel, because then AB and $A'B'$ would be parallel. If the medial lines coincide, then $ABB'A'$ would be a trapezium and we take as center O the intersection of AB and $A'B'$. The hypothesis excludes the case of the trapezium being a rectangle. Because of the medial lines, the triangles OAB and $OA'B'$ are congruent and triangles OAA', OBB' are similar and the rotation angle is the one of signed measure $(AOA') = (BOB')$.

Exercise 2.19. In the preceding figure show that the circles $(AA'O)$ and $(BB'O)$ intersect a second time at the intersection point T of AB and $A'B'$. Conclude that the rotation angle of the preceding exercise is equal to the angle formed by the (extended) two line segments or its supplementary and the rotation center is the other than T intersection of the circles $\{(ATA'), (BTB')\}$.

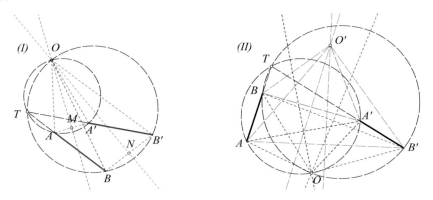

Fig. 2.25: Rotations which map AB onto $A'B'$

Corollary 2.6. *Given two equal and non parallel line segments AB and $A'B'$, there exist exactly two rotations which map the segment AB onto $A'B'$.*

Proof. The first one is the one that maps A onto A' and B onto B' (Theorem 2.15) and the other the one that maps A onto B' and B onto A' (See Figure 2.25-II).

Exercise 2.20. The composition $g \circ f$ of a rotation f and a reflection g, whose axis passes through the center of the rotation, is a reflection (See Figure 2.26).

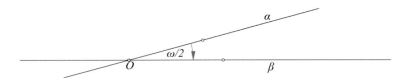

Fig. 2.26: Composition of rotation and reflection through the center

Hint: Express the rotation as a composition $f = g \circ h$ of two reflections, one of which is g.

Exercise 2.21. Show that, if an isometry of the plane fixes exactly one point O, then it coincides with a rotation with this point as the center.

Hint: If f is the isometry and $X \neq O$, then point $X' = f(X)$ will have $|X'O| = |XO|$, in other words it will lie on the circle with center O and radius $|OX|$. Then the medial line ε of XX' passes through point O. Suppose g is the reflection relative to ε. Then the composition $g \circ f$ fixes the points O and X, consequently it fixes all the points of the line OX, therefore it coincides with a reflection h with axis which passes through point O. We then have

$g \circ f = h \Rightarrow f = g \circ h$, in other words f is the composition of two reflections with axes which intersect at O.

Exercise 2.22. Using the conclusions of the three preceding exercises prove Theorem 2.5.

Hint: To show that every isometry f, different from the identity, is the composition of at most three reflections, we consider the fixed points of f. If it has exactly one, then (Exercise 2.21) it is a rotation, therefore a composition of two reflections. If it has two fixed points, then it also has a whole line consisting of fixed points and consequently coincides with a reflection (Theorem 2.7). If it fixes no point, consider an arbitrary point X and its image $X' = f(X)$. The reflection g with axis the medial line of XX' defines a composition $h = g \circ f$ which fixes point X. Therefore h will be either a reflection or a rotation and consequently $f = g^{-1} \circ h$ will be the composition of two or three reflections.

2.5 Congruency or isometry or equality

> The various modes of worship, which prevailed in the Roman world, were all considered by the people as equally true; by the philosopher, as equally false; and by the magistrate, as equally useful.
>
> E. Gibbon, *The Decline and Fall of the Roman Empire*

Isometries of the plane lie at the root of the concept of **congruency** or **isometry** or **equality** between *shapes* of the Euclidean plane, which is defined as follows:

Two shapes Σ and Σ' of the plane are **congruent** or **isometric** or **equal**, if and only if there exists an isometry f, which maps the one to the other ($f(\Sigma) = \Sigma'$).

In the next exercises, we give specific shapes Σ, Σ' and we search for an isometry f, which satisfies the above definition. Most of these exercises have been already expressed in another form in the preceding sections.

Exercise 2.23. Find an isometry, which maps a line α onto a line β (identical to Exercise 2.3).

Exercise 2.24. Find an isometry, which maps a circle α onto a circle β of equal radius (identical to Exercise 2.8).

Exercise 2.25. Given are lines α, β, which intersect and one of the formed angles between them is angle ω. If also the lines α', β' intersect under the same angle, then find an isometry f, which maps line α onto α' and line β onto β'.

Exercise 2.26. Find an isometry, which maps a line segment AB onto another line segment $\Gamma\Delta$ of the same length.

Hint: Theorem 2.15 gives the solution for the general case, it leaves however some special cases, which must be dealt with.

Exercise 2.27. Given are two congruent triangles $AB\Gamma$ and $A'B'\Gamma'$. Find an isometry which maps the one to the other.

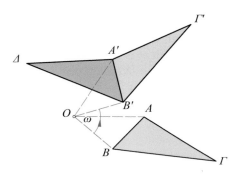

Fig. 2.27: Rotation which maps $AB\Gamma$ to $A'B'\Gamma'$

Hint: Again Theorem 2.15 applied to the line segments AB and $A'B'$, defines a rotation f (or translation) which maps the one line segment to the other and drifts the triangle $AB\Gamma$ to a congruent $A'B'\Delta$ ($f(A) = A'$, $f(B) = B'$, $f(\Gamma) = \Delta$) (See Figure 2.27). The two congruent triangles $A'B'\Delta$ and $A'B'\Gamma'$ have in common the side $A'B'$, therefore they will either be coincident ($\Delta = \Gamma'$), which is exactly the case when the given triangles are similarly oriented, or Δ will be the mirror image of Γ'. In this case by composing with the reflection g relative to $A'B'$ we get the isometry $g \circ f$, which maps $AB\Gamma$ onto (the reversely oriented) $A'B'\Gamma'$.

According to Theorem 2.5, every isometry is the composition of at most three reflections. We saw that compositions of two reflections give rotations or translations. Here we'll examine compositions of three reflections, which lead to the so called *glide reflections* (See Figure 2.28). We call **glide reflection** the composition $g \circ f_\varepsilon$ of a reflection f relative to line ε and a translation g by an oriented segment AB parallel to ε. We can easily see that in this definition the order of the composition is irrelevant, in other words it holds $g \circ f_\varepsilon = f_\varepsilon \circ g$.

Remark 2.7. When point X is on the axis ε of the glide reflection h, then the corresponding image $Z = h(X)$ is also on the axis and XZ is equal and similarly oriented to the line segment AB, which defines the translation of the glide reflection. For every point not lying on the line ε the corresponding image $Z = h(X)$ lies on the opposite side of ε. Consequently if, for a given glide

2.5. CONGRUENCY OR ISOMETRY OR EQUALITY

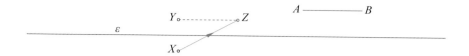

Fig. 2.28: Glide reflection $Z = h(X)$

reflection h, a line ε with $h(\varepsilon) = \varepsilon$ is found (we say: ε is invariant relative to h), then this is the axis of the glide reflection, and for every point X of it and its image $Z = h(X)$, segment XZ is the translation of the glide reflection.

Proposition 2.8. *The composition of a translation f_{AB} by the oriented line segment AB and a reflection f_ε relative to a line ε is a glide reflection (See Figure 2.29).*

Proof. The formulation leaves the order of composition of the two isometries on purpose indeterminate. The proof for both cases is the same. Let us suppose then that we have the order $f_\varepsilon \circ f_{AB}$. If AB is parallel to ε, then no

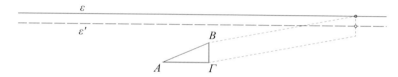

Fig. 2.29: Composition of translation-reflection $f_\varepsilon \circ f_{AB}$

proof is needed. If AB is not parallel to ε, then it can be considered as a hypotenuse of a right triangle $A\Gamma B$ with $A\Gamma$ parallel and ΓB orthogonal to ε. The translation f_{AB} is then written as the composition $f_{AB} = f_{\Gamma B} \circ f_{A\Gamma}$ and, denoting by ε' the parallel-translate of ε by $B\Gamma/2$, the original composition becomes

$$f_\varepsilon \circ f_{AB} = f_\varepsilon \circ (f_{\Gamma B} \circ f_{A\Gamma}) = (f_\varepsilon \circ f_{\Gamma B}) \circ f_{A\Gamma} = f_{\varepsilon'} \circ f_{A\Gamma}.$$

The replacement of the parenthesis relies on Exercise 2.6.

Proposition 2.9. *The composition of three reflections $f = f_{\Gamma A} \circ f_{B\Gamma} \circ f_{AB}$ relative to the sides of a triangle is a glide reflection.*

Proof. We write the composition as $f = f_{\Gamma A} \circ f_{B\Gamma} \circ f_{AB} = (f_{\Gamma A} \circ f_{B\Gamma}) \circ f_{AB} = g \circ f_{AB}$. The transformation $g = f_{\Gamma A} \circ f_{B\Gamma}$ is a rotation and we can rotate the angle $A\Gamma B$ to a position $A'\Gamma B'$ so that $B'\Gamma$ is parallel to AB, while preserving $g = f_{\Gamma A'} \circ f_{B'\Gamma}$ (See Figure 2.30-I). Then

$$f = g \circ f_{AB} = f_{\Gamma A'} \circ f_{B'\Gamma} \circ f_{AB} = f_{\Gamma A'} \circ (f_{B'\Gamma} \circ f_{AB}) = f_{\Gamma A'} \circ h.$$

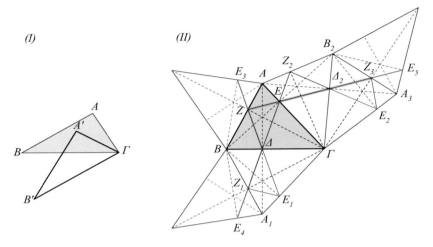

Fig. 2.30: Composition of reflections in the sides of triangle

Here $h = f_{B'\Gamma} \circ f_{AB}$ is a translation by the double of the distance of the parallel segments $B'\Gamma$, AB. The conclusion follows by applying the preceding proposition.

Proposition 2.10. *The composition of three reflections $f = f_{\Gamma A} \circ f_{B\Gamma} \circ f_{AB}$ relative to the sides of a triangle $AB\Gamma$ is a glide reflection relative to the line, which is defined by the side EZ of the orthic triangle which is opposite to the vertex A. The distance of the translation is equal to the perimeter of the orthic triangle ΔEZ.*

Proof. From the preceding proposition we know that f is a glide reflection. The proof follows from remark 2.7 in combination with the properties (Exercise I-2.175) of the orthic triangle ΔEZ of $AB\Gamma$, the basic of which is, that the altitudes of $AB\Gamma$ are bisectors of ΔEZ. This implies that the side ZE of the orthic, which is opposite to vertex A, maps via f onto an equal segment $Z_3 E_5$ on the same line (See Figure 2.30-II). The proof results by considering the points Z, E and following the trajectories of their images through the successive reflections building up f. This is better seen in figure 2.30-II, instead of using a verbal description (also see Exercise I-2.176). In this figure the various triangles result by reflection of $AB\Gamma$ and its orthic on the sides of $AB\Gamma$ and the sides of its reflected images.

Theorem 2.16. *Every isometry is either a reflection or a translation or a rotation or a glide reflection.*

Proof. Combination of Theorem 2.5, of Theorem 2.9, of Proposition 2.6 and of Proposition 2.9.

2.6. HOMOTHETIES

Exercise 2.28. Let f be a glide reflection, which maps a line ε to the line $\varepsilon' = f(\varepsilon)$. Show that the lines $\{\varepsilon, \varepsilon'\}$ are parallel, if and only if, the line ε is parallel or orthogonal to the axis of the glide reflection.

2.6 Homotheties

> The fact then emerges that the overwhelming majority of things that interest and appeal to the more refined and discriminating tastes, to every higher nature, will strike the average person as utterly "uninteresting".
>
> F. Nietzsche, Beyond Good and Evil

Given a number $\kappa \neq 0$ and point O of the plane, we call **homothety** of **center** O and **ratio** κ the transformation which corresponds: a) to point O, itself, b) to every point $X \neq O$ the point X' on the line OX, such that the following signed ratio relation holds (See Figure 2.31):

$$\frac{OX'}{OX} = \kappa.$$

A direct consequence of the definition is, that for every point O the homo-

Fig. 2.31: Homothety

thety of center O and ratio $\kappa = 1$ is the identity transformation. Often, when the ratio is $\kappa < 0$ we say that the transformation is an **antihomothety**. Its characteristic is that point O is between X and X'.

Proposition 2.11. *The composition of two homotheties with center O and ratios κ and λ is a homothety of center O and ratio $\kappa \cdot \lambda$.*

Proof. Obvious consequence of the definition. If f and g are the two homotheties with the same center O and ratios respectively κ and λ, then, for every point X, points $Y = f(X), Z = g(Y)$ and O will be four points on the same line and will satisfy,

$$\frac{OY}{OX} = \kappa, \quad \frac{OZ}{OY} = \lambda \quad \Rightarrow \quad \frac{OZ}{OX} = \frac{OZ}{OY} \cdot \frac{OY}{OX} = \lambda \cdot \kappa.$$

Corollary 2.7. *The inverse transformation of a homothety f, of center O and ratio κ, is the homothety with the same center and ratio $\frac{1}{\kappa}$.*

Remark 2.8. The homothety is a special transformation closely connected with the theorem of Thales and the similarity (I-§ 3.9). The *homothetic* triangles we studied in that section are triangles and images of triangles under homotheties. The angles are preserved but the lengths are multiplied by the ratio of the homothety. This happens because homotheties map lines to lines parallel to them.

Theorem 2.17. *A homothety maps every line ε of the plane to another line ε' parallel to ε.*

Proof. The proof is the same as that of Theorem I-3.12.

Theorem 2.18. *A homothety f preserves the angles and multiplies the distances between points with its ratio. In other words, for every pair of points X, Y and their images $X' = f(X)$, $Y' = f(Y)$ holds $|X'Y'| = \kappa|XY|$ and for every three points the respective angles are preserved $\widehat{Y'X'Z'} = \widehat{YXZ}$.*

Proof. The proof follows directly from the fact that the triangles OXY and $OX'Y'$ are homothetic.

Theorem 2.19. *The composition of two homotheties f and g with different centers O and P and ratios respectively κ and λ, with $\kappa \cdot \lambda \neq 1$, is a homothety with center T on the line OP and ratio equal to $\kappa \cdot \lambda$.*

Proof. The proof is an interesting application of the theorem of Menelaus I-§ 5.16. Let X be an arbitrary point and $Y = f(X)$, $Z = g(Y)$. This defines the triangle OYP and the points X, Z are contained in its sides OY and YP respectively. Let T be the intersection point of ZX with OP. Applying the

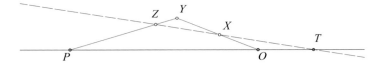

Fig. 2.32: Composition of homotheties with $\kappa\lambda \neq 1$

theorem of Menelaus we have (See Figure 2.32),

$$\frac{XO}{XY} \cdot \frac{ZY}{ZP} \cdot \frac{TP}{TO} = 1 \quad \Rightarrow \quad \frac{TP}{TO} = \frac{XY}{XO} \cdot \frac{ZP}{ZY}.$$

However, for the oriented line segments holds

$$XY = XO + OY \quad \Rightarrow \quad \frac{XY}{XO} = \frac{XO + OY}{XO} = 1 + \frac{OY}{XO} = 1 - \kappa,$$

$$ZY = ZP + PY \quad \Rightarrow \quad \frac{ZY}{ZP} = \frac{ZP + PY}{ZP} = 1 + \frac{PY}{ZP} = 1 - \frac{1}{\lambda} \quad \Rightarrow$$

$$\frac{TP}{TO} = \frac{XY}{XO} \cdot \frac{ZP}{ZY} = (1 - \kappa) \cdot \left(\frac{1}{1 - \frac{1}{\lambda}}\right) = \frac{\lambda \cdot (1 - \kappa)}{\lambda - 1}.$$

2.6. HOMOTHETIES

The last formula shows, that the position of T on the line OP is fixed and independent of X. In addition, the ratio $\mu = \frac{TZ}{TX}$ is calculated, by applying the theorem of Menelaus to the triangle OXT, this time with PY as secant:

$$\frac{PT}{PO} \cdot \frac{ZX}{ZT} \cdot \frac{YO}{YX} = 1 \quad \Rightarrow$$

$$\frac{ZX}{ZT} = \frac{YX}{YO} \cdot \frac{PO}{PT} \quad \Leftrightarrow$$

$$\frac{ZT+TX}{ZT} = \frac{YO+OX}{YO} \cdot \frac{PT+TO}{PT} \quad \Leftrightarrow$$

$$1 - \frac{1}{\mu} = \left(1 + \frac{OX}{YO}\right)\left(1 + \frac{TO}{PT}\right) \quad \Leftrightarrow$$

$$1 - \frac{1}{\mu} = \left(1 - \frac{1}{\kappa}\right)\left(1 - \frac{\lambda-1}{\lambda(1-\kappa)}\right) \quad \Leftrightarrow$$

$$\mu = \kappa \lambda.$$

Theorem 2.20. *The composition of two homotheties f and g with different centers O and P respectively and ratios κ and λ with $\kappa \cdot \lambda = 1$ is a translation by a segment parallel to OP.*

Fig. 2.33: Composition of homotheties with $\kappa \lambda = 1$

Proof. Let X be an arbitrary point and $Y = f(X)$, $Z = g(Y)$. This defines the triangle OYP (See Figure 2.33) and the points X, Z are contained in its sides OY and YP respectively. According to the hypothesis

$$\frac{YX}{YO} = \frac{YO+OX}{YO} = 1 - \frac{1}{\kappa}, \quad \frac{YZ}{YP} = \frac{YP+PZ}{YP} = 1 - \frac{PZ}{YP} = 1 - \lambda = 1 - \frac{1}{\kappa}.$$

The equality of the ratios shows, that the line segment XZ is parallel to OP. From the similarity of triangles YOP and YXZ, follows that

$$XZ = (1-\lambda)OP,$$

therefore XZ has fixed length and direction.

Theorem 2.21. *The composition $g \circ f$ of a homothety and a translation g is a homothety.*

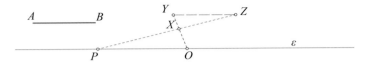

Fig. 2.34: Compositions of homotheties and a translation

Proof. Suppose that the homothety has center O and ratio κ and the translation is defined by the fixed, oriented line segment AB. Let also $X \neq O$ be arbitrary and $Y = f(X), Z = g(Y)$. Suppose finally that P is the intersection of the line XZ and the line ε is the parallel to AB from O (See Figure 2.34). From the similarity of the triangles XYZ and OXP follows that

$$\frac{OP}{AB} = \frac{OP}{YZ} = \frac{OX}{YX} = \frac{OX}{YO+OX} = \frac{1}{\frac{YO+OX}{OX}} = \frac{1}{1-\kappa} \Rightarrow OP = \frac{1}{1-\kappa}AB.$$

It follows that the position of P on ε is fixed and independent of X. Also for the ratio,

$$\frac{PZ}{PX} = \frac{OY}{OX} = \kappa.$$

Therefore the composition $g \circ f$ is a homothety of center P and ratio κ.

Exercise 2.29. Show that the composition $g \circ f$ of a translation f and a homothety g is a homothety.

Remark 2.9. The last theorems and the exercise show that homotheties and translations build a *closed*, as we say, set of transformations with respect to composition. We saw something similar also for rotations and translations.

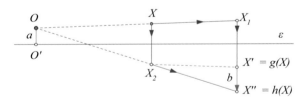

Fig. 2.35: Composition of a homothety and a reflection

Exercise 2.30. Consider a line ε and a point O at distance a from it (See Figure 2.35). Let f_1 be the homothety with center at O and ratio k and f_2 the reflection in ε. Show that the compositions $g = f_2 \circ f_1$ and $h = f_1 \circ f_2$ differ by a translation. In other words, for every point X of the plane it is valid $h(X) - g(X) = b$, where b is a line segment of length $|2a(1-k)|$ and direction orthogonal to ε.

2.7. SIMILARITIES

Exercise 2.31. Given two circles with different radii, show that there exist homotheties which map one to the other. How many are there? What are their centers and ratios?

Exercise 2.32. Given two circles κ and λ, draw a line intersecting them, which forms chords AB, $\Gamma\Delta$, having given lengths [27, p.21].

Exercise 2.33. Show that a shape Σ, with more than one points, for which there is a homothety f, different from the identity, leaving Σ invariant ($f(\Sigma) = \Sigma$), extends to infinity. To find a shape with this property.

Hint: If f leaves Σ invariant, then also the inverse homothety $g = f^{-1}$ will leave it invariant. If $\{O, k\}$ is the center and the ratio of f, then $\{O, \frac{1}{k}\}$ will be respectively the center and ratio of the inverse homothety. Thus, we can suppose $k > 1$. Then if X is an arbitrary point of Σ, the $X' = f(X)$ will satisfy $|OX'| = k|OX|$. Repeating this procedure we find $X'' = f(X')$, with $|OX''| = k^2|OX|$ and after n similar steps, we find points $X^{(n)} = f(X^{(n-1)})$, with $|OX^{(n)}| = k^n|OX|$.

A shape with the aforementioned property is a set of lines through a fixed point O.

2.7 Similarities

> We shall not cease from exploration
> And the end of all our exploring
> Will be to arrive where we started
> And know the place for the first time.
>
> T.S. Eliot, *Little Gidding*

Similarity is called a transformation f of the plane, which multiplies the distances of points with a constant $\kappa > 0$, which is called **ratio** or **scale** of the similarity. By definition then, for every pair of points X, Y a similarity corresponds points $X' = f(X)$, $Y' = f(Y)$, which satisfy

$$|X'Y'| = \kappa \cdot |XY|.$$

This general definition includes the isometries, for which $\kappa = 1$, and the homotheties. Similarities not coincident with isometries, in other words, similarities for which $\kappa \neq 1$ are called **proper** similarities. As we'll see further down (Theorem 2.24), proper similarities are divided into two categories: **direct similarities** or **rotational similarities** and **antisimilarities** or **reflective similarities** [2, p.217].

A *direct similarity* or *rotational similarity* is defined as a composition $g \circ f$ of a rotation f and a homothety g, which shares the same center with f. The rotation angle of f is called **angle of similarity**. An *antisimilarity* is defined as a composition $g \circ f$ of a reflection f and a homothety g with center on the axis of f. The axis of f is called **axis of antisimilarity**. In both categories therefore there exists a point, the center O of the homothety g which is fixed under the transformation. Obviously, proper similarities cannot have also a second fixed point T different from O. For if they had, then for the two points and their images $O' = f(O) = O$, $T' = f(T) = T$ would hold $|OT| = |O'T'|$, while a proper similarity requires $|O'T'| = \kappa|OT|$ with $\kappa \neq 1$. This unique fixed point is called **center** of the proper similarity.

The order of the transformations, which participate in the definition of a proper similarity, is irrelevant because of the following proposition.

Proposition 2.12. *The two transformations, which participate in the definition of a proper similarity, commute ($g \circ f = f \circ g$).*

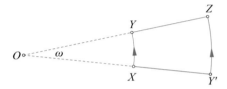

Fig. 2.36: Commutativity of rotation and concentric homothety

Proof. Let us see the proof for the direct similarities, which are compositions $g \circ f$ of rotations f and homotheties g (See Figure 2.36). The proof for antisimilarities is similar. For the proof then, it suffices to observe the orbit of an arbitrary point X under the application of the two transformations. According to $g \circ f$, we first rotate X, about the center O of the rotation, to Y and next we take the homothetic Z of Y. It holds therefore $(XOY) = \omega$ and $\frac{OZ}{OY} = \kappa$, where ω is the angle of rotation of f and κ the homothety ratio of g. According to $f \circ g$, we first take the homothetic Y' of X and next we rotate Y' by ω. It is obvious that the two processes give the same final result, which is the point Z.

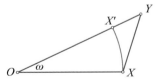

Fig. 2.37: Triangle OXY for direct similarities

Proposition 2.13. *For every direct similarity f with center O and rotation angle, which is not a multiple of π, the triangles OXY with $Y = f(X)$, which result for the different positions of X on the plane, are similar (See Figure 2.37).*

2.7. SIMILARITIES

Proof. Direct consequence of the definition, according to which $\widehat{XOY}$ is the rotation angle ω and the ratio $\frac{|OY|}{|OX|}$ is the ratio κ of the similarity.

Proposition 2.14. *For every antisimilarity f with center O and axis ε and every point X of the plane, for which points $O, X, Y = f(X)$ are not collinear, the angles $\widehat{XOY}$ have the same bisectors, which coincide with ε and its orthogonal ε' at O. Points X of ε and ε' are the only points for which O, X, Y are collinear.*

Proof. Direct consequence of the definition (See Figure 2.37-I), according to which the lines OX, OY are always symmetric relative to ε (See Figure 2.38-I).

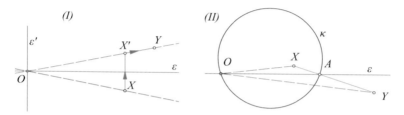

Fig. 2.38: Antisimilarity   ... and Apollonian circles

Exercise 2.34. Show that, if f is an anitsimilarity with center at O and axis ε, then for every point X different from O and its image $Y = f(X)$, holds $|k| = \frac{|OY|}{|OX|} = \frac{|AY|}{|AX|}$, where k is the similarity ratio and A is the intersection point of line XY with line ε. Conclude that the Apollonian circle κ of the segment XY for the ratio $|k|$, passes through points $\{O,A\}$ and points $\{X,Y\}$ are inverse with respect to κ (See Figure 2.38-II).

Exercise 2.35. Show that for every triple of non collinear points X, Y, Z and their images X', Y', Z' through a similarity, triangles XYZ and $X'Y'Z'$ are similar.

Exercise 2.36. Show that a direct similarity maps a triangle $AB\Gamma$ to a similar triangle $A'B'\Gamma'$, which is also similarly oriented to $AB\Gamma$. An antisimilarity reverses the orientation of the triangles.

Exercise 2.37. Show that a similarity maps a line ε to a line ε' and a circle κ to a circle κ'.

Exercise 2.38. Show that two similarities f, g, which are coincident at two different points A and B, they are coincident at every point of the line AB. Conclude then, that the composition of the transformations $g^{-1} \circ f$ is either the identity transformation or a reflection.

Exercise 2.39. Show that two similarities f, g, which are coincident at three non collinear points, they are coincident at every point of the plane.

Next theorem expresses a basic property of similarities.

Theorem 2.22. *For two line segments AB and $A'B'$ of the plane, of different length, there exists a unique direct similarity which maps A to A' and B to B', consequently mapping AB to $A'B'$.*

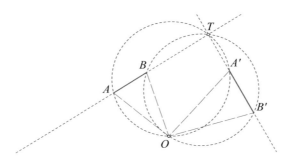

Fig. 2.39: Similarity from two line segments

Proof. Leaving the special cases for the end, let us suppose that the two segments are in general position and the lines they define intersect at a point T (See Figure 2.39). This defines two circles $(AA'T)$ and $(BB'T)$ which intersect not only at T but also at a second point O. The quadrilaterals $TBOB'$ and $TAOA'$ are inscriptible, therefore their angles at O are equal as supplementary to the angle at T. This shows that $(AOA') = (BOB')$ and defines the rotation f of the similarity. From this property follows that the angles of triangles $A'B'O$ and ABO at O are equal as are their angles at A' and A (as internal and opposite external in quadrilateral $AOA'T$). It follows that the two triangles are similar and the similarity ratio is $\kappa = \frac{|A'B'|}{|AB|}$. The requested similarity then is the composition of the rotation f and the similarity with ratio κ and center O.

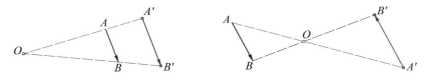

Fig. 2.40: Similarity from two parallel line segments

In the special case where T does not exist, that is when AB and $A'B'$ are parallel and not collinear (See Figure 2.40), then O is the intersection point of AA' and BB'. If AB and $A'B'$ are equally oriented, then the requested similarity is the homothety with center O and ratio $\kappa = \frac{|A'B'|}{|AB|}$. If AB and $A'B'$ are

2.7. SIMILARITIES

inversely oriented, then the requested similarity is the composition of the rotation f by π about O (which coincides with point symmetry relative to O) and the homothety with ratio $\kappa = \frac{|A'B'|}{|AB|}$ relative to O. The reasoning for collinear AB and $A'B'$ is similar, but I leave this case as an exercise.

The uniqueness of this similarity follows from the fact that the arguments can be reversed. If O is the center of a similarity, which maps AB to $A'B'$, then for the angles, $(AOA') = (BOB')$ and further the triangles AOB and $A'OB'$ will be similar. This however means that the quadrilaterals $AOA'T$ and $BOB'T$ are inscriptible in circles and O is the intersection point of the circles (ATA') and (BTB'), as in the preceding case.

Theorem 2.23. *For two line segments AB and $A'B'$ of the plane, of different length, there exists a unique antisimilarity, which maps A to A' and B to B', consequently mapping AB to $A'B'$.*

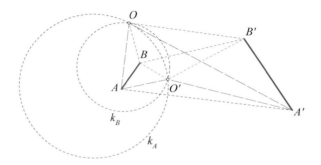

Fig. 2.41: Antisimilarity from two line segments

Proof. We seek an antisimilarity whose ratio $\kappa = \frac{|A'B'|}{|AB|}$ is known. Therefore it suffices to find its center O'. This point will be on the bisectors of the angles $\widehat{AO'A'}$ and $\widehat{BO'B'}$ (Proposition 2.14). These bisectors will intersect the respective sides AA' and BB' of the triangles AOA' and BOB' at points which divide them in ratio κ (See Figure 2.41). Therefore O' will be contained in the two Apollonian circles k_A and k_B, which are respectively the loci of the points which divide segments AA' and BB' in ratio κ. Consequently it will coincide with an intersection point of these circles. A similar property will be valid also for the center O of the direct similarity, which is guaranteed by the preceding theorem. Therefore this, too will be contained in the intersection of k_A and k_B. Consequently the two circles will intersect. From the equality of ratios

$$\frac{|OA|}{|OA'|} = \frac{|OB|}{|OB'|} = \frac{|O'A|}{|O'A'|} = \frac{|O'B|}{|O'B'|} = \frac{|AB|}{|A'B'|},$$

it follows that triangles $O'AB$ and $O'A'B'$ are similar and OAB, $OA'B'$ are also equal.

In the case where the two circles intersect at exactly two points (See Figure 2.41), it is impossible for both pairs of similar triangles to consist of similarly oriented triangles. This, because otherwise we would have two direct similarities with centers at O and O', something which is excluded by the preceding theorem. Therefore one of the two pairs will consist of reversely oriented triangles and consequently one of the two will be an antisimilarity and the other a direct similarity. If the two points O and O' coincide then

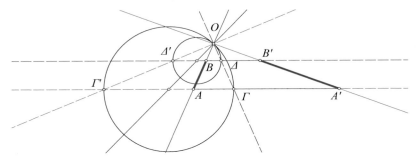

Fig. 2.42: Coincidence of centers of similarity and antisimilarity

AA' and BB' must be parallel (See Figure 2.42). Indeed, then, the bisectors of the angles AOA' and BOB' will coincide and the two circles k_A and k_B will be tangent at O. However the lines AA' and BB' contain the diametrically opposite pairs of points Γ, Γ' and Δ, Δ' respectively, which are defined by the mutually orthogonal bisectors which pass through O. Because of the circle tangency at O, the diameters $\Gamma\Gamma'$ and $\Delta\Delta'$, which are excised by the two orthogonal lines on the circles are parallel, something which proves the claim. In this case the direct similarity has rotation angle $\widehat{AOA'}$ and the antisimilarity has axis line $\Gamma\Delta$.

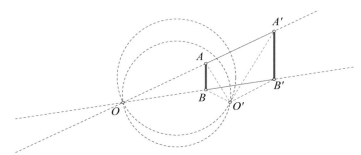

Fig. 2.43: Centers of similarity and antisimilarity when $AB \| A'B'$

In the special case, in which the lines AB and $A'B'$ are parallel, the two Apollonian circles pass through the intersection point O of AA' and BB', which is a homothety center and, consequently, the center of a direct similarity between AA' and BB'. The antisimilarity center coincides in this case also with the other intersection point O' of the two circles (See Figure 2.43).

Exercise 2.40. Complete the proof of the last two theorems, by examining the case where AB and $A'B'$ are on the same line.

Theorem 2.24. *Every proper similarity is a direct similarity, if it preserves the orientation of triangles and an antisimilarity if it reverses the orientation of triangles.*

Proof. Indeed, let X, Y be two different points and $X' = f(X)$, $Y' = f(Y)$ their images by the similarity. Suppose also that f preserves the orientation of triangles and g is the direct similarity, which maps X to X' and Y to Y' (Theorem 2.22). Then the two similarities f and g coincide on the entire line XY (Exercise 2.22). Let Z be a point not on the line XY. The triangle XYZ maps by f to the similar and similarly oriented (to XYZ) triangle $X'Y'Z'$. The same happens with g. It also maps XYZ to a similar and similarly oriented triangle $X'Y'Z''$. Triangles XYZ, $X'Y'Z'$, $X'Y'Z''$ are similar and similarly oriented, and the last two have $X'Y'$ in common. Therefore they either coincide or one is the mirror image of the other. The latter however cannot happen, because then the two triangles would have opposite orientations. Therefore the triangles coincide and consequently $Z' = Z''$, in other words f and g are coincident on three non collinear points, therefore they are coincident everywhere and $f = g$.

The case where the transformation f reverses the orientation of the triangles is proved similarly.

Corollary 2.8. *Every proper similarity has exactly one fixed point.*

Exercise 2.41. Determine the fixed point of a given proper similarity f.

Hint: Use Theorem 2.22 for direct similarities and Theorem 2.23 for antisimilarities ([34, p.74]).

Exercise 2.42. Show that, for two similar but not congruent triangles $AB\Gamma$ and $A'B'\Gamma'$, there exists one unique proper similarity which maps $AB\Gamma$ to $A'B'\Gamma'$.

Hint: Use the similarity (direct or antisimilarity) which maps AB to $A'B'$.

The next two exercises show, that in the definition of the proper similarity it is not necessary to restrict ourselves to homotheties and rotations (resp. reflections) with coincident centers (resp. with homothety center on the the axis of the reflection). Even if the centers are different (resp. the center is not on the axis of reflection), the composition of a homothety and a rotation (resp. reflection) is a proper similarity.

Exercise 2.43. Show that the composition $g \circ f$ of a homothety f with center O and a rotation g with center $P \neq O$ is a direct similarity with rotation angle that of g, ratio that of f and center which is determined by f and g. Show that the same happens also for the composition $f \circ g$.

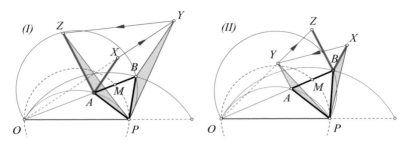

Fig. 2.44: Composition of homothety and rotation

Hint: Let κ be the ratio of the homothety f and ω be the angle of the rotation g. There exists an isosceles triangle PAB, with vertex at the center P of the rotation, whose two other vertices A, B are centers of the similarities $g \circ f$ and $f \circ g$ (respectively (I) and (II) in figure 2.44). This triangle can be constructed using two characteristic properties it has: (i) an apical angle equal to ω and (ii) $B = f(A)$.

Indeed, if such a triangle exists, then B will see the line segment OP under the angle $\frac{\pi-\omega}{2}$ and A will see OP under the angle $\frac{\pi+\omega}{2}$. Both of the latter if $\kappa > 1$. If $\kappa < 1$ the roles of A and B must be reversed. Let us then suppose that $\kappa > 1$ and that $f(A) = B$. Point B is on the intersection of the arc of the points which see OP under angle $\frac{\pi-\omega}{2}$ and of the arc which results through the homothety f from the arc of points which see OP under angle $\frac{\pi+\omega}{2}$. Consequently point B is constructible and from it the isosceles PAB with angle ω at P is also constructible. Then $g(f(A)) = g(B) = A$, therefore point A is a fixed point of $h = g \circ f$.

Let X an arbitrary point, $Y = f(X)$ and $Z = g(Y)$. The angle $(XAZ) = \omega$. Indeed, the triangles PAZ and PBY are congruent, because they have $|PA| = |PB|$ by hypothesis, $|PY| = |PZ|$, since point Z results from Y through a rotation about P and the angles $\widehat{APZ}$, $\widehat{BPY}$ are equal since both added to $\widehat{ZPB}$ give ω. Also, because of the similarity, $\kappa = \frac{|OB|}{|OA|} = \frac{|OY|}{|OX|}$, therefore AX and BY are parallel and $|BY| = |AZ|$. Therefore $\frac{|AZ|}{|AX|} = \kappa$ and the angle between the lines AZ and AX is equal to the angle between AZ and BY, which is ω. Consequently, the correspondence $Z = g(f(X))$ coincides with the composition $g' \circ f'$, where f' is the rotation about A by ω and g' is the homothety relative to A with ratio κ. We have then $g \circ f = g' \circ f'$ and the second composition satisfies the definition of the direct rotation.

The proof of the claim for the other ordering of the composition, that is $f \circ g$ (corresponding to case (II) in figure 2.44) is similar.

Exercise 2.44. Show that the composition $g \circ f$ of a homothety f with center O and a reflection g with axis ε, which does not contain O, is an antisimilarity with axis a line ε', parallel to ε and center the projection P of O on ε'. Show that the same happens also for the composition $f \circ g$.

2.7. SIMILARITIES

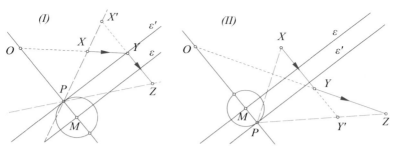

Fig. 2.45: Composition of homothety and reflection

Hint: The key role here is played by the circle with center the projection M of O on the axis of g and radius $r = \frac{\kappa-1}{\kappa+1}|OM|$. The intersection point P of this circle with OM, which is contained between points O and M is proved to be a fixed point of $g \circ f$. Its diametrically opposite is proved to be a fixed point of $f \circ g$ (cases (I) and (II) respectively in figure 2.45). The rest follows easily from the figures, in which X is an arbitrary point of the plane, $Y = f(X)$ (resp. $Y = g(X)$) and $Z = g(Y)$ (resp. $Z = f(Y)$). In the first case $g \circ f = g' \circ f'$, where f' is the homothety with center P and ratio equal to the ratio κ of f and g' is the reflection relative to the line ε', which is parallel to ε and passes through point P. In the second case $f \circ g = f'' \circ g''$, where g'' is the reflection relative to ε', which is parallel to ε and passes through point P and f'' is the homothety with center P and ratio equal to κ. The figures show the trajectory of the arbitrary point X under the application of these new transformations. In the first case point X maps by f' to X', which next, by g' maps to Z. In the second case point X maps by g'' to Y', which, by f'' maps to Z'.

Exercise 2.45. Show that the composition of two direct similarities is a direct similarity with rotation the sum of the rotations and ratio the product of the ratios if their angles sum up to $\omega + \omega' \neq 2k\pi$ and the ratios κ and κ' satisfy $\kappa \cdot \kappa' \neq 1$, otherwise it is a translation.

Hint: Combination of the two preceding exercises and of Proposition 2.7. Write the two similarities in their "normal" form, as compositions of homotheties and rotations with the same center: $g \circ f$ and $g' \circ f'$. Then their composition would be $(g' \circ f') \circ (g \circ f) = (g' \circ f') \circ (f \circ g) = g' \circ (f' \circ f) \circ g$. Apply next Proposition 2.7 to $f' \circ f$ and subsequently to the resulting rotation or translation h apply Exercise 2.43 or Theorem 2.21, etc. ([58, II, p.42]).

Exercise 2.46. State and prove an exercise similar to the preceding one for the composition of two antisimilarities and the composition of an antisimilarity and a direct similarity.

As it happens with isometries and congruence § 2.5, so it happens also with similarity transformations, which are at the root of a general definition of the similarity for plane shapes: Two shapes Σ, Σ' of the plane are called **similar**, when there exists a similarity f which maps one to the other ($f(\Sigma) = \Sigma'$).

Next exercise gives an application of similarity, which produces a relatively simple solution to a complex problem. The problem is related to the construction of triangles on the sides of a given triangle. Given a triangle $AB\Gamma$

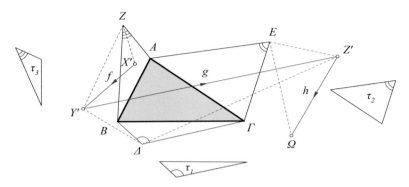

Fig. 2.46: Triangles on the sides of $AB\Gamma$

and three other triangles τ_1, τ_2, τ_3, we choose a side on each one of the three last triangles. Next we construct on the sides of triangle $AB\Gamma$ externally lying triangles $B\Delta\Gamma$, ΓEA, AZB respectively similar to τ_1, τ_2, τ_3, on $AB\Gamma$, so that the respectively similar to the selected sides of the three triangles coincide with the sides of the triangle ([36, p.141]). I call such a construction briefly: a *welding by similarity* of τ_1, τ_2, τ_3 onto $AB\Gamma$. The resulting points Δ, E, Z I call *vertices of the welding* (See Figure 2.46).

Exercise 2.47. Construct the triangle $AB\Gamma$ from the triangles τ_1, τ_2, τ_3 and the respective vertices Δ, E, Z of a welding by similarity on $AB\Gamma$.

Hint: From the given triangles and points Δ, E, Z there are defined respectively three similarities f, g, h. Similrarity f has center Z, angle $\omega_1 = (AZB)$ and ratio $\kappa_1 = \frac{|ZB|}{|ZA|}$. Similarity g has center Δ, angle $\omega_2 = (B\Delta\Gamma)$ and ratio $\kappa_2 = \frac{|\Delta\Gamma|}{|\Delta B|}$. Finally, the similarity h has center E, angle $\omega_3 = (\Gamma EA)$ and ratio $\kappa_3 = \frac{|EA|}{|E\Gamma|}$. The angles and the ratios of the similarities are determined completely from the given triangles τ_1, τ_2, τ_3. We then observe that point A satisfies

$$f(A) = B, \ g(B) = \Gamma, \ h(\Gamma) = A \ \Rightarrow \ (h \circ g \circ f)(A) = A.$$

Point A therefore coincides with the unique fixed point of the composed similarity $h \circ g \circ f$. Consequently point A is determined from the given data (even if its actual construction is somewhat involved). As soon as A is determined, the rest of the vertices of the requested triangle are constructed by applying the similarities: $B = f(A)$ and $\Gamma = g(B)$.

The preceding exercise includes many interesting special cases, which offer themselves for further study, for example, when the three triangles τ_1, τ_2,

2.7. SIMILARITIES

τ_3 coincide or have a more special form (isosceles, equilateral) or when for the corresponding ratios holds $k_1 k_2 k_3 = 1$.

Theorem 2.25. *Let $\{\Delta_1, \Delta_2\}$ be the similarity centers of the direct and indirect (antisimilarity) mapping the segment AB onto $A'B'$ (See Figure 2.47). Let also Θ denote the intersection $AB \cap A'B'$ and consider the circles $\kappa_1 = (B\Theta B')$, $\kappa_2 = (A\Theta A')$, whose second intersection defines Δ_1. The second intersection point $\Delta_2 \neq \Delta_1$ of the Apollonian circles $\{\lambda_1, \lambda_2\}$ of the segments $\{AA', BB'\}$ w.r.t. the ratio $r = AB/A'B'$ defines the center of the antisimilarity mapping AB onto $A'B'$. Let also $\Sigma = \kappa_1 \cap \lambda_1$, $T = \kappa_2 \cap \lambda_2$. The following are valid properties.*

1. *Triangles $\{A\Sigma A', BTB'\}$ are similar.*
2. *Points $\{T, \Sigma, \Theta\}$ are collinear.*
3. *Δ_2 is collinear with $\{T, \Sigma, \Theta\}$.*
4. *The angle $\widehat{\Delta_2 \Delta_1 \Theta}$ is right.*
5. *The lines $\{\Theta\Delta_1, \Theta\Delta_2\}$ are harmonic conjugate w.r.t. $\{AB, A'B'\}$.*

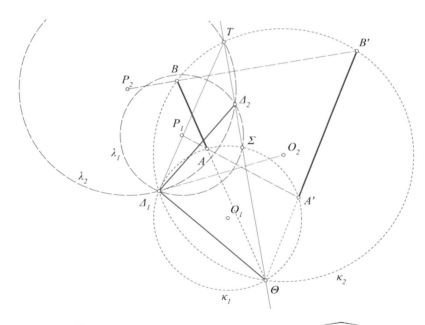

Fig. 2.47: The line ΘT and the right angle $\widehat{\Theta \Delta_1 \Delta_2}$

Proof. Nr-1 is valid because the ratios $\Sigma A/\Sigma A' = TB/TB' = r$ and the angles are equal: $\widehat{\Sigma} = \widehat{T} = \pi - \widehat{\Theta}$.

Nr-2 is valid because $\widehat{A\Theta\Sigma} = \widehat{AA'\Sigma} = \widehat{BB'T} = \widehat{B\Theta T}$.

Nr-3 is valid because the line $T\Sigma$ is characterized by the ratio of distances of its points from the segments $d(X,AB)/d(X,A'B') = r$, satisfied by $\{\Sigma, T\}$. But Δ_2 satisfies also this condition, hence belongs to that line.

Nr-4 follows by an angle chasing argument. $\widehat{\Delta_2 \Delta_1 O_2} = \widehat{\Delta_1 T \Delta_2}$ because $\Delta_1 O_2$ is tangent to λ_2 at Δ_1. This follows from the fact that κ_2 passing through $\{B, B'\}$, which are inverse relative to λ_2 is orthogonal to λ_2. Also $\widehat{O_2 \Delta_1 \Theta} = \frac{1}{2}(\pi - \widehat{\Delta_1 O_2 \Theta}) = \pi/2 - \widehat{\Delta_1 T \Theta}$.

Nr-5 is a consequence of the characterization of their points to have ratio of distances from $\{AB, A'B'\}$: $d(X, AB)/d(X, A'B') = r$.

Remark 2.10. The preceding theorem makes more precise the distinction between the centers $\{\Delta_1, \Delta_2\}$ respectively of the direct similarity and the anti-similarity mapping the segment AB onto $A'B'$: In the right angled triangle $\Theta \Delta_1 \Delta_2$ the right angle is at Δ_1.

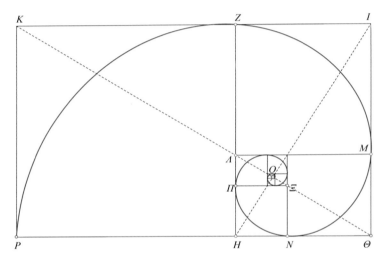

Fig. 2.48: Golden section rectangles and logarithmic spiral

Exercise 2.48. Starting from a "golden" rectangle $KP\Theta I$ and successively subtracting the squares of its small sides, we construct a sequence of other, pairwise similar rectangles (See Figure 2.48). Show that each one of them, for example $M\Theta H\Lambda$, results from its preceding $ZI\Theta H$ through a similarity f with center the intersection point O of $K\Theta$, HI, rotation angle $\frac{\pi}{2}$ and ratio $x = \frac{\sqrt{5}-1}{2}$.

Hint: Simple use of the definitions and the properties of the golden section (I-§ 4.2).

Figure 2.48, shows a curve, called a **logarithmic spiral**, which passes through a vertex of the initial rectangle P, as well as its successive positions $Z = f(P)$, $M = f(Z)$, $N = f(M)$, $\Pi = f(N)$, ..., which result by applying repeatedly the similarity f ([26, p.227]). Similar sequences of points and logarithmic spirals containing them result by starting from any point $P \neq O$ and taking the successive $f(P)$, $f^2(P)$, $f^3(P)$, ..., etc.

This curve also shows up as a pursuit trajectory in *pursuit problems*, like that with 4 bugs initially placed at the vertices of a square. The bugs start pursuing each other, moving at a constant speed. Each time their positions are at the vertices of a square, which gradually shrinks and simultaneously rotates, until they all meet at the center of the square. Figure 2.49, on the left,

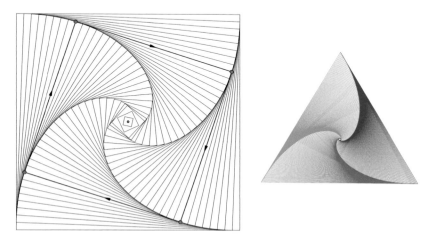

Fig. 2.49: Logarithmic spirals as pursuit curves

shows the positions of the bugs at different moments in time and the corresponding square defined by their positions. The same figure, on the right, shows the corresponding curves for three bugs, which start at the vertices of an equilateral triangle ([54, p.136], [39, p.203], [46, p.109]).

2.8 Inversions

> For in every problem which the pupil himself solves, or theorem which he demonstrates, not having precedingly seen it solved or demonstrated, the same faculties are exercised which, in their higher degrees, produced the greatest discoveries in geometry.
>
> J.S. Mill, *On the Study of Mathematics*

Given a point O on the plane and a constant $\delta > 0$, **inversion** with center O and power δ is called the transformation, which to every point $X \neq O$ corresponds the point Y on the half-line OX, such that the following relation is satisfied:

$$|OX|\cdot|OY|=\delta.$$

δ is called **power** of the inversion, the circle with center O and radius $\sqrt{\delta}$ is called **circle of inversion** and its center **center of inversion**. The definition repeats essentially that of I-§ 4.8. Indeed, these are the same things, which we see here from a different viewpoint. The domain, in which is defined the inversion, is the entire plane except point O. Also, the inversion, contrary to the transformations we saw previously, does not preserve lines except only in the special case where these pass through the center of inversion O. This follows directly from the definition and Proposition I-4.16. Generally speaking, all the properties of I-§ 4.8 are properties of the transformation of inversion. This way, for example, the pairing of two "inverse" points, in which, if Y is the inverse of X, then also X is the inverse of Y, is translated into the basic property of transformation f of the inversion, according to which the transformation coincides with its inverse transformation (we say, it is involutive):

$$f\circ f=e.$$

Using the properties of I-§ 4.8, we prove easily the properties of this transformation, which I formulate as exercises.

Exercise 2.49. The fixed points of the inversion f are precisely all the points of the corresponding circle of inversion of f.

Exercise 2.50. An inversion f is completely determined from two different points X, Y and their images X', Y', provided the four points are concyclic or collinear.

Exercise 2.51. Consider the inverse points $\{A'=f(A), B'=f(B)\}$ of the inversion f relative to the circle $\kappa(O,r)$. Show that the triangles $\{OAB, OA'B'\}$ are similar.

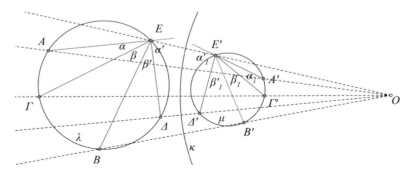

Fig. 2.50: Inversion and cross ratio of points on a circle

Theorem 2.26. *The inversion preserves the cross ratio of four points on a circle or a line.*

2.8. INVERSIONS

Proof. Let us consider the case of four points $\{A,B,\Gamma,\Delta\}$ on a circle λ. If the inversion relative to the circle $\kappa(O,r)$ maps these points respectively to $\{A',B',\Gamma',\Delta'\}$ (See Figure 2.50), then we have to show that the cross ratios are equal: $(AB;\Gamma\Delta) = (A'B';\Gamma'\Delta')$. Taking an arbitrary point E on the circle and its inverse E', it suffices, according to the definition of the cross ratio of four points on a circle (Theorem I-5.32), to show that the two pencils of four lines $E(A,B,\Gamma,\Delta)$ and $E'(A',B',\Gamma',\Delta')$ define the same cross ratio. If R is the radius of λ, then from the sine rule we have

$$(AB;\Gamma\Delta) = \frac{\sin(\alpha)}{\sin(\alpha')} : \frac{\sin(\beta)}{\sin(\beta')} = \frac{|A\Gamma|/R}{|A\Delta|/R} : \frac{|B\Gamma|/R}{|B\Delta|/R} = \frac{|A\Gamma|}{|A\Delta|} : \frac{|B\Gamma|}{|B\Delta|}.$$

Analogously we have $(A'B';\Gamma'\Delta') = \frac{|A'\Gamma'|}{|A'\Delta'|} : \frac{|B'\Gamma'|}{|B'\Delta'|}$. The requested equality $(AB;\Gamma\Delta) = (A'B';\Gamma'\Delta')$ results using the formula of exercise I-4.93 and replacing in the cross ratio

$$|A'\Gamma'| = |A\Gamma| \cdot \frac{r^2}{|OA||O\Gamma|}, \quad |A'\Delta'| = |A\Delta| \cdot \frac{r^2}{|OA||O\Delta|}, \quad \text{etc.}$$

The case of four points on a line, which by the inversion map to points on a circle passing through the center of inversion, results immediately from the definition of the cross ratio of four points on a circle and I leave it as an exercise.

Exercise 2.52. The composition $g \circ f$ of two inversions having the same center is a homothety. The compositions $g \circ f$ and $f \circ g$ of a homothety f with ratio $\kappa \neq 1$ and an inversion g with the same center are two different inversions with the same center.

Exercise 2.53. The compositions $g \circ f$ and $f \circ g$ of two inversions relative to non concentric circles κ and λ is not an inversion. For every point X not on the center-line of the two circles, the five points X, $Y = f(X)$, $Z = g(Y)$, $Y' = g(X)$, $Z' = f(Y')$ lie on a circle μ which is orthogonal to both circles κ and λ (See Figure 2.51-I).

Exercise 2.54. Show that the inscriptible in the circle κ quadrilateral $AB\Gamma\Delta$, is harmonic, if and only if there is an inversion relative to some circle $\lambda(O)$, which leaves κ invariant and maps the quadrilateral to a square $A'B'\Gamma'\Delta'$ (See Figure 2.51-II).

Hint: If the quadrilateral $AB\Gamma\Delta$ results from a square $A'B'\Gamma'\Delta'$ through an inversion relative to the circle $\lambda(O,r)$, as stated in the exercise, then we have (Exercise I-4.93)

$$|A'B'| = |AB| \cdot \frac{\rho^2}{|OA||OB|}, \quad |\Gamma'\Delta'| = |\Gamma\Delta| \cdot \frac{\rho^2}{|O\Gamma||O\Delta|}, \quad \Rightarrow$$

$$|AB||\Gamma\Delta| = |A'B'|^2 \cdot \frac{|OA||OB||O\Gamma||O\Delta|}{r^4}.$$

From the corresponding formula for the product $|B\Gamma||A\Delta|$ follows then that $|AB||\Gamma\Delta| = |B\Gamma||A\Delta|$, which characterizes the harmonic quadrilateral (Corollary I-5.15).

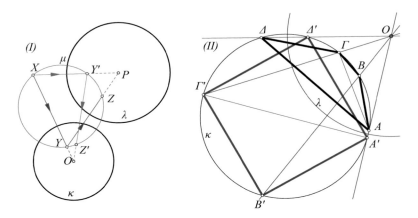

Fig. 2.51: Inversions composition   Harmonic quadrilateral and square

Conversely, if the quadrilateral $AB\Gamma\Delta$ is harmonic, hence inscriptible, then (Exercise 4.189) there is a point O, such that the second intersection points $\{A', B', \Gamma'\}$ of the lines $\{OA, OB, O\Gamma\}$ with the circle κ form an isosceles right triangle $A'B'\Gamma'$. Then for the diametral point Δ' of B', using the same formula and the harmonicity hypothesis for $AB\Gamma\Delta$, we show that $\{\Delta, \Delta', O\}$ are collinear. The requested circle λ has its center at O and is orthogonal to circle κ.

2.9 The hyperbolic plane

> ... in astronomy, as in geometry, we should employ problems, and let the heavens alone if we would approach the subject in the right way and so make the natural gift of reason to be of any real use.
>
> *Plato, Republic VII-530*

One of the more widespread and important uses of inversions is that which results through their perception as a sort of "reflections" on the so called Poincare model. This is simply the interior of a circle κ, which, using suitable definitions, can be considered as the set of points of the plane of a non-euclidean geometry, referred to as *geometry of Bolyai-Lobatsevsky* ([48]) or *hyperbolic geometry*.

2.9. THE HYPERBOLIC PLANE

To build this model we start from one, otherwise arbitrary, circle κ (See Figure 2.52), the *interior* of which we call the **hyperbolic plane** and we refer to it briefly as the *H*-**plane** and to its points as *H*-**points**. We do not consider the points of the circle κ itself as normal points, rather as *H*-**points at infinity** or **limit points** or **points at the horizon**. The circle itself then is called often the **line at infinity** or **horizon** of the hyperbolic plane.

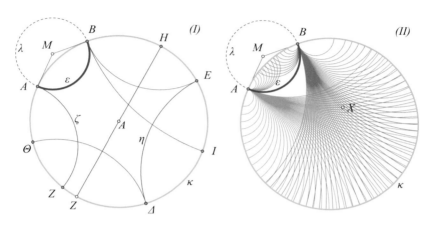

Fig. 2.52: The Poincare disk: a model of hyperbolic geometry

A special characteristic of this model, is that for us, who are looking at it *from outside*, the model is finite, since it is constrained in the interior of a circle. With the definitions, however, and the rules (axioms) we accept below, for its lines, its angles and its distances, we create a system (a geometry), in which, the beings which subject to these rules, have the feeling that they live in a plane which extends to infinity. These beings then, can never reach a point of the horizon (the circle κ).

This model relies on the existence of a distance function $d(A,B)$ of two *H*-points ([49, p.260], [29, p.130], [37]), for which all the reflections relative to lines passing through the center of κ, as well as all the inversions f_λ relative to circles λ, orthogonal to circle κ, are **H-isometries**, that is, they satisfy

$$d(f_\lambda(A), f_\lambda(B)) = d(A,B), \quad \text{for every pair of } H\text{-points } (A,B).$$

We call these special reflections/inversions *H*-**reflections**. It is proved, that for this distance, the corresponding "lines" (called *geodesics of the metric* $d(X,Y)$), in other words the curves which realize the shortest path (of length $d(A,B)$) between two *H*-points (by analogy to the lines of the euclidean plane), are either (i) arcs of circles λ which intersect the circle κ orthogonally, or (ii) diameters of the circle κ. It is also proved, that every other *H*-isometry f of the *H*-plane, in other words, every other mapping of the *H*-plane to itself, which preserves the distance

$$d(f(A), f(B)) = d(A,B), \quad \text{for every pair of } H\text{-points } (A,B),$$

can be written as a composition

$$f = f_k \circ f_{k-1} \circ \ldots \circ f_2 \circ f_1, \quad \text{where the } f_i \text{ are } H\text{-reflections, for } i = 1, \ldots, k.$$

We call these maps *H*-**isometries**. Two shapes Σ, Σ' of the *H*-plane are considered congruent (we say *H*-congruent) when there exists a *H*-isometry f, such that $f(\Sigma) = \Sigma'$.

Figure 2.52-I shows some typical *H*-lines. Each one of them has two points at the horizon, which, it is true that they do not belong to the *H*-line and are not accessible by the *H*-citizens, yet, for us however, the *euclideans* or *outsiders*, suffice to determine the whole line. As it is seen from ε and ζ in the figure, two *H*-lines can have a common point at the horizon. However, because this point is not included in the set of normal *H*-points, lines ε and ζ don't have common points and they are like the parallel lines of euclidean geometry. Somewhat "more" parallel are the *H*-lines ε and η, which do not have *H*-points in common, not even horizon points in common.

Every *H*-line defines two points at the horizon and is determined completely by them. Contrary to this, on the euclidean plane, every line defines a point at the horizon (corresponding point at infinity) and this point is not sufficient by itself for the line's complete determination. On the euclidean plane, two lines can have a common point at the horizon (parallel lines) like on the *H*-plane. However, on the euclidean plane, if two lines do not have a common point at the horizon (at infinity), then, necessarily, they will intersect. Contrary to this again, on the *H*-plane there exist lines (like ε and η in figure 2.52-I), which have no common points at the horizon and yet they do not intersect.

Immediately then, we can distinguish two kinds of parallels on the *H*-plane: (i) those which have exactly one common point at the horizon (like ε and ζ) and which we call **limiting parallels** and (ii) those which have neither a *H*-point nor a horizon point in common and we call **ultraparallels** or **hyperparallels** (like ε and ζ). Figure 2.52-II shows examples of limiting parallels defined by the limiting points of the *H*-line ε.

Exercise 2.55. Given a *H*-line $\varepsilon = AB$ and a *H*-point Γ not lying on it, show that there exist exactly two limiting parallels ΓA, ΓB of ε, passing through point Γ.

Hint: The two lines are defined by the member-circles, which pass through point Γ, of the two circle pencils which are tangent to λ at A and B (Exercise I-4.62) (See Figure 2.53-I).

2.9. THE HYPERBOLIC PLANE

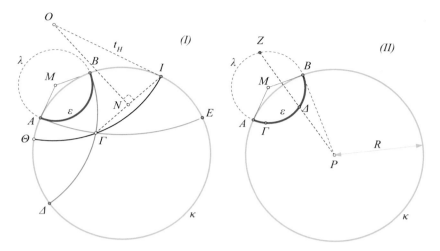

Fig. 2.53: The two limiting parallels of ε through Γ,   H-line through Γ, Δ

Exercise 2.56. Given a H-line $\varepsilon = AB$ and a H-point Γ not on it, show that there exist infinite many hyperparallels $I\Theta$ of ε passing through point Γ.

Hint: Every circle/diameter passing through point Γ, orthogonal to κ and not intersecting the arc AB, defines one such H-line. Figure 2.53-I shows the way to construct one such H-line $I\Theta$, which is defined from point Γ and a point of the horizon I. Point I must lie on the same side of AB as Γ. The center of the circle, which defines the arc $I\Theta$, lies at the intersection of the medial line of ΓI and the tangent to κ at I.

If we define as **parallel** two H-lines which do not intersect, then, the two preceding exercises show that, *from a point Γ, not lying on the line AB, we can draw infinitely many lines parallel to it*. Among these parallels exactly two limiting parallels are included, while all the rest are hyperparallels.

Exercise 2.57. Given two H-points of the H-plane or/and its horizon, there exists exactly one H-line, which passes through them.

Hint: If both the given Γ, Δ are H-points, then the requested circle (carrier of the arc AB) is constructed easily (See Figure 2.53-II). Indeed, from the orthogonality of the requested λ and of κ, follows that the second intersection point Z of $P\Delta$ with λ will satisfy $|PZ||P\Delta| = R^2$, where $\kappa(P,R)$ is the circle, whose interior defines the H-plane. Point Z then is found from the given data and λ is defined from the three points Γ, Δ, Z. The other cases, according to which one of Γ, Δ or both lie at the horizon, are equally easy.

As with euclidean geometry, so here also the notion of a **H-line segment** AB is defined to include all the H-points **between** the H-points A and B. With our euclidean eyes, these are the H-points of the arc AB between A and B.

Finally, as the **angle** of two H-lines we consider the angle between the corresponding circles, which represent these two "lines" (I-§ 4.8). With these few definitions, we can see that properties of circles, circle pencils and inverses translate to geometric properties of the H-plane. A typical related example is that of the next theorem, which characterizes all the H-circles. The latter are again defined by analogy to the euclidean circles, as the geometric loci of the H-points A, which lie at a fixed distance $d(A,O) = r$ from a fixed H-point O. Point O is then called the H-center of the circle and r is called the H-radius.

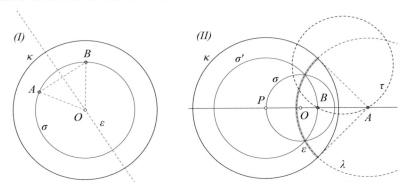

Fig. 2.54: H-circles σ, σ'

Theorem 2.27. *A euclidean circle $\sigma(O,r)$, which is contained entirely in the H-plane, is also a H-circle.*

Proof. If the center O of the euclidean circle σ coincides with the center of κ, the proof is obvious (See Figure 2.54-I). Indeed, then the medial line ε of the chord AB of two points of σ will pass through the center of κ and will define a H-reflection f (coincident with the euclidean reflection relative to ε) with $f(A) = B$ and $f(O) = O$, therefore $d(B,O) = d(f(A), f(O)) = d(A,O)$. Keeping A on σ fixed and varying B on σ, we conclude that for every point B of σ, holds $d(B,O) = d(A,O)$, which is fixed. σ therefore, is simultaneously a H-circle.

In the case where $\sigma(O,r)$ has a center different from the center P of κ, a non-intersecting pencil $\mathscr{D}$ is defined, which is generated from the circles κ and σ. This pencil has two limit points A, B on the center-line O, P of the two circles, of which one (A) is external to both circles and the other is internal to both. We consider then the circle $\lambda(A)$, with center point A, which is orthogonal to κ (See Figure 2.54-II). The inversion f_λ relative to this circle is a H-reflection which maps σ to a circle $\sigma' = f_\lambda(\sigma)$ concentric of κ. Then will also hold $\sigma = f_\lambda(\sigma')$. Because f_λ is a H-isometry and σ' is, according to the first part of the proof, a H-circle, circle σ will also be a H-circle.

Remark 2.11. The first part of the proof shows that the concentric with κ circles $\sigma(O,r)$ are simultaneously H-circles. We must, however, note that on

2.9. THE HYPERBOLIC PLANE

the H-circle, we measure the distance of its points from the center using the function $d(A,B)$, which gives a different value from the euclidean radius r ([38, p.402]). Indeed, in this case the radius $r_H = d(O,A)$ of the H-circle is $r_H = \left|\ln\left(\frac{R-r}{R+r}\right)\right|$, where R is the radius of κ and $y = \ln(x)$ is the function of the natural logarithm, which is studied in elementary calculus ([55, p.31]).

The second part of the proof constructs the circle σ' which is "congruent" to σ relative to the H-distance $d(A,B)$, but not relative to the euclidean distance. The two circles obviously have different euclidean radii, however the corresponding hyperbolic radii are equal. We must also note that the center of the H-circle is in this case different from the euclidean center O and coincides with the point B (See Figure 2.54-II).

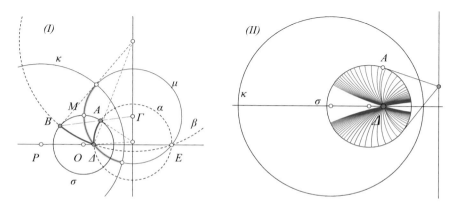

Fig. 2.55: Y-center Δ and the Y-radii of H-circles

Exercise 2.58. Show that the center of a H-circle σ, non-concentric with κ, is the internal limit point Δ of the pencil $\mathcal{D}$, generated by σ and κ (See Figure 2.55-I).

Hint: Consider the orthogonal to $\mathcal{D}$ pencil $\mathcal{D}'$. Every circle-member α of $\mathcal{D}'$ defines a H-line passing through point Δ. Let A be an intersection point of α and σ. Define analogously the intersection point B of a circle-member β of the pencil $\mathcal{D}'$. All the points of the H-circle σ are constructed like A and B. The line segment AB defines an intersection point Γ on the radical axis of $\mathcal{D}$. The circle μ of the orthogonal pencil $\mathcal{D}'$ with center point Γ defines a H-reflection f_μ with $f_\mu(A) = B$. Figure 2.55-II shows the center Δ of the H-circle σ as well as some of its radii ΔA, for A varying on the circle.

Exercise 2.59. Let λ be a circle contained entirely in κ. Show that the circles λ and κ generate a circle pencil $\mathcal{D}$, whose members that are contained in κ represent concentric H-circles, in other words circles with the same H-center.

Exercise 2.60. Construct the middle of a given H-line segment AB. Also, double AB towards B to produce segment $A\Gamma$, such that B becomes the middle of $A\Gamma$.

Hint: The H-line segment AB is an arc of a circle ε orthogonal to κ (See Figure 2.56-I). Let Γ be the intersection point of the line AB with the radical axis δ of the circles κ and ε. The circle $\mu(\Gamma)$, which is orthogonal to κ intersects AB at its middle M. Here we mean, of course the H-middle, for which holds $d(M,A) = d(M,B)$ and not the euclidean middle.

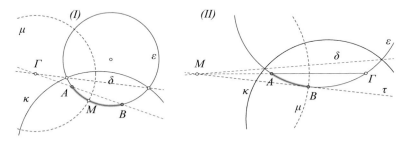

Fig. 2.56: The H-middle M of AB Doubling: $d(A,\Gamma) = 2d(A,B)$

For the doubling of AB consider the intersection M of the tangent τ of ε at B and the radical axis δ of κ and ε (See Figure 2.56-II). Define the circle μ which is simultaneously orthogonal to κ, ε and take the inverse Γ of A relative to μ. In the case where AB is part of a diameter of κ, the corresponding constructions are easier. Figure 2.57-I shows the way to construct the middle

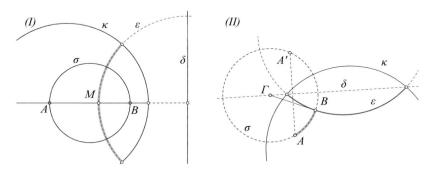

Fig. 2.57: The H-middle M of AB The orthogonal AB from A to ε

in such a case.

Exercise 2.61. Construct the orthogonal AB relative to H-line ε from a point A not on it, as well as from a point B contained in it.

2.9. THE HYPERBOLIC PLANE

Hint: In both cases the orthogonal is an arc of a circle σ simultaneously orthogonal to κ and to the circle which defines ε (See Figure 2.57-II). Its center lies on the radical axis δ of circles κ and ε. Also, in the first case the requested circle will pass through the symmetric A' of A relative to δ. In this case, then, the problem is reduced to Exercise I-4.62. In the second case, the center of σ is the intersection point Γ of δ and the tangent to ε at B.

Exercise 2.62. Construct the H-medial line of a H-line segment AB.

Hint: It is the circle μ of the first part of Exercise 2.60.

Exercise 2.63. Show that given two H-points A, B, there exists exactly one H-reflection f_δ relative to a H-line δ, which maps A to B and B to A.

Hint: δ is the H-medial line of the H-line segment AB.

Exercise 2.64. Given a point B on the H-plane, find a H-line ε, such that the corresponding H-reflection f_ε maps point B to the center A of the circle κ. Show that, through this mapping, a H-circle with H-center at B gets mapped onto a H-circle with center A.

Hint: It is the circle λ used in figure 2.57-I.

Exercise 2.65. Show that the H-reflection of the preceding exercise maps the pencil of H-lines which pass through B, to the pencil of H-lines which pass through A, which coincide with diameters of the circle κ.

Exercise 2.66. Show that a euclidean rotation about the center A of the circle κ, which represents the H-plane, is a H-isometry.

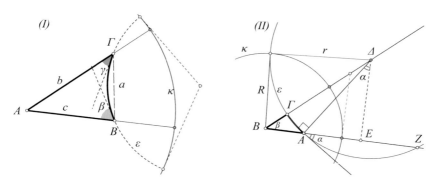

Fig. 2.58: Sum of angles of Y-triangles   Right Y-triangles

Exercise 2.67. Show that hyperbolic triangles have sum of angle measures less than π.

Hint: According to Exercise 2.64, an arbitrary *H*-triangle can be placed on the *H*-plane in such a way, that one of its vertices, *A* say, coincides with the center of the circle κ which represents the hyperbolic plane (See Figure 2.58-I). Then the angle of the triangle at *A* coincides with the euclidean angle and its sides *AB*, *AΓ* coincide with corresponding euclidean line segments. The third side *BΓ* is a circle arc ε orthogonal to κ and the *H*-lines, which are the angles between the corresponding line segments *AB*, *AΓ* and the arc *BΓ* are strictly less than the euclidean angles *ABΓ* and *BΓA*. Consequently if α, β, γ are the measures of the *H*-angles, it holds

$$\alpha + \beta + \gamma < \alpha + |\widehat{AB\Gamma}| + |\widehat{A\Gamma B}| = \pi.$$

Exercise 2.68. Show that a right *H*-triangle is determined uniquely from its two acute angles. Consequently two right *H*-triangles, which have respective acute angles equal, are *H*-congruent, in other words there exists a *H*-isometry which maps the one onto the other.

Hint: Place the right triangle, so that one of its vertices *B* with acute angle, coincides with the center of κ (See Figure 2.58-II). Then its sides *BΓ*, *BA* coincide with corresponding euclidean line segments. Assuming a right angle at *Γ*, the various right triangles with the same angle β and the other acute angle α variable, result through the circles ε of a pencil $\mathscr{D}$. These circles are all orthogonal to κ, their centers are contained in the line *BΓ* and each one of them defines the second side *ΓA* of the right triangle. It suffices to show, that the different angles α correspond to different circles ε, that is, different respective radii *r*. *R* is the radius of the circle κ. The orthogonality of the circles κ and ε corresponds to $r^2 + R^2 = |B\Delta|^2$ (See Figure 2.58-II). Also, the next formulas follow through the power of *B* relative to ε and some calculations.

$$|BE|^2 - r^2\sin^2(\alpha) = R^2, |BE| = r\cos(\alpha)\cot(\beta) \Rightarrow \cos^2(\alpha) = \left(1 + \frac{R^2}{r^2}\right)\sin^2(\beta).$$

The last formula relates the other acute angle α of the right *H*-triangle *ABΓ* with the radius *r* of the euclidean circle ε which represents the second orthogonal side *AΓ* of the triangle. Among other things it shows also that different angles $\alpha \neq \alpha'$ result from corresponding different radii $r \neq r'$.

Exercise 2.69. Show that a hyperbolic triangle is determined completely from its angles.

Hint: Place the triangle, so that one of its vertices *A* with the greater angle, coincides with the center of κ (See Figure 2.59-I). Then its sides *AB* and *AΓ* coincide with corresponding euclidean line segments. The altitude *AΔ* from *A* also coincides with a euclidean line segment which divides *ABΓ* into two *H*-right triangles, to both of which the conclusion of the preceding exercise is applied.

Exercise 2.70. Show that for every integer $\nu \geq 3$ there exist infinitely many *regular H-polygons*, each of which is characterized by the measure of the angle formed by two consecutive sides. Construct the regular pentagon for which this angle is right.

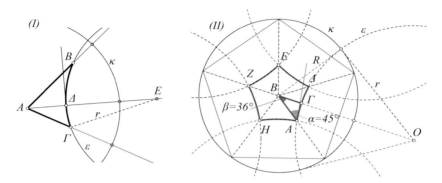

Fig. 2.59: Y-triangle   Regular Y-pentagon with right angle

Hint: In the circle κ we inscribe a regular ν-gon. We construct the H-orthogonal triangle $AB\Gamma$ (See Figure 2.59-II), with $\beta = \frac{\pi}{\nu}$ and $\alpha \leq \frac{\pi}{2}$, as in Exercise 2.68. We also construct the symmetric $B\Gamma\Delta$ of $AB\Gamma$ relative to $B\Gamma$. $AB\Delta$ is H-isosceles, with apical angle $2\beta = \frac{2\pi}{\nu}$. By rotating $BA\Delta$ about B ν times successively we create the polygon $A\Delta EZ...$, which is characterized by the H-angle 2α. As we saw in the aforementioned exercise, different α lead to different right triangles $AB\Delta$ and through them to different polygons $A\Delta EZ....$

For the construction of the regular H-pentagon with a right angle, it suffices to perform the preceding process for $\nu = 5$ (See Figure 2.59-II). The formula of the aforementioned exercise in this case gives

$$\cos^2(45°) = \left(1 + \frac{R^2}{r^2}\right)\sin^2(36°) \Leftrightarrow r^2\left(\frac{2}{x\sqrt{5}} - 1\right) = R^2 \Leftrightarrow \frac{r^2}{R^2} = \sqrt{5}.$$

Here $x = \frac{\sqrt{5}-1}{2}$ is the ratio of the golden section (see also Exercise I-4.13) and r is the radius of the circle ε which is orthogonal to κ and defines the side of the regular pentagon.

Exercise 2.71. Show that for two arbitrary hyperparallel H-lines ε, ζ, there exists exactly one H-line η, which is simultaneously orthogonal to both.

Hint: The circle η, which represents the corresponding H-line, must be simultaneously orthogonal to κ, ε and ζ (See Figure 2.60-I). It is therefore the uniquely defined circle with this property and center the radical center P of the three circles (Exercise I-4.22).

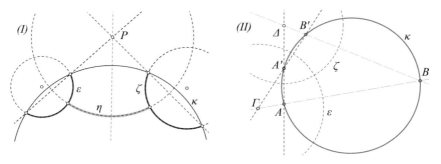

Fig. 2.60: Common orthogonal   Horizon's correspondent points

Exercise 2.72. Given are two points A, B on the horizon as well as two other points A', B' of the horizon, on the same side of the line AB. Show that there exists a H-reflection f_ε relative to an H-line ε, such that $f_\varepsilon(A) = B$ and $f_\varepsilon(A') = B'$. Also show that there exists a H-reflection f_ζ relative to H-line ζ, such that $f_\zeta(A) = A'$ and $f_\zeta(B) = B'$.

Hint: The circle ε, which represents the corresponding H-line, is the orthogonal to κ with center the intersection point $\Gamma = (AB, A'B')$ (See Figure 2.60-II). Respectively, the circle ζ, which represents the corresponding H-line, is the orthogonal to κ with center the intersection point $\Delta = (AA', BB')$.

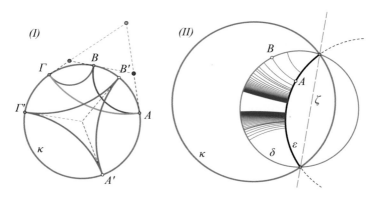

Fig. 2.61: Asymptotic triangles   Distance line δ of the H-line ε

Joining three points A, B, Γ of the horizon with H-lines, creates the so called **asymptotic triangles** of the hyperbolic plane (See Figure 2.61-I). Depending on the position of the three points on the horizon, these "triangles" take on different shapes. However, as the next exercise reveals, from the perspective of hyperbolic geometry, all these triangles are "congruent".

2.9. THE HYPERBOLIC PLANE

Exercise 2.73. Show that every asymptotic triangle $AB\Gamma$ is H-isometric with a special asymptotic triangle $A'B'\Gamma'$, whose vertices form an euclidean equilateral triangle.

Hint: Use Exercise 2.72, first to map isometrically $AB\Gamma$ to a triangle $A''B'\Gamma'$, where B', Γ' are two vertices of an equilateral. Next, find a H-reflection interchanging $\{B',\Gamma'\}$ and $\{A'',A'\}$.

As we saw already, the euclidean circles, which intersect κ orthogonally define H-lines and the circles which are entirely contained in the H-plane define H-circles. In the two other cases of euclidean circles, which intersect κ, correspond curves of the H-plane with noteworthy properties. The first category consists of the so called **distance lines**. This category results by choosing a H-line ε and raising at its points A and to the same side, a H-line segment AB of fixed H-length $d(A,B) = \sigma$ (See Figure 2.61-II). The endpoints B of all these line segments are contained in a circle δ, which intersects the horizon at the same points Γ, Δ with ε. This property shows, that a pencil of intersecting circles with basic points two points Γ, Δ of the horizon, defines through exactly one of its members a H-line, while all its other members define distance lines from this H-line.

The second category of noteworthy curves of the hyperbolic plane, which is defined by (parts of) euclidean circles, is that of **horocycles**. These curves are defined from the members of a tangential pencil $\mathscr{D}$ of circles which are tangent to κ at a point A of the horizon (See Figure 2.62). Point A is called

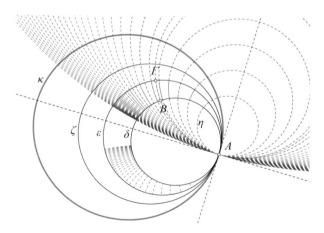

Fig. 2.62: Horocycles δ, ε, ζ

limit point of the horocycles. Two horocycles δ, ε, with common limit point, are orthogonal to every line which passes through A and excise on each such line a H-line segment $B\Gamma$ of fixed H-length $d(B,\Gamma)$. The corresponding image in euclidean geometry is that of a pencil $\mathscr{D}$ of parallel lines, whose direction defines the point at infinity (euclidean horizon) A. This defines then the pencil $\mathscr{D}'$ of the parallel lines, which are orthogonal to the lines of $\mathscr{D}$. In hyperbolic geometry the corresponding image is that of the pencil $\mathscr{D}$ of the limiting parallels which "pass through" the point of the horizon A. The corresponding pencil $\mathscr{D}'$, which intersects the lines of $\mathscr{D}$ orthogonally, is not any more a pencil of H-lines, rather a pencil of horocycles, which are curves with properties different from those of H-lines. The horocycles may be considered also as limiting cases of **distance lines**, whose two points on the horizon tend to coincide.

Exercise 2.74. For a given H-line ε and a corresponding distance line δ of it. Show that the segments AB of the H-lines which are orthogonal to ε and are excised by ε and δ are equal.

Hint: Two such segments AB, $A'B'$ are mapped to each other through a H-reflection relative to the circle of the orthogonal pencil to the one generated by the circles δ and ε.

Exercise 2.75. Given are two horocycles δ, ε with the same limit point A. Show that the segments AB of the H-lines which they excise from H-lines η, which pass through point A, have fixed length $d(A,B)$.

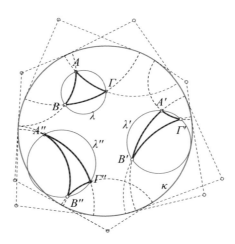

Fig. 2.63: Circumscribed "circles"(?) of H-triangles

An important difference between H-triangles and the corresponding euclidean ones is that the former may not always have a circumcircle (See Figure 2.63). As we saw precedingly, the H-circles coincide with the euclidean ones which are contained completely in the interior of κ. This implies that,

2.9. THE HYPERBOLIC PLANE

if a *H*-triangle admits a circumcircle, that is a *H*-circle, which passes through all three of its vertices, this circle will coincide with a euclidean circle contained entirely in the interior of κ. However, we can easily construct *H*-triangles $AB\Gamma$, whose euclidean circles λ, which are defined by their three vertices, lie in the interior, or are tangent to, or even intersect κ at two points (See Figure 2.63). We conclude, then, that in the hyperbolic plane there exist three categories of triangles: (i) the ones which have a circumcircle, (ii) the ones which have a circumscribed horocycle ($A'B'\Gamma'$ in 2.63) and finally, (iii) the ones which have a circumscribed distance line ($A''B''\Gamma''$ in figure 2.63). With the last category, the inconceivable for euclidean geometry happens: there exists a *H*-line ε, such that all three vertices of the triangle are on the same side of ε and share the same distance from it.

The **angle of parallelism**, which was introduced by Lobatsevsky and, traditionaly, is denoted by $\phi = \Pi(x)$, is a special function with fundamental importance in *H*-trigonometry ([48, p.7], [51, p.221], [31, p.41]). Here x denotes

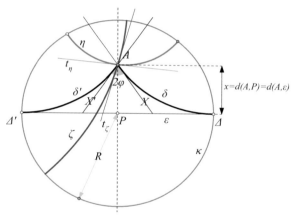

Fig. 2.64: Angle of parallelism $\phi = \Pi(x)$

the hyperbolic distance $d(A,P)$ (not the euclidean $|AP|$) from a *H*-line ε. ϕ denotes the angle $\phi = \widehat{PAX}$, whose one side is the *H*-orthogonal AP on ε and its other side is a limiting parallel δ with a point on the horizon Δ common with a respective point on the horizon of ε (See Figure 2.64). As we saw in the beginning of the section, the *H*-lines through A which do not intersect ε are infinite in number and are all hyperparallel to it, except two: δ, δ'. The last two are those which have exactly one common point Δ on the horizon with ε. Angle ϕ then defines at A two supplementary angles of measure 2ϕ and $(\pi - 2\phi)$. Of whatever *H*-lines ζ, their tangents t_ζ at A fall inside the first angle, these *H*-lines intersect ε. Of whatever *H*-lines η, their tangents t_η at A fall inside the second angle, these do not intersect ε. An easy calculation relying on the figure 2.64, shows that, measuring in our model with euclidean means, the following formula holds

$$\tan\left(\frac{\phi}{2}\right) = \frac{R - |AP|}{R + |AP|}.$$

This formula, among other things, also reveals the fact that the angle of parallelism decreases and tends to 0, as A moves away from the line ε. A further study of the angle between parallels and its applications to the H-trigonometry would require the knowledge of elementary transcendental functions which transcends the purposes of this lesson.

Exercise 2.76. Show that, for every pair (ε', A') from a H-line and one H-point, there exists a H-reflection f_δ relative to H-line δ, such that $f_\delta(\varepsilon') = \varepsilon$ and $f_\delta(A') = A$, where (ε, A) are those of figure 2.64 and the H-distance of A from ε is equal to the corresponding of A' from ε'.

Hint: If P' is the trace of the H-orthogonal from A' to ε, then there exists a H-reflection f_δ which maps P' onto P (Exercise 2.63).

2.10 Archimedean tilings

> They are not taught, or not sufficiently taught, chamber music and ensemble techniques and, moreover, they are not stimulated to love music as such, instead of loving only themselves and their careers.
>
> George Szell, Quoted in "Szigeti on the Violin" p. 43

With the term **tiling** we mean a covering of the plane with mutually non overlapping tiles. It is allowed for adjacent tiles to have in common either

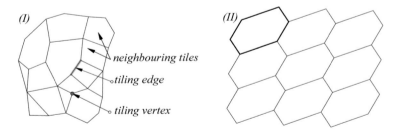

Fig. 2.65: General tiling   Tiling with congruent tiles

an entire side or a vertex, which are called respectively **tiling edge** and **tiling vertex**. We suppose that we have at our disposal an infinite supply of tiles and that the tiling must cover the plane entirely, without leaving gaps between tiles. There are many kinds of tilings, with tiles of many different shapes (See Figure 2.65-I) or/and with tiles of the same shape (See Figure

2.10. ARCHIMEDEAN TILINGS

2.65-II). Next theorem shows which are the possible **regular** tilings with tiles of one only shape (which are called **monohedral**), when this shape is a regular polygon.

Theorem 2.28. *The only possible tilings of the plane with equal tiles, in the shape of a regular polygon, are for the figure of the equilateral triangle, the square and the regular hexagon (See Figure 2.66).*

Proof. The fact that the aforementioned shapes may be used for tiling, follows directly in the case of the equilateral and the square. Indeed, we see that in these cases, using 9 tiles we can make a similar tile having a side three times larger, which we can name tile of *second order*. Next, in the same way, we can using 9 tiles of second order make tiles of *third order* and so on and so forth. The tiling with hexagons reduces to the tiling with triangles, since each hexagon decomposes into 6 equilateral triangles. The fact that

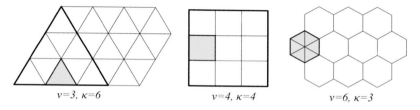

Fig. 2.66: The only possible tilings with regular tiles

these are the only possible tilings, follows by considering the number κ of the tiles with a common vertex. If the tile is a regular v-gon, then its angle is $\omega_v = \frac{v-2}{v}180°$ (Corollary I-2.19). This means that the tiles around a vertex must cover exactly the entire horizon around the point, consequently

$$360° = \kappa \cdot \omega_v = \kappa \cdot \frac{v-2}{v}180° \quad \Leftrightarrow \quad \kappa = \frac{2v}{v-2}.$$

It is easily seen that κ will be the same for each vertex, and the last equation implies a strong restriction illustrated by the following table

v	3	4	5	6
$\kappa = \frac{2v}{v-2}$	6	4	3.333...	3

in which κ, v must be integers. The table shows that the only acceptable pairs of values (v, κ) are $(3,6)$, $(4,4)$ and $(6,3)$. For $v = 7, 8, ...$ the corresponding κ, which is written as $\kappa = 2 + \frac{4}{v-2}$, is not acceptable since it has a value less than 3.

Exercise 2.77. Show that every triangle $AB\Gamma$ defines a tiling of the plane.

Exercise 2.78. Show that every quadrilateral $AB\Gamma\Delta$ defines a tiling of the plane.

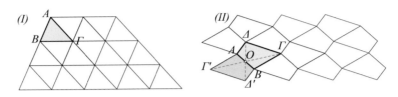

Fig. 2.67: Tilings with arbitrary triangle (I) and arbitrary quadrilateral (II)

Hint: Let O be the middle of the side AB. Consider the symmetric $AB\Delta'\Gamma'$ of $AB\Delta\Gamma$ relative to O (See Figure 2.67-II). This forms a symmetric hexagon $A\Delta\Gamma B\Delta'\Gamma'$ which, according to the next exercise, defines a tiling ([47]).

Exercise 2.79. Show that every hexagon $AB\Gamma\Delta EZ$, which is symmetric relative to a point O, defines a tiling of the plane.

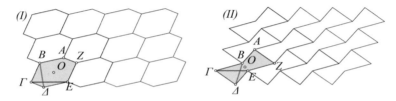

Fig. 2.68: Tilings with convex (I) and non-convex (II), symmetric hexagon

Hint: The parallel translations by the oriented line segments of the "small diagonals", like ΓE, ΔB (See Figure 2.68), applied to the tile $p = AB\Gamma\Delta EZ$ define its neighboring tiles.

The more general problem, of the *monohedral* tiling, has been studied extensively. It remains though open and it is not at all trivial to determine if a given polygon p can or cannot tile the plane. It has been proved that it is not possible to construct a method (algorithm) which, in a finite number of steps, can decide if a specific polygon tiles the plane or not ([40, p.21], [50]). In the very interesting and rich in results and figures book of Gruenbaum and Shephard [40] various categories of tilings are investigated and classified. From time to time though, a new tiling is discovered, which is not among the already known ones, like, for example, the monohedral tiling of figure 2.69, which was discovered by Marjorie Rice, a housewife from California, without special knowledge of Mathematics ([52]). The tiling in the figure differs from regular ones also in that each vertex is not surrounded by the same number of tiles (defining the so called order of the tiling). Here, there are vertices surrounded by 3 or 4 tiles.

If we allow regular tiles, but with different shapes, then it is proved that, besides the preceding monohedral ones (with tiles of one and only shape), there exist 8 more tilings, which are called **semiregular** ([45, p.123]). The

2.10. ARCHIMEDEAN TILINGS

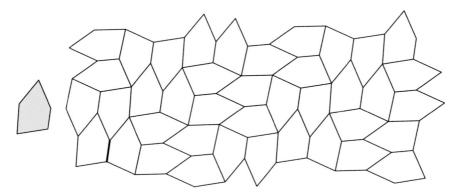

Fig. 2.69: Rice pentagonal tiling

eleven tilings with one or more regular tiles are called **Archimedean** tilings of the plane ([40, p.63]). In Archimedean tilings we require that their vertices are surrounded always by the same number of tiles κ, which are regular polygons. Also, for their arrangement around a vertex we require to be the same for each vertex. We call κ **order** of the tiling. For the distinction of these

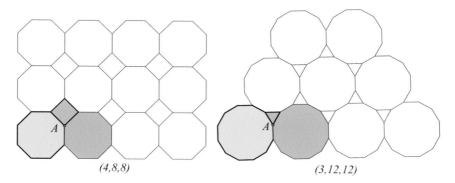

Fig. 2.70: Two Archimedean tilings of order 3 (2 different tiles)

tilings we use exactly the integers which give the number of the sides of the tiles and the order under which they appear around each vertex. This way, for example, the symbol (4,8,8) (or also $4 \cdot 8^2$), in figure 2.70, corresponds to a tiling where each vertex is common to a square and two octagons. Any cyclic permutation of the numbers, like for example (8,8,4) or (8,4,8), denotes the same tiling.

Next theorem shows that all tilings with three regular tiles around each vertex (that is of order 3) are those of figures 2.70 and 2.71.

Theorem 2.29. *There are four tilings with three regular tiles around each vertex (of order 3). These are: (6,6,6), (4,8,8), (3,12,12) and (4,6,12).*

Proof. Leaving aside for now the details of the fact that the four mentioned are indeed tilings, we'll prove that they are the only ones. The proof for this is a generalization of the preceding proof. Indeed, let us suppose that the three tiles around each vertex are regular with number of sides respectively

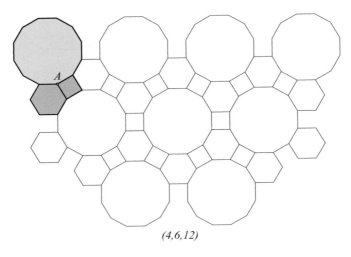

(4,6,12)

Fig. 2.71: The unique Archimedean tiling of order 3 (3 different regular tiles)

$$v_1 \leq v_2 \leq v_3.$$

The angles around each vertex must sum up to

$$360° = \frac{v_1 - 2}{v_1}180° + \frac{v_2 - 2}{v_2}180° + \frac{v_3 - 2}{v_3}180° \Leftrightarrow$$

$$2 = \frac{v_1 - 2}{v_1} + \frac{v_2 - 2}{v_2} + \frac{v_3 - 2}{v_3} \Leftrightarrow$$

$$\frac{1}{2} = \frac{1}{v_1} + \frac{1}{v_2} + \frac{1}{v_3}.$$

This equation must hold true for integers greater than 3. If these integers are all the same $v_1 = v_2 = v_3 = v$, we see immediately that the solution is $v = 6$, that is, the case (6,6,6) of the tiling with hexagons. If two exactly from the tiles are congruent v_1, $v_2 = v_3$, the equation becomes

$$\frac{1}{2} = \frac{1}{v_1} + \frac{2}{v_2} \Leftrightarrow v_2 = \frac{4v_1}{v_1 - 2}.$$

A small table with integers gives the values of the right side of the equation.

2.10. ARCHIMEDEAN TILINGS

v_1	3	4	5	6	7
$v_2 = \frac{4v_1}{v_1-2}$	12	8	6.66...	6	5.6

The integer values 12, 8 are the only acceptable ones, as $v_2 = 6$ corresponds to (6,6,6) which, for now, we have excluded and the rest are non integral or give $v_2 < v_1$. Interchanging the role of v_1, v_2 in the table, we see that there are no tilings with two congruent tiles with number of sides less than that of the third tile, in other words case (v_1, v_1, v_2) with $v_1 < v_2$.

An additional restriction results for tilings with three tiles around the vertex with symbol $(3, v_1, v_2)$. Then it must be $v_1 = v_2$. Indeed, these tilings have at

Fig. 2.72: Restriction $(3, v_1, v_2) \Rightarrow v_1 = v_2$

each vertex a triangular equilateral tile and from figure 2.72 follows that at vertex A two of the angles must be the same.

Finally, if all three integers are different, then for the smaller one, $v_1 \leq 6$. Otherwise the sum of the angles of the three tiles around a vertex would be greater than $360°$. Also for the greater one, $v_3 \geq 6$. Otherwise the sum of the angles around a vertex would be less than $360°$. This restriction, in combination with the preceding general equation, implies

$$\frac{1}{v_1} + \frac{1}{v_2} = \frac{1}{2} - \frac{1}{v_3} \geq \frac{1}{2} - \frac{1}{6} = \frac{1}{3} \Rightarrow v_2 \leq \frac{3v_1}{v_1 - 3}.$$

The corresponding table for the first few integers

v_1	4	5	6	7
$v_2 \leq \frac{3v_1}{v_1-3}$	12	7.5	6	5.2

shows that the only acceptable values of v_1 are $v_1 = 4, 5$ with corresponding values $v_2 \leq 12$ and $v_2 \leq 7$. Examining

$$\frac{1}{v_1} + \frac{1}{v_2} + \frac{1}{v_3} = \frac{1}{2} \text{ with } v_1 < v_2 < v_3,$$

and these additional restrictions, we see that the only acceptable solution is (4,6,12)

Figure 2.73 shows the two Archimedean tilings of order 4. Finally, the next figures show the three Archimedean tilings of order 5.

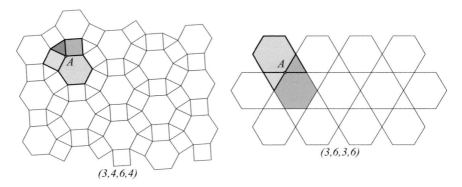

Fig. 2.73: Two Archimedean tilings of order 4

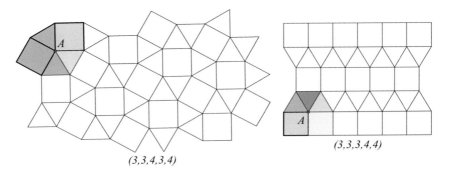

Fig. 2.74: Two Archimedean tilings of order 5

Theorem 2.30. *There are 3 regular and 8 semiregular tilings of the plane, whose kinds are given by the table*

order				
3	(6,6,6)	(4,8,8)	(3,12,12)	(4,6,12)
4	(4,4,4,4)	(3,4,6,4)	(3,6,3,6)	
5	(3,3,4,3,4)	(3,3,3,4,4)	(3,3,3,3,6)	
6	(3,3,3,3,3,3)			

*For one of them only, (3,3,3,3,6), the symmetry relative to some axis produces a different tiling (called **enantiomorphic**), a tiling which cannot be placed on top of the other so that they coincide.*

I am not going to proceed with the proof of this theorem. This may be done in a similar way to the proof of the preceding theorem. Let us note however, that the order κ of an Archimedean tiling cannot be greater than 6. This because regular polygons have, all of them, angles greater than or equal

2.10. ARCHIMEDEAN TILINGS

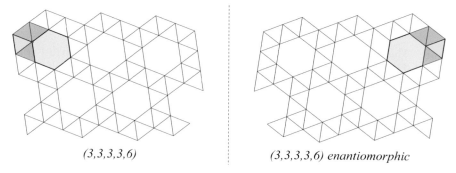

(3,3,3,3,6)   (3,3,3,3,6) enantiomorphic

Fig. 2.75: The Archimedean tiling (3,3,3,3,6) and its enantiomorphic tiling

to 60° and six such angles around a vertex cover the 360° of the horizon. For the proof, consequently, in the case of order 4, it suffices to "solve" the corresponding equation for integers

$$360° = \frac{v_1-2}{v_1}180° + \frac{v_2-2}{v_2}180° + \frac{v_3-2}{v_3}180° + \frac{v_4-2}{v_4}180° \Leftrightarrow$$
$$\frac{1}{v_1} + \frac{1}{v_2} + \frac{1}{v_3} + \frac{1}{v_4} = 1,$$

and for the case of order 5 the corresponding one, which is

$$\frac{1}{v_1} + \frac{1}{v_2} + \frac{1}{v_3} + \frac{1}{v_4} + \frac{1}{v_5} = \frac{3}{2}.$$

We see easily that in the first equation all the integers will satisfy $3 \leq v_i \leq 12$ while in the second $3 \leq v_i \leq 6$. The equations then may be solved by "trial and error" for the first 10^4 quadruples of integers for the first equation and for the first 4^5 pentuplets for the second. The numbers are fairly small and the solution of the two equations may be found with the use of a small computer program.

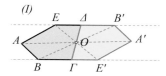
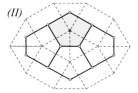

Fig. 2.76: Pentagons which allow tilings

Exercise 2.80. Show that every pentagon $AB\Gamma\Delta E$, possesing two parallel sides, defines a tiling of the plane.

Hint: There exists a side of the pentagon which has its neighboring sides parallel. The symmetry relative to the middle O of this side defines a neighboring pentagon which along with the initial one forms a hexagon symmetric relative to the point O (See Figure 2.76-I).

Exercise 2.81. Show that the dual tile of the Archimedean tiling (3,3,4,3,4), that is, the tile which has as vertices the centers of mass of the tiles around one vertex of this tiling, defines a tiling with pentagons.

Hint: Figure 2.76-II shows, with dotted lines, some of the tiles of the (3,3,4,3,4) tiling. The pentagonal tiling is defined from the centers of the two squares and the three triangles which surround a vertex of the tiling. Four such tiles form a symmetric hexagon. These tiles and the corresponding tiling are called *Cairo*, as they are often encountered on old streets of that city. In the book [40] are discussed thoroughly the preceding tilings, as well

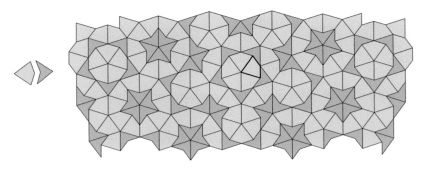

Fig. 2.77: A Penrose tiling

as a plethora of other categories of tilings. I note one more example, that of Penrose ([57, p.176], [30]), of a tiling with two tiles (see remark I-4.11), which allow the formation of infinitely many different tilings. In these also, however, there is no restriction of order. A vertex of the tiling may be surrounded by 3, 4, or 5 tiles (See Figure 2.77). The next exercise constructs a tile like that of Rice, which was mentioned precedingly. This tile satisfies the specifications of figure 2.78-I, where capital letters label the measures of the corresponding angles and small latin caps label the side lengths ([52, p.35]).

Exercise 2.82. Starting from a line segment AB of arbitrary length x construct four circles (See Figure 2.78-II). First the two with radius x: $\kappa(A,x), \lambda(B,x)$ and the other two after the selection of an arbitrary point E on κ. If Γ' is the second intersection point of λ with AE, circle μ is defined as the circle $\mu(\Gamma',x)$ and circle ν as the circle with diameter EB. Let Δ be the second intersection point of the circles μ and ν, let also M be the middle of $B\Gamma$ and

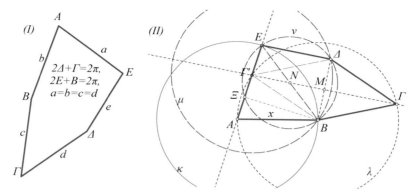

Fig. 2.78: Specification of the Rice tile and construction of one such tile

Γ be the second intersection point of $M\Gamma'$ with the circle λ. Show that the pentagon $AB\Gamma\Delta E$ fulfills the specifications of the Rice tile (See Figure 2.78-I).

Hint: By construction $B\Gamma'\Delta\Gamma$ is a rhombus and the lengths of the sides are $a = b = c = d = x$. The relation between angles $2E + B = 2\pi$ results by projecting point B at Ξ on $A\Gamma'$. By construction $B\Xi E\Delta$ is inscriptible, therefore E is supplementary to $\Xi B\Delta$, which is half of $AB\Gamma$. It holds then $E + \frac{B}{2} = \pi$, which is equivalent to the specification. The other specification is proved more simply from the trapezium $\Delta E\Gamma'\Gamma$, in which the angles Δ and $\Gamma'\Gamma\Delta$ are supplementary. It follows $\Delta + \frac{\Gamma}{2} = \pi$, which is again equivalent to the specification.

2.11 Comments and exercises for the chapter

> It may be so, there is no arguing against facts and experiments.
>
> I. Newton, *Memoirs of the life vol. 2*

Exercise 2.83. Two balls X, Y are placed on a billiard table. Determine the trajectory of ball X, which reflected in the four walls will hit next ball Y. Examine also the case where the ball after the reflections in the four walls of the rectangle returns to its original position (See Figure 2.79).

Hint: ([32, p.7]) Y_1 is the reflected of Y relative to side $A\Delta$. Y_2 is the reflected of Y_1 relative to $\Gamma\Delta$. X_1 is the reflected of X relative to $B\Gamma$. X_2 is the reflected

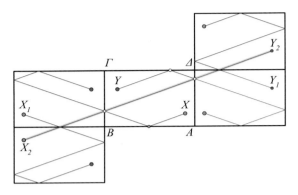

Fig. 2.79: Billiard trajectory

of X_1 relative to AB. The segment X_2Y_2 determines the reflection points. For the second part see Exercise I-3.45.

Figure 2.80-I shows the construction of a polygonal line which connects the contact points of four circles α, β, γ, δ, which are pairwise tangent externally respectively at points A, B, Γ, Δ. The construction process of the polygonal line is the following:

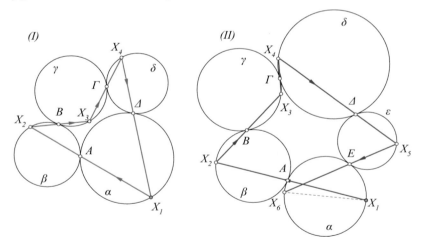

Fig. 2.80: Chains of pairwise tangent circles

1. We start from an arbitrary point X_1 of the circle α and we join it with the contact point A of α, β,
2. The line $X_1 A$ intersects again β at point X_2, which we join with the contact point B of β, γ,

3. The line X_2B intersects again γ at point X_3, which we join with the contact point Γ of γ, δ,
4. The line $X_3\Gamma$ intersects again δ at point X_4, which we join with the contact point Δ of δ, α,
5. The line $X_4\Delta$ intersects again κ at point X_5 which coincides with X_1.

The preceding process can be extended to as large chains of circles $\alpha, \beta, \gamma, ...$, pairwise tangent, as we please. Figure 2.80-II shows a similar polygonal line for a chain of five circles. A difference which shows up between the figures and happens more generally for similar chains of ν circles is noteworthy.

Exercise 2.84. Show that in the preceding construction, when ν is even, then the last point $X_{\nu+1}$ (point $X_5 = X_1$ in figure 2.80-I) coincides with the start point X_1. When ν is odd, then the last point (X_6 in figure 2.80-II) coincides with the diametrically opposite of the starting point.

Hint: ([58, p.31]). Point X_2 is the image $X_2 = f_A(X_1)$ of the homothety f_A with center A and ratio λ_A equal to the ratio of the radii of circles α and β: $\lambda_A = -\frac{r_\beta}{r_\alpha}$. Similarly, point X_3 is the image $X_3 = f_B(X_2)$ of the homothety with center B and ratio $\lambda_B = -\frac{r_\gamma}{r_\beta}$. Point X_4 is the image $X_4 = f_\Gamma(X_3)$ of the homothety with center Γ and ratio $\lambda_\Gamma = -\frac{r_\delta}{r_\gamma}$. Finally point X_5 is the image $X_5 = f_\Gamma(X_4)$ of the homothety with center Δ and ratio $\lambda_\Delta = -\frac{r_\alpha}{r_\delta}$. Totally then point X_5 is the image of the composition of transformations

$$X_5 = (f_\Delta \circ f_\Gamma \circ f_B \circ f_A)(X_1).$$

According to Theorem 2.19, the composition of the homotheties is a homothety with ratio λ equal to the product of the ratios

$$\lambda = \lambda_\Delta \cdot \lambda_\Gamma \cdot \lambda_B \cdot \lambda_A = \left(-\frac{r_\alpha}{r_\delta}\right) \cdot \left(-\frac{r_\delta}{r_\gamma}\right) \cdot \left(-\frac{r_\gamma}{r_\beta}\right) \cdot \left(-\frac{r_\beta}{r_\alpha}\right) = 1.$$

Note however that the homothety with ratio $\lambda = 1$ is the identity transformation, hence also the conclusion for $\nu = 4$. The proof for even $\nu = 2\kappa$ is similar.

In the second case, with the five circles, point $X_6 = f(X_1) = (f_E \circ f_\Delta \circ f_\Gamma \circ f_B \circ f_A)(X_1)$, is again the image of a homothety f with ratio which is calculated similarly and is equal to $\lambda = -1$. Such a homothety, however, is coincident with the symmetry relative to a center, which must coincide with the center of the circle α since this circle remains invariant by f (in other words f maps it to itself). From this follows also the conclusion for odd ν.

Exercise 2.85. Given the similar and similarly oriented triangles $AB\Gamma$ and $A'B'\Gamma'$, whose respective sides intersect at points $E = (AB, A'B')$, $Z = (B\Gamma, B'\Gamma')$, $H = (\Gamma A, \Gamma'A')$, show that (See Figure 2.81-I):

1. Circles $(AA'E)$, $(BB'E)$, $(\Gamma\Gamma'Z)$ pass through a common point Δ.

2. Point H lies in both circles $(AA'E)$, $(\Gamma\Gamma'Z)$.
3. Point Δ is a similarity center of the two triangles.
4. The three centers of circles $(AA'E)$, $(BB'E)$, $(\Gamma\Gamma'Z)$ define a triangle $A_0 B_0 \Gamma_0$ similar to $AB\Gamma$.

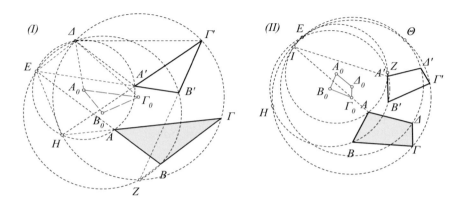

Fig. 2.81: Similar triangles   Similar quadrilaterals

Hint: The second point of intersection Δ of the circles $(AA'E)$, $(BB'E)$ defines the center of the similarity f, which maps $f(A) = A'$, $f(B) = B'$ (See Figure 2.81-I). That similarity, because of the preservation of angles and ratios, maps the triangle $AB\Gamma$ to $A'B'\Gamma'$. This proves (1), (2) and (3). For (2) also, it suffices to observe that angles $\widehat{BEB'}$ and $\widehat{BZB'}$ are equal.

(4) reduces to the fact that $(\Delta A_0 \Gamma_0, \Delta A'\Gamma')$, $(\Delta B_0 A_0, \Delta B' A')$, $(\Delta \Gamma_0 B_0, \Delta \Gamma' B')$ are pairs of similar triangles. For the first pair, for example, this follows from the fact that the angles $\widehat{\Delta A'\Gamma'}$ and $\widehat{\Delta A_0 \Gamma_0}$ are equal, because the first is equal to $\widehat{\Delta EH}$, which, because of the inscriptible quadrilateral $\Delta EHA'$, is half of the (non-convex) $\widehat{\Delta A_0 H}$, which is exactly equal to $\widehat{\Delta A_0 \Gamma_0}$. Consequently, triangles $\Delta A_0 \Gamma_0$, $\Delta A'\Gamma'$ have an angle equal and the angles $\widehat{A'\Gamma'\Delta} = \widehat{A\Gamma_0 H}/2 = \widehat{\Delta \Gamma_0 A}$, etc.

Exercise 2.86. Formulate and prove properties similar to these of the preceding exercise for quadrilaterals (See Figure 2.81-II).

Exercise 2.87. Given is a rectangular strip $AB\Gamma\Delta$. Find at which points and using what angles you must fold it, so that the created square ΔZMN, is the maximal possible (See Figure 2.82).

Exercise 2.88. Given is some rectangular strip $AB\Gamma\Delta$. Find at which points and using what angles you must fold it, so that the created shape $\Delta EZH\Theta$, is the maximal possible regular pentagon (See Figure 2.83).

2.11. COMMENTS AND EXERCISES FOR THE CHAPTER

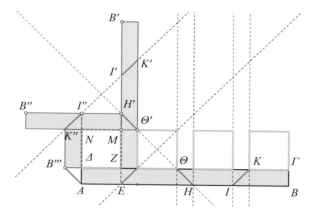

Fig. 2.82: Fold in square

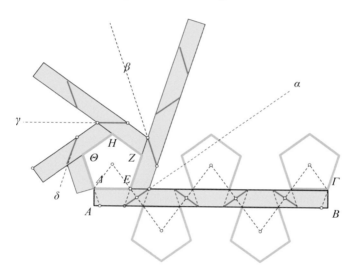

Fig. 2.83: Folding into regular pentagon

Exercise 2.89. Generalize the preceding exercises and find the way to fold a rectangular strip of dimensions $a \times b$, so that the resulting n-gon $AB\Gamma\Delta...$ in its interior, is the maximal possible (See Figure 2.84).

Exercise 2.90. Given the two similar and similarly oriented parallelograms $AB\Gamma\Delta$ and $A'B'\Gamma'\Delta'$, consider the intersection points of respective sides $\Theta = (AB, A'B')$, $H = (\Gamma\Delta, \Gamma'\Delta')$, $E = (B\Gamma, B'\Gamma')$ and $Z = (A\Delta, A'\Delta')$. Show that $\{ZE, H\Theta\}$ pass through the center of similarity I of the parallelograms (See Figure 2.85-I).

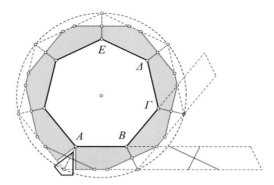

Fig. 2.84: Folding into a regular *n*-gon

Hint: $\widehat{\Theta IZ} = \widehat{\Theta AZ}$ and, because of the parallel lines, the last angle is equal to $\widehat{\Theta BE}$, etc.

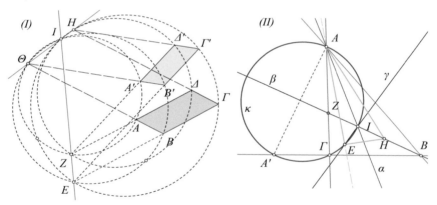

Fig. 2.85: Similar parallelograms   Triangle of given bisectors

Exercise 2.91. Let α, β, γ be three different lines passing through the point *I*. Construct a triangle *AEH*, which has these lines as internal angle bisectors. Also show, that if two of the lines are orthogonal, then there is no solution (See Figure 2.85-II).

Hint: If *A* is arbitrary point of α and *Z* is arbitrary point of β, consider the line *AZ*, the reflection of *AB* relative to α and the reflection *BΓ* of *AB* relative to β. The three lines *AZ*, *AB* and *BΓ* define a triangle *ABΓ*. For fixed *A* and variable *Z* on β the vertex *Γ* of this triangle traces circle κ, passing through points *A*, *I* and the mirror image *A'* of *A* relative to β. The requested triangle results from the intersection point *E* of this circle with line γ.

Exercise 2.92. From an arbitrary point Δ of the side AB of triangle $AB\Gamma$ we draw successively antiparallels ΔE, EZ, ZH, $H\Theta$, ΘI, respectively, towards the sides $B\Gamma$, AB, $A\Gamma$, $B\Gamma$, AB. Show that $I\Delta$ is also antiparallel to $A\Gamma$.

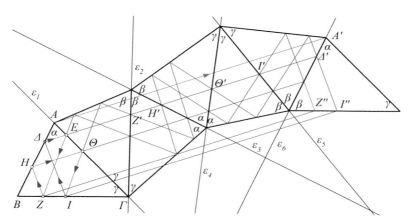

Fig. 2.86: Inscribed in $AB\Gamma$ hexagons with antiparallel sides

Hint: Consider the composition $f = f_6 \circ f_5 \circ f_4 \circ f_3 \circ f_2 \circ f_1$, where f_i is the reflection relative to line ε_i for $i = 1, 2, ..., 6$ (See Figure 2.86). This composition is a translation by the oriented segment AA' and the polygon, because of antiparallel lines, forms equal angles with the sides. $f(HZ)$, $f(I\Delta)$ are parallel, therefore also HZ, $I\Delta$ are parallel and consequently also $I\Delta$ is antiparallel to $A\Gamma$.

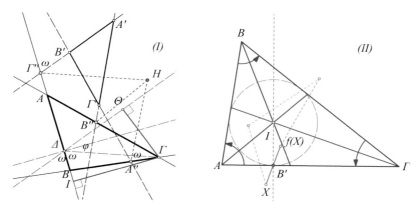

Fig. 2.87: Similarity by reflections   Rotations w.r. to triangle-angles

Exercise 2.93. Given a triangle $AB\Gamma$ and a point H, we project H to the sides of the triangle respectively at points $\{A'', B'', \Gamma''\}$ so that the angles of the lines $(HA'', \Gamma B)$, $(HB'', A\Gamma)$, $(H\Gamma'', BA)$ are all equal to the fixed angle ω (See

Figure 2.87-I). Show that the triangle $A'B'\Gamma'$, which is formed from the reflections of the lines $\{HA'', HB'', H\Gamma''\}$ relative to the corresponding sides $\{B\Gamma, \Gamma A, AB\}$ is similar to $AB\Gamma$. Also show that the center of similarity coincides with H and the similarity ratio is $\lambda = 2\cos(\omega)$.

Hint: Show first that the similarity ratio λ is independent of the position of H. Subsequently calculate this ratio by placing H on a special position, like for example, $H = \Gamma$.

Exercise 2.94. Consider the rotations $\{f_1, f_2, f_3\}$ about the vertices $\{A, B, \Gamma\}$ of the triangle $AB\Gamma$ by angles equal to the respective positively oriented angles $\{\alpha, \beta, \gamma\}$ of the triangle (See Figure 2.87-II). Show that their composition $f = f_3 \circ f_2 \circ f_1$ is a point-symmetry with respect to the contact point B' of the inscribed circle with side $A\Gamma$.

Hint: Since the angle-sum $\alpha + \beta + \gamma = \pi$, the composition f of the rotations is certainly (Proposition 2.7) a rotation by π or a half-turn i.e. a point symmetry. Show that B' remains constant under f, hence it is the center of symmetry.

Last exercise can be generalized, initially, for convex polygons with ν sides $A_1 A_2 \ldots A_\nu$. Thus, considering the angles of the polygon to be positively oriented, we can define the composition of rotations

$$f = f_\nu \circ f_{\nu-1} \circ \cdots \circ f_2 \circ f_1,$$

where each f_i is the rotation about A_i by the angle ω_i of the polygon at A_i. Since the sum of the polygon-angles $\omega_1 + \cdots + \omega_\nu = (\nu - 2)\pi$, by the general theorems for rotations, we conclude that f, for odd ν will coincide with a point symmetry and for even ν will coincide with a translation.

It is easy to see that f leaves the line $\varepsilon = A_\nu A_1$ invariant, i.e. maps ε to itself. Consequently, in the case of the point-symmetry, the center of the symmetry will be on ε and in the case of the translation, the oriented segment of translation will be parallel to ε. Next exercises discuss the simplest cases of such compositions.

Exercise 2.95. Consider the cyclic quadrilateral $AB\Gamma\Delta$, and the composition $f = f_4 \circ f_3 \circ f_2 \circ f_1$ of the reflections with respect to the sides $\{AB, B\Gamma, \Gamma\Delta, \Delta A\}$ having respective lengths $\{a, b, c, d\}$. Show that f is a translation parallel to the oriented segment $2E\Delta$ (See Figure 2.88-I)

of length $\dfrac{ac+bd}{R}$ and slope to the diagonal $B\Delta$: $\widehat{B\Delta E} = \dfrac{\pi}{2} - \widehat{B}$,

where R is the circumradius.

Hint: Separate the composition in pairs $f = (f_4 \circ f_3) \circ (f_2 \circ f_1)$. The first two reflections $g = (f_2 \circ f_1)$ define a rotation about B by an angle equal to the double of $\widehat{B}$.

This rotation can be represented also by two other reflections $g = f'_2 \circ f'_1$, where f'_1 is the reflection on BA' and f'_2 is the reflection in the diagonal $B\Delta$.

2.11. COMMENTS AND EXERCISES FOR THE CHAPTER

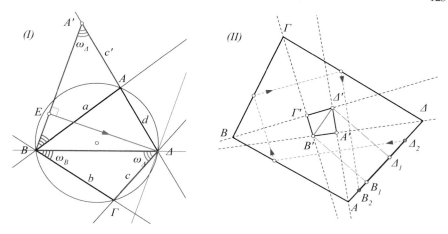

Fig. 2.88: 4 reflections composition   4 rotations composition

For this, it suffices for BA' to make with $B\Delta$ the same angle $\widehat{B}$. The second pair $h = f_4 \circ f_3$ represents also a rotation, which, can be tto represented by the pair of reflections $h = f'_4 \circ f'_3$, where $f'_3 = f'_2$ is the reflection on ΔB and f'_4 is the reflection on the line which forms with $B\Delta$ an angle equal to $\widehat{\Delta}$. Because of the hypothesis $\widehat{B} + \widehat{\Delta} = \pi$, the reflection-axes of $\{f'_1, f'_4\}$ are parallel and the composition

$$f = (f_4 \circ f_3) \circ (f_2 \circ f_1) = (f'_4 \circ f'_3) \circ (f'_2 \circ f'_1) = f'_4 \circ f'_1,$$

is the translation by the double of the distance of these parallels. The rest follows from simple calculations. Noticable is the symmetry of the expression in terms of the lengths of sides, which, by Ptolemy's theorem, can be represented also through the product of the diagonals.

Exercise 2.96. Let $f = f_4 \circ f_3 \circ f_2 \circ f_1$ be the composition of the rotations about the vertices, correspondingly, $\{A, B, \Gamma, \Delta\}$ of the quadrilateral $AB\Gamma\Delta$ by the respective positively oriented angles of the quadrilateral. Show that f is a translation by the oriented segment $\Delta_2 B_2$ of the side ΔA. This segment is the double of the projection $\Delta_1 B_1$ on ΔA of the diagonal of a quadrilateral $A'B'\Gamma'\Delta'$ (See Figure 2.88-II). This quadrilateral is cyclic and is formed by the intersection points of the inner bisectors of the angles of the given quadrilateral $AB\Gamma\Delta$.

Hint: The rotation f_1 about A can be represented as a composition of two reflections $f_1 = h_1 \circ g_1$. g_1 is the refelction with respect to the bisector AA' of $\widehat{A}$. h_1 is the reflection in the side AB. Similarly, the rotation f_2 about B can be represented as a composition $f_2 = h_2 \circ g_2$, where $g_2 = h_1$ is, again, the reflection on the side AB and h_2 is the reflection on the bisector $B\Gamma'$ of $\widehat{B}$. Then the composition of the rotations

$$f_2 \circ f_1 = (h_2 \circ g_2) \circ (h_1 \circ g_1) = h_2 \circ g_1,$$

is the composition of the two reflections in the two consecutive sides $A'\Delta'$, $\Delta'\Gamma'$ of the quadrilateral $A'B'\Gamma'\Delta'$. A similar argument shows that the composition of the rotations about Γ and Δ coincides with the composition of the reflections in $B'\Gamma'$ and $B'A'$. Apply the preceding exercise to the inscriptible quadrilateral $A'B'\Gamma'\Delta'$.

Exercise 2.97. The tile of type p_n is defined by a regular n-gon q. For even n we consider the symmetric q' of q relative to the second greater diagonal AB and delete from q the common with q' part. What remains is the tile p_n, like the p_8 in figure 2.89-I. For an odd n we do the same procedure relative to a maximal diagonal AB, from which results p_n, like the p_{15} in figure 2.89-II. Show that these non-convex tiles can tile the plane in concentric circular rings, like the p_8 in 2.89-I and the p_{15} in figure 2.89-III. Compute the number of tiles contained in each ring.

Exercise 2.98. Show that a convex quadrilateral $AB\Gamma\Delta$ is circumscriptible if and only if, the composition $f = f_4 \circ f_3 \circ f_2 \circ f_1$ of the rotations by the positive oriented angles of the quadrilateral, represents the identity transformation.

Exercise 2.99. On the sides $\{AB, \Gamma A\}$ of the triangle $AB\Gamma$ we take respectively points $\{E, \Delta\}$ such that $\{AE = x \cdot AB, \Gamma\Delta = x \cdot \Gamma A\}$. Show that the segment ΔE becomes minimal, when it is orthogonal to the median AM of the triangle. Compute the minimal length of ΔE and the value of x for which this is obtained.

Hint: Draw $\Gamma\Delta'$ parallel, equal and equal oriented to ΔE. Show that Δ' lies on the median AM of the triangle. The minimal ΔE is the altitude from Γ of $A\Gamma A'$, where A' the symmetric of A relative to M.

Exercise 2.100 (Hjelmslev's theorem). Prove that for a line α and its isometric image $\beta = f(\alpha)$ by an isometry f of the plane, the middpoints of all the segments XX', where $X' = f(X)$, are either collinear on the same line γ or coincident to a point O ([34, p.47]).

Hint: From theorem 2.15 follows that there are two types of isometries between two lines: rotations and translations preserving the orientation, and reflections and glidereflections reversing it. Examine in each case what happens with the middles of XX'. In the case of rotation apply exercise I-3.95. Show that for glidereflections all these middles are on its axis.

Fig. 2.89: Tilings with tiles of type p_n

References

26. J. Aarts (2011) Plane and Solid Geometry. Springer, Heidelberg
27. A. Adler (1906) Theorie der Geometrischen Konstruktionen. Goeschensche Verlagshandlung, Leipzig

28. Ch. Baloglou (2001) Scattered drops of Geometry (in Greek). Chromotyp, Thessaloniki
29. A. Beardon (1983) The Geometry of Discrete Groups. Springer, Berlin
30. N. Bruijn, 1981. Algebraic theory of Penrose non-periodic tilings of the plane I, II, *Proceedings of the AMS*, 84:39-66
31. H. Carslaw (1916) The elements of non-euclidean plane geometry and trigonometry. Longmans, London
32. E. Catalan (18582) Theoremes et problemes de Geometrie Elementaire. Carilian-Coeury, Paris
33. J. Coolidge (1980) A history of the Geometrical Methods. Oxford University Press, Oxford
34. H. Coxeter (1961) Introduction to Geometry. John Wiley and Sons Inc., New York
35. H. Coxeter, L. Greitzer (1967) Geometry Revisited. Math. Assoc. Amer. Washington DC
36. R. Deaux (1956) Introduction to the Geometry of Complex Numbers. Dover, New York
37. H. Eves, V.Hoggatt, 1951. Hyperbolic Trigonometry Derived from the Poincare Model, *The American Mathematical Monthly*, 58:469-474
38. H. Eves (1963) A survey of Geometry. Allyn and Bacon, Inc., Boston
39. L. Graham (1959) Ingenious Mathematical Problems and Methods. Dover Publications, New York
40. B. Gruenbaum, G. Shephard (1987) Tilings and Patterns. W.H. Freeman and company, New York
41. T. Heath (1931) A manual of Greek Mathematics. Oxford University Press, Oxford
42. J. Hadamard (1905) Lecons de Geometrie elementaire I, II. Librairie Armand Colin, Paris
43. R. Lachlan (1893) Modern Pure Geometry. Macmillan and Co., London
44. T. Lalesco (1952) La Geometrie du triangle. Librairie Vuibert, Paris
45. G. Martin (1983) Transformation Geometry. Springer, Heidelberg
46. P. Nahin (2007) Chases and Escapes, The Mathematics of Pursuit and Evasion. Princeton University Press, Princeton
47. I. Niven, 1978. Convex Polygons that Cannot tile the Plane, *The American Mathematical Monthly*, 85:785-792
48. A. Papadopoulos (2010) Nikolai I. Lobachevsky, Pangeometry. European Mathematical Society, Zürich
49. A. Pressley (2001) Elementary Differential Geometry. Springer, Heidelberg
50. R. Robinson, 1971. Undecidability and nonperiodicity of tilings of the plane, *Inventiones Mathematicae*, 12:177-209
51. A. Rosenfeld (1988) A history of Non-Euclidean Geometry. Springer, Heidelberg
52. D. Schattschneider, 1978. Tilling the Plane with Congruent Pentagons, *Mathematics Magazine*, 51:29-44
53. D. Singer, 1995. Isometries of the Plane, *The American Mathematical Monthly*, 102:628-631
54. H. Steinhaus (1983) Mathematical Snapshots, 3rd Edition. Oxford University Press, Oxford
55. G. Thomas, R. Finney (2012) Calculus. Ginn and Company, Addisson Wesley, New York
56. O. Veblen, J. Young (1910) Projective Geometry vol. I, II. Ginn and Company, New York
57. D. Wells (1991) Dictionary of Curious and Interesting Geometry. Penguin Books
58. I. Yaglom (1962) Geometric Transformations I, II, III. Mathematical Association of America, New York

Chapter 3
Lines and planes in space

3.1 Axioms for space

> The eye of understanding is like the eye of the sense; for as you may see great objects through small crannies or levels, so you may see great axioms of nature through small and contemptible instances.
>
> F. Bacon, Sylva Sylvarum

For the study of shapes of the space we need the abstract notion of the **plane** and some additional axioms to these we met in the preceding chapters. Axioms which describe the basic properties of planes, as well as, properties of lines in space and their relation to planes.

Axiom 3.1 *Three non-collinear points define precisely one plane. Every plane ε contains infinitely many points and, additionally, there also exist infinitely many points not contained in the plane ε.*

Axiom 3.2 *If a line has two common points with a plane ε, then it is contained in the plane ε.*

Axiom 3.3 *If two planes have a point in common, then they also have an entire line in common.*

Axiom 3.4 *Every plane ε separates the space into two parts, which have no points in common and are called* **half spaces**. *A line segment, which has its endpoints in different half spaces intersects the plane at exactly one point. A line segment, which has both its endpoints in the same half space, does not intersect the plane.*

Planes play the same role in space as lines on the plane. In plane geometry we isolate one of the planes of space and we examine the shapes which are contained in that plane. In **Space Geometry** or **stereometry** or **solid geometry** we examine shapes, like for example, the cube and the sphere, which extend in space and can not be confined on a single plane.

Theorem 3.1. *If two different planes intersect, then their intersection coincides with a line (See Figure 3.1).*

Proof. According to the last axiom their intersection includes at least a line AB. It cannot include however points not lying on the line AB. This because two points on the line, for example, points A, B and a point X not lying on it, define exactly one plane. Therefore if X belonged to both planes then these, according to the first axiom, would have to be coincident, contrary to the hypothesis.

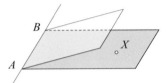

Fig. 3.1: Intersecting planes

Theorem 3.2. *Two different intersecting lines define one and only plane which contains them both.*

Proof. Consider the point of intersection O of two lines α and β. Also consider two points X and Y, the first belonging to α and the second to β (See

Fig. 3.2: Plane of two intersecting lines

Figure 3.2). According to the first axiom this defines exactly one plane ε which contains the three points A, X and Y. According to the second axiom the plane will contain both lines.

Proposition 3.1. *A line ε and a point X not lying on it define exactly one plane which contains them.*

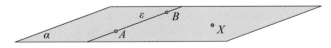

Fig. 3.3: Plane of line and point

Proof. Consider two points A and B of ε (See Figure 3.3). The plane α, which contains the three points A, B and X also contains the entire line ε (Axiom 3.2). Each plane which contains ε and X will contain also points A, B and X, therefore it will coincide with α (Axiom 3.1).

3.1. AXIOMS FOR SPACE

Proposition 3.2. *If two different lines ε and ζ are contained in a plane, then this plane is unique.*

Proof. The same as the preceding, considering two points A, B of ε and a point X of ζ etc.

Parallel planes are called two planes which do not intersect (See Figure 3.4). A line is called **parallel** to a plane when it does not intersect it. A plane is called **parallel** to a line when it does not intersect it. Two lines are called

Fig. 3.4: Parallel planes

parallel (in space) when (1) they are contained in a plane α and (2) they are parallel lines in that plane.

Two lines α and β not contained in a plane are called **non coplanar** or **skew lines** (See Figure 3.5).

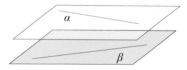

Fig. 3.5: Non coplanar or "skew" lines

I clarify first a subtle point, which has to do with two parallel lines in space. The definition of parallels in space and on the plane has a subtle difference. While on the plane two parallels are defined as two non-intersecting lines, in space this property is insufficient, as can be seen with two *skew* lines, which, while it is true that they do not intersect, they are not parallel. Parallels in space are, by assumption, contained in the same plane and their definition, as parallels in space, reduces to the definition of them being parallel on their containing plane (See Figure 3.6).

Fig. 3.6: Unique parallel postulate in space

Theorem 3.3. *From a point A not lying on the line ε, one and only line ζ parallel to ε can be drawn.*

Proof. On the plane α, which is defined by the line ε and the point A, consider the unique parallel ζ to ε (Axiom of parallels I-1.15). Any other plane, which contains ε and point A will coincide with α, therefore the parallel from A to ε will be unique.

Corollary 3.1. *For every point X not lying on the line ε, the plane which contains point X and line ε also contains the parallel ζ to ε from X (See Figure 3.7).*

Fig. 3.7: Plane of two parallel lines

Proof. Consider the unique plane α which contains ε and X. In this plane the parallel ζ to ε from X can be defined. Every plane which contains ε and X will coincide with α and the parallel of ε from X will coincide with ζ.

Proposition 3.3. *If a line ε is not parallel to a plane α and is not contained in it, then it intersects it at exactly one point (See Figure 3.8).*

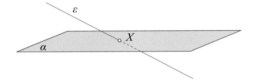

Fig. 3.8: Line intersecting plane

Proof. Obviously ε must intersect α. Otherwise it would be parallel to it. Further, if it intersected it at more than one points then, according to the second axiom, it would have to be entirely contained in α, contrary to the assumption. Therefore ε intersects α at exactly one point.

Proposition 3.4. *If the line ε intersects plane α at a point, then also every parallel ζ to it will intersect the plane at one point (See Figure 3.9).*

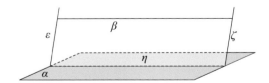

Fig. 3.9: Parallel lines $\{\varepsilon, \zeta\}$ intersecting plane

3.1. AXIOMS FOR SPACE

Proof. Consider the plane β which contains the parallels ε and ζ (See Figure 3.9). It will intersect plane α along a line η. On plane β lines ε and ζ are parallel and ε intersects η. Therefore line ζ also intersects η at a point (Corollary I-1.20), which also happens to be a point of α.

Construction 3.1 *From a point A not lying on the plane α, draw a parallel line ζ to the plane α.*

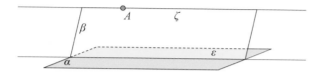

Fig. 3.10: Construction of a line parallel to a plane

Construction: Consider an arbitrary line ε of the plane α and define the plane β which contains the line ε and the point A (See Figure 3.10). On this plane draw the parallel ζ of ε from A. ζ cannot intersect α, because if it did, then the common point Σ of ζ with α would also be a point of ε, which is contradictory.

Proposition 3.5. *If the plane β passes through line XY and intersects plane α, relative to which line XY is parallel, then the intersection ε of α and β is a line parallel to XY (See Figure 3.11).*

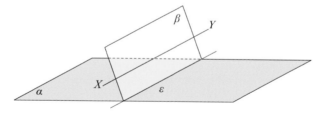

Fig. 3.11: Plane through a line intersecting a parallel plane

Proof. If line ε is not parallel but intersected XY at a point Σ, then Σ would also be on XY and on α, therefore XY would not be parallel to α, contrary to the hypothesis. This contradiction shows that ε is parallel to XY.

Theorem 3.4. *A line XY is parallel to a plane α, if and only if α contains a line ε parallel to XY.*

Proof. If line XY is parallel to the plane α (See Figure 3.12), consider a point Z of α and the plane β, which contains line XY and point Z. According to the preceding proposition, plane β would intersect α along a line parallel to XY. Conversely, suppose that line XY is parallel to line ε of the plane α and suppose that line XY intersects the plane α at a point Σ. According to the definition, the lines XY and ε are contained in plane β and line ε coincides with the intersection of α and β. Also point Σ will be contained in the intersection of α and β, therefore it will be a point of ε. This is contradictory, since line ε was supposed parallel to XY. Thus, line XY cannot intersect α.

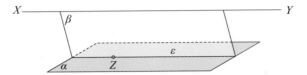

Fig. 3.12: Line parallel to plane

Proposition 3.6. *Two different intersecting planes, which are both parallel to line ε, intersect along a line ζ parallel to ε (See Figure 3.13).*

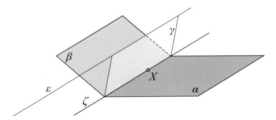

Fig. 3.13: Intersection of planes parallel to line

Proof. Suppose that planes α and β are parallel to the line ε and intersect along line ζ. Let then X be a point of ζ. According to Proposition 3.5 plane γ, which is defined from ε and point X, will intersect α and β along lines parallel to ε and pass through point X. This implies that these two lines will coincide and will also coincide with the intersection of α and β.

Exercise 3.1. Suppose that lines α and β are different and parallel as well as parallel to plane ε. Show that the plane of α and β either is parallel to plane ε or intersects it along a line γ parallel to α and β.

Exercise 3.2. Suppose that planes α and β pass through the line ε, which is parallel to plane ζ. Also suppose that α and β intersect the plane ζ. Show that their intersections are two parallel lines of ζ.

Exercise 3.3. Consider all the planes α, which pass through the line ε. Let also ζ be a plane not containing ε and not parallel to it (See Figure 3.14). Show that the lines which are defined as intersections of the planes α and ζ all pass through a fixed point of ζ. How is this point defined from the given data?

3.2. PARALLEL PLANES

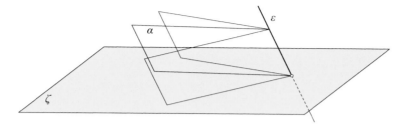

Fig. 3.14: Planes α through line ε and their intersection with ζ

3.2 Parallel planes

> In other endeavors it comes after they are completed. In philosophy, on the contrary, delight and knowledge are combined, because pleasure doesn't follow learning, but they happen simultaneously.
>
> *Epicurus, Address*

Proposition 3.7. *If two planes α and β are parallel, then every line of one is parallel to the other plane (See Figure 3.15).*

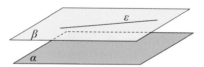

Fig. 3.15: Line parallel to plane

Proof. If the line ε of α intersected β at X, then α and β would have X as a common point, contrary to the assumption. Thus, line ε cannot intersect β.

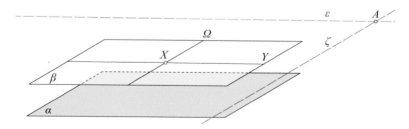

Fig. 3.16: Plane of two parallels to plane

Proposition 3.8. *The plane β of two intersecting lines, which are parallel to plane α, is parallel to α (See Figure 3.16).*

Proof. Suppose that the plane β of the intersecting lines XY and $X\Omega$ intersects the plane α along line η. We'll show that this is contradictory. To this, consider a point A outside both planes and draw from A the parallel ε to XY and the parallel ζ to $X\Omega$. Because ε is parallel to XY it intersects neither β nor α (Proposition 3.4, Theorem 3.4). Therefore, according to Proposition 3.6 the intersection η of α and β will be parallel to ε. Similarly we also show for ζ, that it will be parallel to η. From point A then we'll have two parallels to ζ, which is contradictory and shows that α and β do not intersect.

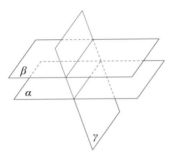

Fig. 3.17: Plane intersecting parallel planes

Proposition 3.9. *A plane γ intersecting two parallel planes α and β, intersects them along parallel lines (See Figure 3.17).*

Proof. The two lines-intersections of γ with α and β cannot intersect, because then the planes would, too. Also they are contained in plane γ, therefore, by definition, they are parallel.

Theorem 3.5. *From a point X, not lying on the plane α, one and only one plane β parallel to α can be laid (See Figure 3.18).*

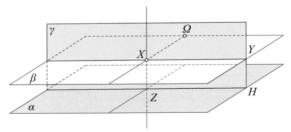

Fig. 3.18: Plane from point, parallel to an other plane

Proof. From Proposition 3.8 follows that there exists a plane β through X parallel to α. This plane is unique. Indeed, if there was also a second plane β' through X then β and β' would intersect along a line η passing through

3.2. PARALLEL PLANES

X, which without loss of generality we may suppose different from XY. Consider then a point Z on α and consider the plane γ which contains points X, Y and Z. This intersects plane β along XY, β' along a line XY' (not drawn) and α along line ZH. Because the planes are parallel, the lines XY and XY' will be two different parallels from point X to line ZH of plane γ (Proposition 3.9). This is contradictory and shows the uniqueness of the plane β parallel to α from X.

Corollary 3.2. *A plane γ, which intersects plane β, will also intersect every other plane α parallel to β.*

Proof. Suppose that the plane γ intersects β but not its parallel α. Then from one point X of the intersection of β and γ can be drawn two different planes parallel to α: β and γ, something contradictory according to Theorem 3.5. Therefore if γ intersects β it must also intersect α.

Corollary 3.3. *For every line ε parallel to plane β, there exists one exactly plane α which contains ε and is parallel to β.*

Proof. From a point X of ε draw a parallel ζ to β different from ε. According to Proposition 3.8 the plane α of ε and ζ is parallel to β. According to Theorem 3.5 α is unique with this property.

Proposition 3.10. *If the line ε intersects plane α, then it will intersect every other plane β parallel to α (See Figure 3.19).*

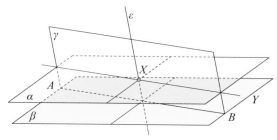

Fig. 3.19: Line intersecting parallel planes

Proof. Suppose that the line ε intersects plane α at point X but does not intersect its parallel plane β. Draw an arbitrary plane γ containing ε and intersecting α along the line XY. Because planes α, β are parallel and γ intersects α it will also intersect β (Corollary 3.2) and in fact by parallel lines XY and AB (Proposition 3.9). It follows, that on plane γ we have from point X two parallels to AB: line XY and line ε, for which we supposed that it does not intersect β, therefore also AB. This contradiction shows that if the line ε intersects α, then it must also intersect β.

Corollary 3.4. *If the line ε is parallel to the plane α, then it is also parallel to every plane β parallel to α.*

Corollary 3.5. *From a point X not lying on the plane β are drawn infinitely many lines parallel to plane β. All these lines are contained in the unique plane α which passes through X and is parallel to β.*

Proposition 3.11. *Parallel planes excise on parallel lines intersecting them equal line segments (See Figure 3.20).*

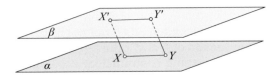

Fig. 3.20: Parallels between parallels

Proof. Suppose that the parallel planes α and β excise on parallel lines the segments XX' and YY'. Then $XX'YY'$ is a parallelogram, therefore XX' and YY' are equal.

Proposition 3.12. *If plane β is parallel to α and plane γ is parallel to β, then plane γ is also parallel to α.*

Proof. If planes α and γ were not parallel, then from their intersection, which is a line ε not intersecting β, therefore parallel to it, we would have two different planes parallel to β, which is contradictory (Corollary 3.3). Consequently α and γ must be parallel.

Exercise 3.4. Show that two circles contained in two different planes α and β have at most two common points.

3.3 Angles in space

> What is easy and obvious is never valued; and even what is in itself difficult, if we come to the knowledge of it without difficulty, and without any stretch of thought or judgment, is but little regarded.
>
> *D. Hume, A Treatise of Human Nature*

We measure angles in space the same way we measure them on the plane. Two different and intersecting lines α and β define exactly one plane ε (Theorem 3.2) and on this plane we measure the angles formed by the lines the same way we do for the angles of that plane (I-§ 1.4) (See Figure 3.21). In particular, two lines in space are called **orthogonal**, when they are contained in the same plane ε and they are orthogonal lines of this plane ε.

3.3. ANGLES IN SPACE

Remark 3.1. In the next section we'll see that the notion of angle in space is extended to include non coplanar (skew) lines, which, by definition, are not parallel yet they do not intersect.

Fig. 3.21: Angle in space

Theorem 3.6. *Angles with parallel and equally oriented sides are equal.*

Proof. Suppose that the two angles $\widehat{XOY}$ and $\widehat{X'O'Y'}$ have respective sides parallel (See Figure 3.22). Then they are either contained in different or co-

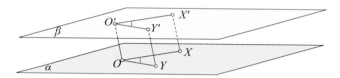

Fig. 3.22: Angles with parallel sides

incident planes. If they are contained in the same plane, then according to Corollary I-1.30 they form equal angles. If they are contained in different planes then these planes are parallel (Proposition 3.8). In this case define on their sides the points such that $|OX| = |O'X'|$ and $|OY| = |O'Y'|$. Then $OXX'O'$ and $OYY'O'$ are parallelograms and consequently XX' and YY' are parallel, therefore $XYY'X'$ also is a parallelogram. It follows that the triangles OXY and $O'X'Y'$ have equal respective sides, therefore they are congruent and consequently their angles at O and O' are equal.

Corollary 3.6. *The two planes XOO' and YOO' intersect along line OO'. Then two parallel planes α and β, which intersect OO', intersect the planes XOO' and YOO' along lines forming equal angles ($\widehat{XOY} = \widehat{X'O'Y'}$ in figure-3.22).*

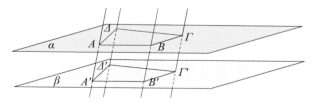

Fig. 3.23: Congruent polygons defined through a pencil of parallel lines

Corollary 3.7. *Let $AB\Gamma\Delta...$ be a polygon on the plane α. Consider also a pencil of parallel lines through its vertices. On every plane β, parallel to α the pencil defines a polygon $A'B'\Gamma'\Delta'...$ congruent to $AB\Gamma\Delta...$ (See Figure 3.23).*

Proof. According to Proposition 3.11, $ABB'A'$, $B\Gamma\Gamma'B'$, ..., are all parallelograms, consequently the polygons have respective sides equal $|AB| = |A'B'|$, $|B\Gamma| = |B'\Gamma'|$, ... According to Theorem 3.6 the polygons also have respective angles equal.

Fig. 3.24: Line orthogonal to line from point

Theorem 3.7. *From a point A in space, not lying on the line ε, exactly one line AB orthogonal to it can be drawn (See Figure 3.24).*

Proof. Point A and line ε define a plane ζ and we apply to it theorem I-1.8.

Point B of ε, which is defined by the preceding proposition, is called the **projection** of the point A on ε. This definition generalizes to space the corresponding notion of projection onto a line in plane geometry (I-§ 1.12).

Exercise 3.5. Show that, if the lines κ and λ are parallel and are contained respectively in intersecting planes α and β, then they are also parallel to the line η of the intersection of the two planes.

Exercise 3.6. Show that, if three lines pass through the same point and intersect a fourth at different points, then all four lines are on the same plane.

Exercise 3.7. Show that a circle contained in a plane α and plane β, different from α, have at most two common points.

Exercise 3.8. Show that if the respective sides of the triangles $AB\Gamma$ and $A'B'\Gamma'$ intersect, then the lines AA', BB' and $\Gamma\Gamma'$ either pass through the same point O or they are parallel (See Figure 3.25).

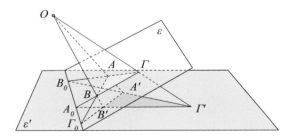

Fig. 3.25: Desargues in space

Hint: The intersection points A_0, B_0, Γ_0 of the pairs of respective sides of the triangles $(AB, A'B')$, $(B\Gamma, B'\Gamma')$ and $(\Gamma A, \Gamma'A')$ are contained in the line which

3.4. SKEW LINES   139

is the intersection of planes ε and ε' of the triangles. Suppose AA' and BB' intersect at O. Point O is then contained in the plane of BB' and $\Gamma\Gamma'$, as well as in the plane of AA' and $\Gamma\Gamma'$. Therefore it will also be contained in the intersection of these planes, which is $\Gamma\Gamma'$. The proof for the case where AA' and BB' are parallel is similar.

Remark 3.2. Last exercise shows, that the theorem of Desargues (I-§ 5.18) follows from figure 3.25 through a process of rotation of the plane ε about the line containing points $\{A_0, B_0, \Gamma_0\}$. Indeed, if by rotating ε about this line, we simultaneously consider also the corresponding point O, then in the limiting position where ε will coincide with ε' point O will take the position of a point of ε' and the figure of the theorem of Desargues will result.

3.4 Skew lines

> To kill an error is as good a service as, and sometimes even better than, the establishing of a new truth or fact.
>
> *Charles Darwin*

As it was already mentioned § 3.1, **skew** or **non coplanar** are called two lines in space for which there is no plane containing them both.

Fig. 3.26: Plane through line β parallel to the skew line α

Theorem 3.8. *Let $\{\alpha, \beta\}$ be two skew lines. Then, there exists exactly one plane ε_β containing β and parallel to α (See Figure 3.26).*

Proof. From a point X of β draw the parallel α_1 to α. The plane ε_β, which contains β and α_1 is parallel to α (Theorem 3.4). If there was another plane ε containing β and parallel to α, then their intersection, which is the line β, would be parallel to α (Proposition 3.6), contradicting the assumption.

Proposition 3.13. *Given two skew lines $\{\alpha, \beta\}$, there exist exactly two parallel planes $\{\varepsilon_\alpha, \varepsilon_\beta\}$, such that ε_α contains α and ε_β contains β (See Figure 3.27).*

Proof. Consider the planes ε_α and ε_β, which are defined in the preceding proposition. These are parallel, because if they intersected, then their intersection would also be parallel to α and to β (Proposition 3.5), which is

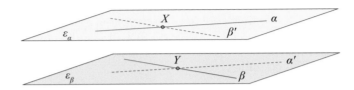

Fig. 3.27: The parallel planes $\{\varepsilon_\alpha, \varepsilon_\beta\}$ of two skew lines $\{\alpha, \beta\}$

contradictory. The preceding proposition shows also that there is no other pair of parallel planes ε' and ε'' containing α and β respectively. Because if there was, and it was for example ε' different than ε_α, then we would have two planes through α parallel to β, hence a contradiction to the preceding proposition.

Last proposition shows the characteristic property of two skew lines. They are simply two non parallel lines contained in two parallel planes. The angle of line α and of the parallel β' of β from an arbitrary point X of α, as it was defined in Theorem 3.8, is independent of the position of the point X. According to Theorem 3.6, it is also equal to the angle between the lines β and α', where α' is a parallel to α from an arbitrary point Y of β. This angle is called the **angle of skew lines**. In the case where this angle is right the two lines α and β are called **skew orthogonal**.

Exercise 3.9. Show that, if the lines $\{\alpha, \alpha'\}$ are contained in two parallel planes $\{\varepsilon, \varepsilon'\}$ and are not parallel, they are skew.

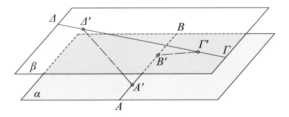

Fig. 3.28: Skew lines $\{A'\Delta', B'\Gamma'\}$ from points on skew lines $\{AB, \Gamma\Delta\}$

Exercise 3.10. Let $\{A', B'\}$ be two different points on a line AB and $\{\Gamma', \Delta'\}$ two different points on the skew line $\Gamma\Delta$ of AB. Show that the lines $A'\Delta'$ and $\Gamma'B'$ are skew (See Figure 3.28).

Hint: It suffices to show that the lines $A'\Delta'$ and $\Gamma'B'$: (i) do not intersected and : (ii) they are not parallel. (i) If they intersected, then the intersection point, X say, would also lie on planes $B'\Gamma'\Delta'$ and $A'\Gamma'\Delta'$, which intersect along line $\Gamma'\Delta'$, therefore X would be on $\Gamma'\Delta'$. Similarly it follows that X

3.5. LINE ORTHOGONAL TO PLANE

would lie on $A'B'$, something which is contradictory. (ii) If lines $A'\Delta'$ and $B'\Gamma'$ were parallel, then there would exist a plane ε, which would contain both of them. Then however plane ε would contain also lines $A'B'$ and $\Gamma'\Delta'$, something which is also contradictory.

Figure $A'B'\Gamma'\Delta'$ of the preceding exercise is called **skew quadrilateral** and obviously it is the generalization of the plane quadrilateral in space. A common property it shares with plane quadrilaterals, which, by the way, is proved the same way (Proposition I-2.7), is that of the next exercise.

Exercise 3.11. *The middles of the sides of a skew quadrilateral form a parallelogram.*

Exercise 3.12. *The line segments which join the middles of opposite sides in a skew quadrilateral intersect at their middle.*

3.5 Line orthogonal to plane

> Facts, however, will ultimately prevail; we must
> therefore take care that they be not against us.
>
> Francis Bacon

A line ζ is called **orthogonal to plane** ε, when ζ intersects ε at a point O and is simultaneously orthogonal to two different lines α and β of ε which pass through point O. We say then also that the **plane ε is orthogonal to line ζ**.

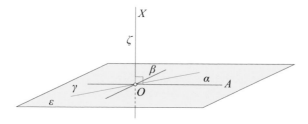

Fig. 3.29: Line orthogonal to plane

The picture of orthogonality between line and plane (See Figure 3.29) is completed by the next propositions which show that a plane orthogonal to a line OX is generated by a a line OA which is rotated around the axis OX remaining always orthogonal to it.

Theorem 3.9. *Line XY, intersecting plane ε at O and being orthogonal to it, is also orthogonal to every line γ of ε passing through point O (See Figure 3.30).*

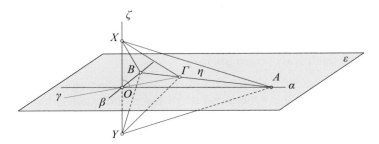

Fig. 3.30: Orthogonality of XY to all lines $O\Gamma$

Proof. Suppose that line $\zeta = XY$ is orthogonal to lines α and β of plane ε, which pass through point O. Let also γ be an arbitrary line through O, different from α and β. We'll show that XY is also orthogonal to γ. For this we consider an arbitrary point Γ of γ and we draw the auxiliary line η through Γ, which intersects α and β at points A and B respectively. Let us also take X, Y on ζ at equal distances from O. Then the triangles XOA, YOA are right with equal orthogonal sides, therefore congruent and consequently AX and AY will be equal. Similarly we show also that BX and BY are equal. This implies that the triangles ABX and ABY are congruent (SSS-criterion). Consequently triangles $B\Gamma X$ and $B\Gamma Y$ will be congruent (SAS-criterion), therefore ΓX and ΓY will be equal, triangle $X\Gamma Y$ will be isosceles and because point O is the middle of XY, $O\Gamma$ will be orthogonal to XY.

Corollary 3.8. *If the plane ε is orthogonal to XY at a point O, then every line OA orthogonal to XY at O is contained in the plane ε (See Figure 3.30).*

Proof. If line OA is orthogonal to XY at the point O, this means that on the plane ζ, which contains XY and OA, these two lines are orthogonal. If OA was not contained in plane ε, then the plane ζ would intersect plane ε by line OA', according to the preceding proposition, also orthogonal to XY, therefore on plane ζ we would have two orthogonals to XY from O. This contradiction shows that OA is contained in ε.

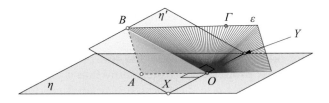

Fig. 3.31: Medial plane of XY

3.5. LINE ORTHOGONAL TO PLANE

Proposition 3.14. *The geometric locus of space points A, which are equidistant from two points X and Y, is a plane ε orthogonal to the line segment XY at its middle O (See Figure 3.31).*

Proof. Consider an arbitrary plane η containing the segment XY and the medial line (Corollary I-1.7) OA of XY on that plane. Similarly define also the medial line OB of XY for a second plane η' containing XY. The two lines OA and OB define a plane ε, which according to Theorem 3.9 will have all its lines $O\Gamma$, which pass through point O orthogonal to XY, therefore will be coincident with the medial lines of XY in the corresponding planes which contain XY and Γ. The plane ε then, which is defined from OA and OB, is contained in the geometric locus. Conversely, if a point Γ is contained in the locus, then $O\Gamma$ will be orthogonal to XY at O and, according to Corollary 3.8, will be contained in ε, therefore point Γ will be contained in plane ε. This completes the proof of the fact that plane ε coincides with the aforementioned geometric locus.

The plane, which is defined from the preceding proposition, is called **medial plane** of the line segment XY.

Proposition 3.15. *If line α is orthogonal to plane ε then, every line β parallel to α will be orthogonal to ε. And conversely, two lines α and β orthogonal to plane ε will be parallel (See Figure 3.32).*

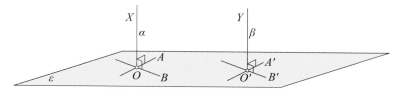

Fig. 3.32: orthogonals to ε are parallel

Proof. Suppose that the orthogonal $\alpha = XO$ intersects ε at point O and OA, OB are two lines of ε, to which α is orthogonal (by definition). Let now O' be the intersection point of the parallel $\beta = O'Y$ of α with ε. From O' draw lines $O'A'$, $O'B'$ respectively parallel to OA and OB. According to Theorem 3.6 angles $\widehat{XOA}$ and $\widehat{YO'A'}$ are equal and angles $\widehat{XOB}$ and $\widehat{YO'B'}$ are also equal. This implies that line $\beta = YO'$ is orthogonal to $O'A'$ and $O'B'$, therefore, according to definition, orthogonal to plane ε.

Conversely, if lines α and β are orthogonal to ε but β is not parallel to α, then draw from O' the parallel α' to α. According to the first part of the proof line α' will be orthogonal to plane ε, therefore the plane of α' and β will intersect plane ε along a line γ which will be orthogonal to α' and β also, something contradictory. The contradiction shows that β must be parallel to α.

Construction 3.2 *From a point A not lying on the plane ε construct the orthogonal to it (See Figure 3.33).*

Construction: Consider two parallel lines α and β of plane ε and on the corresponding planes defined by α, β and point A draw the orthogonals AX and AY to α and β from A. Form the triangle AXY and draw its altitude AB to XY. AB is the requested orthogonal from A to the plane ε.

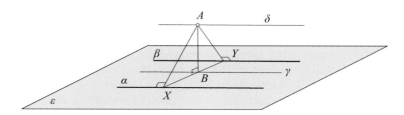

Fig. 3.33: Construction of the orthogonal from a point

To show this it suffices to find two lines of ε orthogonal to AB at B. Obviously one is XY. The other is the parallel γ of α and β from B. The fact that this is also orthogonal to AB is seen as follows. Suppose δ is the parallel of α and β from A. δ is orthogonal to AX and to AY, therefore it is orthogonal to their plane. γ is parallel to δ, therefore it itself is also orthogonal to the plane AXY Proposition 3.15. According to Theorem 3.9 line γ will also be orthogonal to BA, therefore AB will be orthogonal to γ.

Proposition 3.16. *From a point A not lying on the plane ε only one orthogonal to it can be drawn.*

Proof. The preceding construction shows that there exists one such orthogonal from A. If there were a second one AB', then the triangle ABB' would have at B and B' right angles, therefore the sum of its angles would be greater than π. This contradiction shows that there cannot be another orthogonal AB' from A.

For every point A, not lying on the plane ε, the corresponding point B, at which the orthogonal from A intersects ε, is called **projection of the point** A **onto the plane** ε (See Figure 3.33). For every point A of the plane ε we consider that its projection on ε coincides with the point itself. More generally, the set of projections X' of the points X of a shape Σ in space onto the plane ε is a shape Σ' of the plane ε, which is called **projection of the shape** Σ onto the plane. The length $|AB|$ of the orthogonal from point A to its projection B on the plane ε is called **distance of the point from the plane**.

Exercise 3.13. Show that the projection on the plane ε of a line α, non-orthogonal to the plane, is a line α' of ε and the ratio between three points of α is equal to the ratio between their projections on α'.

3.5. LINE ORTHOGONAL TO PLANE

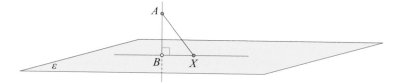

Fig. 3.34: Distance of a point from a plane

Proposition 3.17. *The distance $|AB|$ of a point A from the plane ε is less than the distance $|AX|$, where X is a point of ε different from the projection B of A on ε.*

Proof. For every such point X, the corresponding triangle ABX is right with orthogonal sides AB, BX and hypotenuse AX (See Figure 3.34), therefore $|AX| > |AB|$.

If ε and ε' are parallel planes then the distance $|AB|$ of a point A of ε from ε' is independent of the position of A on ε (Proposition 3.11) and is called **distance** of the two planes ε and ε'. Similarly, if the line α is parallel to the plane ε, A is a point of the line α and B its projection onto ε, the distance $|AB|$ is independent of the position of A on the line α and is called **distance** of the parallel line from the plane. Given two skew lines the distance between the parallel planes which contain them (Proposition 3.13) is called **distance of the skew lines**.

Proposition 3.18. *Line α is orthogonal to the plane ε, if and only if it is orthogonal or skew orthogonal to two non parallel lines β and γ of the plane ε (See Figure 3.35).*

Fig. 3.35: Skew orthogonality of α to lines β, γ of ε

Proof. If line α is orthogonal to ε, then it is orthogonal to every line which passes through its trace (Theorem 3.9) and consequently it is skew orthogonal to every line β of the plane ε, since every such line is parallel to one which passes through its trace. Conversely, if α is orthogonal or skew orthogonal to two lines β, γ intersecting at point O of the plane ε, then its parallel α' from O will be orthogonal to β, γ therefore orthogonal to their plane ε. Consequently the parallel α to α' will also be orthogonal to the plane ε (Proposition 3.15).

Construction 3.3 *Construct the orthogonal to plane ε from a point O on ε.*

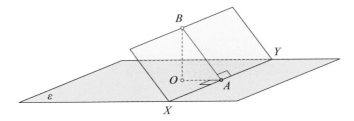

Fig. 3.36: Construction of the orthogonal from a point O of the plane

Construction: Consider an arbitrary line XY of ε not containing O (See Figure 3.36). From O draw the orthogonal OA to line XY. Consider next an arbitrary plane passing through XY and on this plane draw AB also orthogonal to XY. If angle $\widehat{OAB}$ is right rotate the plane about XY so that it becomes acute. On the plane of points $\{O, A, B\}$ draw the orthogonal OB to OA from point O of it. This is the requested orthogonal. This because, by construction, OB is orthogonal to OA but also (skew) orthogonal to XY. Therefore OB is orthogonal to two lines of the plane ε, hence orthogonal to it.

Proposition 3.19. *From a point O of plane ε only one orthogonal to it may be drawn.*

Proof. The preceding construction defines an orthogonal OB from O. If there existed a second OB' then the plane OBB' would intersect plane ε by line α containing point O from which we would have two orthogonals to α contained in plane OBB'. This is contradictory and shows that there exists only one orthogonal from O to ε.

Corollary 3.9. *There exists exactly one plane ε orthogonal to line XY at its point O.*

Proof. Consider an arbitrary plane α containing XY and a line of this plane η, orthogonal to XY at ts point O. Consider next the plane β, which contains η and XY, as well as line ζ orthogonal to β at its point O. Plane γ which is defined from the lines η and ζ is orthogonal to XY at its point O. According to Corollary 3.8 this plane will contain every line orthogonal to XY at its point O, therefore it is unique.

Corollary 3.10. *From a point A not lying on the line ε a unique plane orthogonal to ε can be drawn (See Figure 3.37).*

Proof. Consider the plane α, which contains line ε and point A. Define on plane α the projection B of A onto line ε. The requested plane ζ is the orthogonal to ε through B. This, because plane ζ contains the line BA, therefore also point A (Corollary 3.8). Also, every other plane ζ', which contains A and is

3.5. LINE ORTHOGONAL TO PLANE

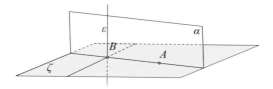

Fig. 3.37: Plane through a point orthogonal to a line

orthogonal to ε, must pass through B, since, otherwise, we would have on the plane of ε and A two orthogonals from A to ε. Plane ζ', consequently, coincides with plane ζ (Corollary 3.9).

Theorem 3.10. *Planes α, β orthogonal to the same line ε or to parallel lines ε, ε' are either coincident or parallel.*

Proof. To begin with, if the planes are orthogonal, respectively, to the parallel lines ε, ε', then (Proposition 3.15) both will be orthogonal to each one of them. It suffices then to examine the case where both planes are orthogonal to the same line ε. If the planes are not coincident, then we must show that they are parallel. Indeed, had they not been, then they would have common a whole line η. Suppose X is a point of η, which, let us suppose is not contained in line ε. Then from X we would have two different planes orthogonal to ε, something contradictory (Corollary 3.10). The fact that X, intersection point of η, α and β, may be chosen so that it may not be contained in ε is obvious. This, because, otherwise line η would coincide with ε, which is contradictory (why?)

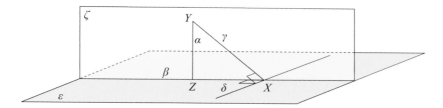

Fig. 3.38: of three orthogonals

Theorem 3.11 (Theorem of three orthogonals). *Three lines α, β, γ form triangle XYZ and a fourth one δ passes through vertex X of the triangle. Then (See Figure 3.38):*

1. *If line α is orthogonal to the plane of $\{\beta, \delta\}$ and line δ is orthogonal to β, then δ is also orthogonal to γ.*
2. *If line α is orthogonal to the plane of $\{\beta, \delta\}$ and line γ is orthogonal to δ, then β is also orthogonal to δ.*

3. If line δ is orthogonal to the plane of $\{\beta, \gamma\}$ and line α is orthogonal to β, then line α is also orthogonal to the plane of β, δ.

Proof. All three properties are consequences of Proposition 3.18.

(1): If α is orthogonal to the plane ε of $\{\beta, \delta\}$ and δ is orthogonal to β, then δ is orthogonal to β and skew orthogonal to α, consequently δ is orthogonal to the plane of α, β, therefore also to every line which passes through X, such as γ.

(2): If α is orthogonal to the plane of $\{\beta, \delta\}$ and γ is orthogonal to δ, then δ is orthogonal to γ and skew orthogonal to α, therefore orthogonal to their plane as well as to every line of their plane which passes through X, such as β.

(3): If δ is orthogonal to the plane of $\{\beta, \gamma\}$ and α is orthogonal to β, then α is orthogonal to β and skew orthogonal to δ, therefore also to their plane.

Proposition 3.20. *There exists exactly one line XY simultaneously orthogonal to two non coplanar (skew) lines α and β (See Figure 3.39).*

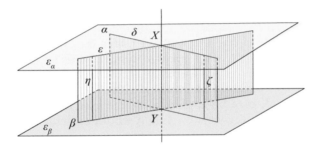

Fig. 3.39: Common orthogonal of two non coplanar lines

Proof. Consider the two parallel planes ε_α, ε_β which contain respectively the skew lines α and β (Proposition 3.13). Also from an arbitrary point of α draw the orthogonal ζ to these planes and from an arbitrary point of β draw the orthogonal η again to these planes. This forms two new planes, δ, which contains α and ζ and ε, which contains β and η. These two planes are parallel to a line orthogonal to the two planes (Theorem 3.4), therefore their intersection θ will be parallel to the orthogonal to the two planes (Proposition 3.6), consequently it will also be orthogonal to the two planes. Line θ meets line α at a point X, because its parallel ζ meets it by assumption. Similarly line θ meets β also at a point Y, therefore θ coincides with XY. XY is the unique line simultaneously orthogonal to both α and β. This, because if there were a second $X'Y'$ orthogonal to both α and β, then $X'Y'$ would be also orthogonal to the parallel planes ε_α, ε_β, consequently it would be parallel to XY (Proposition 3.15). This however, would have as consequence line α, which contains points X, X' and β, which contains points Y, Y' to be on the same plane, which is contradictory.

3.5. LINE ORTHOGONAL TO PLANE

The line XY, which is guaranteed by the preceding proposition, is called (**common perpendicular**) of the skew lines α and β.

Exercise 3.14. Show that the medial planes of the sides of the triangle $AB\Gamma$ intersect along a line passing through the circumcenter O of the triangle and orthogonal to its plane. Also show that for a point X on this line and every point Σ of the circumcircle κ of $AB\Gamma$ the length $X\Sigma$ is constant when Σ varies on κ (See Figure 3.40-I).

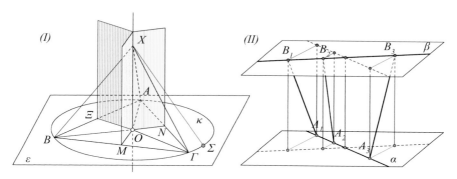

Fig. 3.40: Orthogonal line from circumcenter, Proportionals on skew lines

Hint: Let OX be the intersection of the medial planes of segments AB and $B\Gamma$. Every point X of this line will have $|XA| = |XB|$, as belonging to the medial plane of AB and $|XB| = |X\Gamma|$ as belonging to the medial plane of $B\Gamma$. From these two, follows $|XA| = |X\Gamma|$, therefore X is on the medial plane of $B\Gamma$. Line OX is skew orthogonal to AB and $B\Gamma$, therefore Proposition 3.18 orthogonal to ε. This shows the first part of the exercise. For the second, consider a point Σ of the circumcircle κ and the triangle $XO\Sigma$. This triangle is right at O and congruent to $XO\Gamma$, as having respectively equal orthogonals with it, therefore $|X\Sigma| = |X\Gamma|$.

Exercise 3.15. Given two skew lines $\{\varepsilon, \zeta\}$, when there is a plane containing ζ and orthogonal to ε?

Exercise 3.16. From the points X of a line β, which is skew to α, we draw lines XA orthogonal to α. When does it happen that these lines are contained in a plane?

Exercise 3.17. Construct a line β parallel to a given line α and intersecting two circles $\{\kappa, \kappa'\}$ contained in two different parallel planes $\{\varepsilon, \varepsilon'\}$.

Exercise 3.18. On the skew lines $\{\alpha, \beta\}$ we take points $A_1, A_2, \ldots$ and $B_1, B_2, \ldots$ which define proportional segments $\left\{ \frac{|A_1A_2|}{|B_1B_2|} = \frac{|A_2A_3|}{|B_2B_3|} = \ldots \right\}$ (See Figure 3.40-II). Show that the lines $\{A_1B_1, A_2B_2, \ldots\}$ are parallel to a fixed plane.

Exercise 3.19. Let $\{\alpha, \beta\}$ be two skew lines. From an arbitrary point A of α we draw the segment AB orthogonal to β, with B on β. To find the geometric locus of the middle M of AB.

3.6 Angle between line and plane

> When the five kinds of grain were brought to maturity, the people all obtained a subsistence. But men possess a moral nature; and if they are well fed, warmly clad, and comfortably lodged, without being taught at the same time, they become almost like the beasts.
>
> *Mencius, Teng Wen Gong I*

In space a new possibility arises: the **angle between a line and a plane**. A line XY, which intersects plane ε at a point O, defines various angles through the planes which contain XY. Indeed, every point A of the plane ε defines the plane which contains XY and point A. This intersects plane ε along a line OA and, consequently, defines an angle $\widehat{XOA}$ (See Figure 3.41). Of all angles created this way, there exists one of minimum measure. This angle is called angle between line XY and plane ε. The existence of this angle is guaranteed by the following proposition.

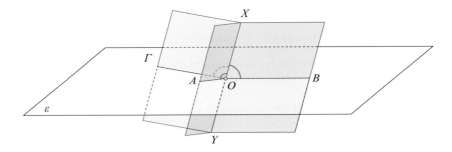

Fig. 3.41: Angle between line XY and plane ε

Theorem 3.12. *Let the line ζ intersect non orthogonally the plane ε at the point O, let also X be a point of ζ different from O. For every plane XOZ containing ζ consider the angle XOZ formed from ζ and the intersection OZ of this plane with ε. Among all these angles the minimal one is $\widehat{XOY}$, where Y is the projection of point X on the plane ε (See Figure 3.42).*

Proof. Consider the circle of ε with center O and radius OY (See Figure 3.42). Every plane which contains ζ intersects the plane along a line OZ passing through point O and defining two points on the circle. Let Z be one of them. Triangle XYZ is right with hypotenuse XZ greater than the orthogonal XY. Consequently, triangles XOZ and XOY have their sides adjacent to their vertex O respectively equal and the opposite side XZ greater than the opposite side XY. From Corollary I-1.17 follows that the angle $\widehat{XOZ}$ will be greater than $\widehat{XOY}$.

3.6. ANGLE BETWEEN LINE AND PLANE

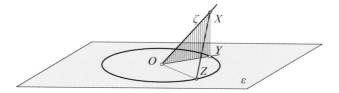

Fig. 3.42: Definition of the angle between line and plane

Remark 3.3. According to this definition of the angle between line and plane, a line which forms a right angle with the plane must coincide with an orthogonal to the plane, in the sense given in § 3.5. That this is indeed the case, follows from the observation, that if the angle $\widehat{XOZ}$ is minimal, then its supplementary is maximal, among all the angles, which result by rotating OZ around point O on the plane ε (See Figure 3.42). This way, if the minimum is a right angle, then it will coincide with the maximum and consequently all the angles $\widehat{XOZ}$ will be right.

Exercise 3.20. Show that two parallel lines form equal angles with the plane ε.

Exercise 3.21. Two planes α and β intersect along the line AB. A line OX is orthogonal to AB at its point O and is contained between the two planes. Suppose δ is the plane which passes through point O and is orthogonal to AB. Show that every other plane γ, which contains OX, intersects planes α and β by an angle $\widehat{\Delta OE}$, which is greater than angle $\widehat{TOY}$ defined by the intersection of planes $\{\alpha, \beta\}$ with plane δ (See Figure 3.43).

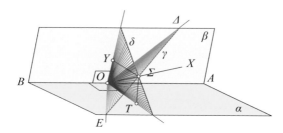

Fig. 3.43: Minimal angle between two planes

Hint: To begin with, plane δ, being orthogonal to AB at O, will also contain every line orthogonal to AB at O, therefore also OX. Also, for an arbitrary point Σ of OX, δ will contain the lines ΣT and ΣY, where T and Y are the projections of Σ to α and β respectively. From the preceding proposition, the angle $\widehat{\Sigma OT}$ is smaller than $\widehat{\Sigma OE}$. Analogously angle $\widehat{\Sigma OY}$, is smaller than $\widehat{\Sigma O\Delta}$, therefore also $\widehat{TOY} = \widehat{\Sigma OT} + \widehat{\Sigma OY} < \widehat{EOX} + \widehat{XO\Delta} = \widehat{EO\Delta}$.

3.7 Theorem of Thales in space

> Concepts, classes, and species, on the other hand, can become its objects only very indirectly; and so the vulgar and uncultured have no thought or desire for universal truths, whereas the genius overlooks and ignores what is individual.
>
> A. Schopenhauer, On Philosophy and its Methods

Theorem 3.13. *Parallel planes excise from lines intersecting them proportional segments (See Figure 3.44).*

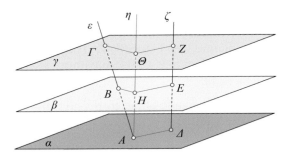

Fig. 3.44: Lines intersected by parallel planes

Proof. Let the lines ε and ζ intersect the parallel planes α, β, γ at points respectively $\{A, B, \Gamma\}$ and $\{\Delta, E, Z\}$. we'll show that

$$\frac{|AB|}{|B\Gamma|} = \frac{|\Delta E|}{|EZ|}.$$

For this, consider the parallel line η to ζ from point A, intersecting the planes β and γ respectively at points H and Θ. According to Proposition 3.11,

$$|\Delta E| = |AH|, |EZ| = |H\Theta| \Rightarrow \frac{|\Delta E|}{|EZ|} = \frac{|AH|}{|H\Theta|}.$$

According to the theorem of Thales applied to the plane of the lines ε, η (Theorem I-3.13) we have

$$\frac{|AB|}{|B\Gamma|} = \frac{|AH|}{|H\Theta|}.$$

Last relation combined with the preceding one leads to the proof.

Theorem 3.14. *Line segments, defined by a point pencil on parallel planes, are proportional (See Figure 3.45).*

3.7. THEOREM OF THALES IN SPACE

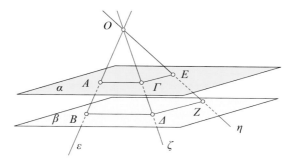

Fig. 3.45: Point pencil and parallel planes

Proof. Let the lines ε, ζ and η, passing through the point O, intersect the planes α and β respectively at points $\{A, \Gamma, E\}$ and $\{B, \Delta, Z\}$. we'll show that

$$\frac{|A\Gamma|}{|\Gamma E|} = \frac{|B\Delta|}{|\Delta Z|}.$$

According to Corollary I-3.16, applied to the plane of the lines ε, ζ, we have

$$\frac{|A\Gamma|}{|B\Delta|} = \frac{|O\Gamma|}{|O\Delta|}.$$

Also applying the same corollary to the plane of the lines ζ, η we have

$$\frac{|O\Gamma|}{|O\Delta|} = \frac{|\Gamma E|}{|\Delta Z|}.$$

Combining these two relations, we get the requested one

Remark 3.4. The preceding proposition shows that the theorem of Thales in the form of Theorem I-3.14, which holds for point pencils of the plane, can be extended also to hold for point pencils in space.

Corollary 3.11. *Let $AB\Gamma\Delta...$ be a polygon on the plane α and point O not lying on that plane. On every plane β parallel to α the point pencil through O defines a polygon $A'B'\Gamma'\Delta'...$ similar to $AB\Gamma\Delta....$ And conversely, if the polygons $AB\Gamma...$ and $A'B'\Gamma'...$ of the planes α and β are similar and have respective sides parallel, then the lines AA', BB', $\Gamma\Gamma'$, ..., which join respective (we often say homologous) vertices, either pass through a common point O, or they are parallel and the two polygons are congruent (See Figure 3.46).*

Proof. The corollary says that the lines OA, OB, $O\Gamma$, ... , which join point O with the vertices of the polygon, define on a parallel plane β to α a polygon $A'B'\Gamma'...$ similar to $AB\Gamma....$ The similarity here means, as in the case of two

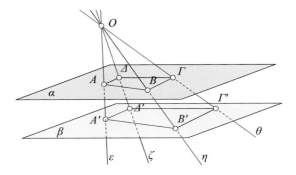

Fig. 3.46: Similar polygons on parallel planes

polygons of the same plane (I-§ 3.10), that the sides of the two polygons are proportional

$$\frac{|A'B'|}{|AB|} = \frac{|B'\Gamma'|}{|B\Gamma|} = \frac{|\Gamma'\Delta'|}{|\Gamma\Delta|} = ... = \kappa,$$

where κ is a fixed constant, and the corresponding angles are equal. The equality of ratios follows from the similarity of the triangles in the pairs $(OA'B', OAB)$, $(OB'\Gamma', OB\Gamma)$, $(O\Gamma'\Delta', O\Gamma\Delta)$, ..., which in turn follows from the preceding proof. The equality of angles follows from the fact that the sides of the respective angles, like $\widehat{A'B'\Gamma'}$ and $\widehat{AB\Gamma}$ are parallel (Theorem 3.6). The converse follows using the method of theorem I-3.17 and is left to the reader.

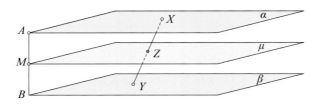

Fig. 3.47: Middle parallel plane of planes α and β

Proposition 3.21. *The geometric locus of the middles M of the line segments XY, which have their endpoints on two parallel planes α and β, is a plane μ, which is equidistant from α and β (See Figure 3.47).*

Proof. Consider a segment AB with endpoints at α and β and middle M. From M draw a parallel plane μ to α and β. According to Theorem 3.13, every other line segment XY would have an intersection point Z with μ, such that $\frac{|ZX|}{|ZY|} = \frac{|MA|}{|MB|} = 1$. Therefore point Z would be the middle of XY. Consequently every point of μ belongs to the geometric locus. For the converse,

3.7. THEOREM OF THALES IN SPACE

that is that each point of the locus is contained in μ, consider an arbitrary segment XY and its middle Z and draw from Z the parallel plane to α and β. Again, according to Theorem 3.13, this will intersect AB at its middle, therefore it will coincide with μ.

The plane which is guaranteed by the preceding proposition is called **middle parallel** plane of α and β.

Exercise 3.22. Show that the geometric locus of the middles M of the line segments XY, which have their endpoints on two skew lines α and β, is a plane ε parallel and equidistant from α and β.

Hint: Consider the two parallel planes η, ζ, which contain respectively lines α and β (Proposition 3.13). Show that the requested locus is the middle parallel plane of η and ζ.

Exercise 3.23. A line segment XY of fixed length has its endpoints on two orthogonal skew lines α and β. Find the geometric locus of its middle Z.

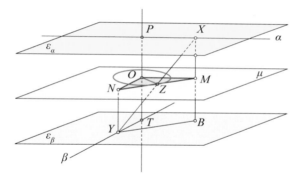

Fig. 3.48: Geometric locus with non coplanars

Hint: (See Figure 3.48) Consider two parallel planes $\varepsilon_\alpha, \varepsilon_\beta$, which contain respectively lines α and β (Proposition 3.13). Project point X onto the plane ε_β at point B. Even though X and Y are variable, the right triangle XBY remains self-congruent, since the hypotenuse XY is of fixed length and $|XB|$ is equal to the distance between the planes. Consequently YB is of fixed length, call it λ. Therefore its projection MN on the middle parallel plane μ of $\varepsilon_\alpha, \varepsilon_\beta$ is also of the same fixed length λ. If PT is the common orthogonal of the two skew orthogonals and O is the point where it meets μ, triangle ONM is right at O. Further point Z is the middle of MN, since $XMYN$ is a parallelogram and its diagonals XY and MN are bisected at Z. Therefore OZ is also the median of the right triangle to the hypotenuse and consequently is of length $\lambda/2$. Therefore Z is contained in the circle κ with center O and radius $\lambda/2$.

Conversely, if Z is a point of the circle κ, then the projections of N', M' onto ON and OM respectively, define the direction $N'M'$ and that of its parallel MN, since M' and N' are the middles of OM and ON respectively. Therefore

MN is constructed by drawing the parallel to $M'N'$ from Z. Line MN and its parallel PT from Z define plane v passing through Z, which intersects α and β at two points X and Y respectively. A reverse process to that of the first part of the proof shows that points X, Z and Y are collinear and Z is the middle of XY.

Exercise 3.24. The isosceles triangle AOB has its apex O on the plane ε and its altitude OM is orthogonal to the plane. Show that for every line XOY of ε passing through O the angles $\widehat{XOA} = \widehat{YOB}$.

Exercise 3.25. The lines $\{\alpha, \beta\}$ make the same angle with the plane ε. Show that the segments on these lines which are cut by ε and a parallel to it plane ε', are equal in length.

Exercise 3.26. Let $\{\alpha, \beta\}$ be two planes and consider all line segments XY with endpoints respectively on these two planes. Show that the geometric locus of points Z of XY which divide them in a constant ratio $ZX/ZY = k$ is a plane parallel to $\{\alpha, \beta\}$.

Exercise 3.27. Let $\{\alpha, \beta\}$ be two parallel planes and A a point of α. Which are the points X alpha α, such that the ratio of distances $XA/XB = \kappa$, from A and the plane β, is constant?

3.8 Comments and exercises for the chapter

> ... life can be compared to a piece of embroidered material of which everyone, in the first half of his time, comes to see the top side, but in the second half the reverse side. The latter is not so beautiful, but is more instructive because it enables one to see how the threads are connected together..
>
> A. Schopenhauer, *On the different periods of life*

Exercise 3.28. Given a line ε and a point A not lying on it, show that there exists a unique line α in space, through A and orthogonal to ε.

Exercise 3.29. Let A be a point not lying on the plane ε. Construct a plane α passing through A, orthogonal to ε and such that the intersection of ε and α is parallel to a given line of ε.

Exercise 3.30. Let A be a point not lying on the plane ε. Construct a plane α passing through A, intersecting ε under a given angle ω and such that the intersection of ε and α is parallel to a given line of ε.

Exercise 3.31. Show that, if lines α and β are skew orthogonal and lines α' and β' are parallel respectively to α and β, then α' and β' are also either orthogonal (if they intersect) or skew orthogonal.

3.8. COMMENTS AND EXERCISES FOR THE CHAPTER

Exercise 3.32. Construct a line α passing through a given point and skew orthogonal to given lines in space β and γ.

Exercise 3.33. Find all lines η intersecting two non coplanar lines α and β in space under equal angles.

Exercise 3.34. Let A, B be two different points not contained in the plane ε. Find a point Γ on ε, such that the sum of the distances $|A\Gamma| + |\Gamma B|$ is minimized.

Exercise 3.35. Let α and β be skew lines and ε_α, ε_β be the parallel planes which contain them. Show that for every point X not contained in these planes, there exists exactly one line through X which intersects both skew lines.

Hint: Consider the intersection of the planes $\{X, \alpha\}$, $\{X, \beta\}$.

Exercise 3.36. Three lines in space α, β and γ are pairwise skew. Show that there exist infinitely many different lines δ which intersect simultaneously all three given lines.

Exercise 3.37. Show that if line OP forms the same angles with three different lines OX, OY, OZ of the plane ε, then it is orthogonal to ε at O.

Exercise 3.38. Show that there cannot exist more than three lines in space which are pairwise skew orthogonal.

Exercise 3.39. Show that ν pairwise intersecting lines in space, either all pass through a common point or they are all coplanar.

Hint: First show the proposition for $\nu = 3$. Subsequently use this case for the proof of the general proposition.

Exercise 3.40. Let $\{\alpha, \beta\}$ be parallel planes and A, B be two points not lying on them and such that A lies on the other side of α than β and B lies on the other side of β than α. Find points A' on α and B' on β, such that $A'B'$ is orthogonal to the planes and the sum of distances $|AA'| + |A'B'| + |B'B|$ is minimized.

Exercise 3.41. If the projections of some points on a plane ε are collinear on that plane, then these points are contained in the same plane ζ.

Exercise 3.42. Show that the geometric locus of the points P, which are equidistant from the sides of angle $\widehat{XOY}$ is the plane ζ which is defined by the bisector of the angle and the line ε, which is orthogonal to the plane of the angle at point O (See Figure 3.49).

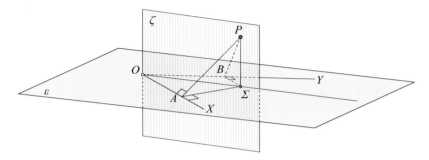

Fig. 3.49: Plane of equal distances from sides of angle

Hint: If P is a point of ζ, Σ its projection onto ε and A, B the projections on OX and OY respectively (See Figure 3.49), then Σ is on the bisector of $\widehat{OXY}$ and the right triangles $P\Sigma A$ and $P\Sigma B$ are congruent. Also, from the Theorem of three orthogonals 3.11, follows that PA and PB are respectively orthogonal to OX and OY, therefore point P is a point of the geometric locus. Conversely, if P is a point of the locus, suppose that PA and PB are the equal orthogonals respectively to OX and OY. Again the right triangles $P\Sigma A$ and $P\Sigma B$ are congruent, and from the theorem of the three orthogonals, ΣA and ΣB are orthogonal respectively to OX and OY, therefore Σ is on the bisector of $\widehat{XOY}$.

Exercise 3.43. Show that the geometric locus of the points P, for which the lines OP form equal angles with the sides of angle $\widehat{XOY}$ is the plane ζ which is defined from the bisector of the angle and the line which is orthogonal to its plane ε at point O (See Figure 3.49).

Exercise 3.44. Show that if a plane ε passes through one diagonal of a parallelogram, then the distances from ε of the endpoints of the other diagonal are equal.

Exercise 3.45. Given is a plane ε and one of its lines η. Find the geometric locus of the points X in space, which are at a fixed distance d from η and at a fixed distance d' from ε. For which d, d' there is no solution?

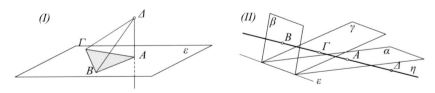

Fig. 3.50: Angle variation   Constant cross ratio

3.8. COMMENTS AND EXERCISES FOR THE CHAPTER

Exercise 3.46. Given is a plane ε and a line η, orthogonal to the plane at vertex A of the isosceles triangle $AB\Gamma$ of the plane (See Figure 3.50-I). Show that, for every point $\Delta \neq A$ of the line η, the angle $\widehat{BA\Gamma}$ is greater than $\widehat{B\Delta\Gamma}$.

Exercise 3.47. The three planes $\{\alpha,\beta,\gamma\}$ are fixed and pass through the same line ε (resp. are parallel). The variable line η intersects these planes respectively at points $\{A,B,\Gamma\}$. Find the locus of a fourth point Δ of the line η, which defines a constant cross ratio $k = (AB\Gamma\Delta)$ (See Figure 3.50-II).

Exercise 3.48. Given are three pairwise skew lines α, β and γ. Construct a line δ, parallel to γ and intersecting α and β.

Hint: From a point A of α draw the parallel γ' to γ and define the plane $\varepsilon_1 = \alpha\gamma'$ which contains them. Similarly, from point B of β draw the parallel γ'' to γ and define the plane $\varepsilon_2 = \beta\gamma''$ which contains them. The intersection of the planes ε_1, ε_2 is the requested line.

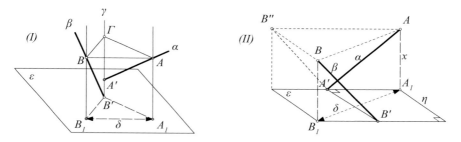

Fig. 3.51: Construction of segment of given length

Exercise 3.49. Given are two skew lines α and β and a plane ε which intersects both. Construct a line segment AB with its endpoints respectively on α and β, which is parallel to ε and has a given length δ (See Figure 3.51-I).

Hint: Planes ε_α, ε_β, which contain respectively α, β and are orthogonal to ε intersect, in general, along a line γ orthogonal to ε. Triangle $AB\Gamma$, formed by the intersections of α, β, γ with a plane parallel to ε, can be calculated, as a function of the distance $x = |A'\Gamma|$, such that $|AB| = \delta$. Examine also the case where ε_α, ε_β are parallel. In this case ε may be considered as containing the line segment η of the minimal distance η of the two skew lines (See Figure 3.51-II). The rectangle with diagonal A_1B_1, constructed through the projections of $\{A,B\}$ on ε, is constructible from the given data. Then, the triangle $A'AB''$ of the projections onto ε_α is constructible, since its angles and the side AB'' are determined from the given data.

Exercise 3.50. For two different points $\{A,B\}$ of space, consider all the planes or/and lines α through A and on each such plane/line the projection B_α of B. Show that all these points $\{B_\alpha\}$ have the same distance $r = |AB|/2$ from the middle M of AB.

Exercise 3.51. The planes $\{\alpha, \beta\}$, intersect along the line ε and contain corresponding points $\{A, B\}$. Show that line AB forms with the planes the same angle, if and only if the points are at equal distance from line ε.

Hint: Project A on line ε at the point A' and to plane β at the point A'' and analogously project point B to $\{B', B''\}$ onto $\{\varepsilon, \alpha\}$. The equality of angles of AB with the planes is equivalent with the equality of the orthogonal triangles $\{A''BA, B''AB\}$. This is equivalent with the equality of $\{|AA''|, |BB''|\}$, which in turn, is equivalent with the equality of the triangles $\{AA'A'', BB'B''\}$.

Exercise 3.52. Show that the medial planes of the sides of a skew quadrilateral pass, all four, through a common point.

Exercise 3.53. Show that if two opposite sides of a skew quadrilateral are equal, then their projections on the line which joins the middles of the two other sides are also equal.

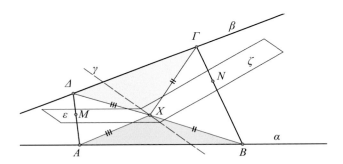

Fig. 3.52: Points X with equal distances from skew lines $\{\alpha, \beta\}$

Exercise 3.54. Show that there are infinite many lines γ, whose points X are equidistant from two given skew lines $\{\alpha, \beta\}$.

Hint: Take two arbitrary but equal segments $\{AB, \Gamma\Delta\}$ respectively on $\{\alpha, \beta\}$. Consider the medial planes $\{\varepsilon, \zeta\}$ of the segments respectively $A\Delta$, $B\Gamma$. The intersection line γ of two such planes does the work (See Figure 3.52). The triangles $\{XAB, X\Delta\Gamma\}$ and their altitudes from X are equal.

Exercise 3.55. The skew quadrilateral $AB\Gamma\Delta$ has fixed the vertices $\{A, B, \Gamma\}$ and Δ moving on a line ε not contained in the plane of $AB\Gamma$. Locate the position of Δ on ε such that the parallelogram of the middles of the sides of the quadrilateral is a rectangle.

Chapter 4
Solids

4.1 Dihedral angles

> Narrative is linear, but action has breadth
> and depth as well as height and is solid.
>
> *Thomas Carlyle*

Two planes α and β, which intersect, separate space into four parts, called **quadrants** (See Figure 4.1-I). They also define shapes analogous to the angles which are defined by two intersecting lines of the plane. Each pair of half

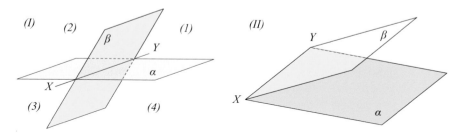

Fig. 4.1: Quadrants   Dihedral

planes which are parts of different planes defines a **convex dihedral angle** or simply **dihedral** (See Figure 4.1-II). The line XY of the intersection of α and β is called **edge of the dihedral**. The half planes (as well as the planes also) which define the dihedral are called **faces of the dihedral**.

To the notion of the dihedral corresponds, in plane geometry, that of the usual angle. This is emphasized also with the next definition which gives the way we measure the magnitude of a dihedral angle. The **measure of the dihedral** is defined by considering a plane orthogonal to the edge. This plane intersects the faces along lines OA and OB and as measure of the dihedral we

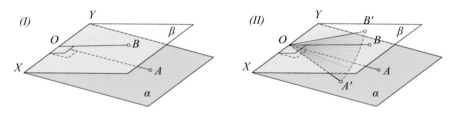

Fig. 4.2: Dihedral angle measure $\widehat{AOB} < \widehat{A'OB'}$

define the measure of the angle $\widehat{AOB}$ (See Figure 4.2-I). This way the convex dihedral corresponds to the convex angle $\widehat{AOB}$ with measure less than π (in radians) and the non-convex dihedral corresponds to the non-convex angle $\widehat{AOB}$ with measure greater than π.

We must note that the measure of the dihedral is independent of the position of O on its edge XY. If we draw the orthogonal to XY plane at another point O' of the edge, then the corresponding angle $\widehat{A'O'B'}$ will have its sides parallel and equally oriented to those of $\widehat{AOB}$, consequently the two angles will be equal (Theorem 3.6).

Theorem 4.1. *The measure $\widehat{AOB}$ of a dihedral is less from that of every other angle $\widehat{A'OB'}$, formed by intersecting the dihedral with a plane which is not orthogonal to its edge (See Figure 4.2-II).*

Proof. The proof of this is identical with that of the exercise 3.21.

We call two dihedral angles **congruent** when they have the same measure. More generally, by transferring the concepts from plane angles to the dihedrals, we talk about **right dihedral, obtuse dihedral, acute dihedral, complementary dihedral, supplementary dihedral, orthogonal dihedral,** etc. Properties similar to these of angles also hold. It suffices to re-examine the properties of plane angles and substitute in them the phrase *side of angle* with the phrase *face of dihedral*. The next proposition gives an example of these analogies.

Proposition 4.1. *Dihedral angles, whose respective faces are parallel planes are either congruent or supplementary (See Figure 4.3).*

Proof. Suppose that the faces α', β' of one dihedral are parallel planes respectively to the faces α, β of the other. Then their edges are parallel. This can be seen by extending one face, for example β' until it intersects α along the line ε. According to Proposition 3.9, ε and XY will be parallel as excised by the parallel planes β and β' through α. Similarly also ε and $X'Y'$ will be parallel as excised by the parallel planes α and α' through β. Consequently the edges XY and $X'Y'$ will be parallel lines. Therefore a plane γ orthogonal

4.1. DIHEDRAL ANGLES

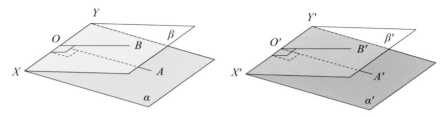

Fig. 4.3: Congruent dihedral angles

to the edge XY which defines the angle $\widehat{AOB}$ of the one dihedral will also be orthogonal to the edge $X'Y'$ of the other dihedral (Proposition 3.15) and will define the angle $\widehat{A'O'B'}$ of the other dihedral. Because of the parallel planes angles $\widehat{AOB}$ and $\widehat{A'O'B'}$ will have parallel sides, therefore they will either be congruent or supplementary (Corollary I-1.30).

We call two planes which form a right dihedral angle, **orthogonal planes**.

Theorem 4.2. *If a line AB is orthogonal to the plane α, then every other plane β which contains AB will be orthogonal to plane α.*

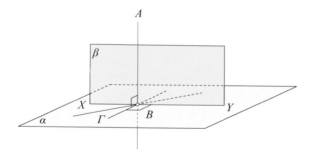

Fig. 4.4: orthogonal planes

Proof. Line AB (with B on α) is orthogonal to every line of α which passes through B (Theorem 3.9), therefore it will be orthogonal also to line $B\Gamma$, which is orthogonal to the intersection XY of the planes and is contained in α (See Figure 4.4). This shows that the angle of the dihedral between α and β is right.

Proposition 4.2. *For two planes α and β, intersecting along line XY, the points in space Γ, which are equidistant from the planes lie on two other planes γ, δ, which pass through XY, form congruent dihedrals with α and β and are orthogonal to each other (See Figure 4.5).*

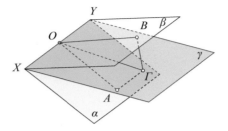

Fig. 4.5: Bisector plane of dihedral

Proof. Let the point Γ be equidistant from planes α and β. Draw the orthogonals ΓA and ΓB to them. Plane $AB\Gamma$ is orthogonal to XY, because XY, as a line of plane β, is orthogonal to $B\Gamma$ and, as line of plane α, is orthogonal to $A\Gamma$. It follows, that angle $\widehat{AOB}$, where O is the point of intersection of XY with plane $AB\Gamma$, is the dihedral angle between α and β which contains Γ. Also the right triangles $O\Gamma A$ and $O\Gamma B$ are congruent, having the hypotenuse $O\Gamma$ common and the orthogonals ΓA and ΓB equal by assumption. It follows that plane γ, which contains Γ and the line XY forms equal angles with the planes α and β. Similarly we show that if Γ belongs to one of the other quadrants defined by the two planes, it is contained either in γ or in plane δ which forms a right angle with γ (bisects the external dihedral).

In the preceding proposition we have again a property of the angles of the plane (Corollary I-1.36) which is transferred to dihedral angles. The planes, defined by the preceding proposition, are called **bisecting planes** or **bisectors** of the dihedral.

Exercise 4.1. Find the geometric locus of the points X in space, which are at a given distance δ from the faces of a dihedral angle.

Theorem 4.3. *A line α, which is not contained in and is not orthogonal to the plane ε, is contained in exactly one plane ζ orthogonal to ε (See Figure 4.6).*

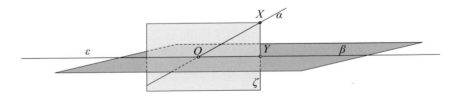

Fig. 4.6: Plane through line orthogonal to other

Proof. From an arbitrary point X of α draw the orthogonal XY to ε. The plane ζ which contains α and XY is orthogonal to ε (Theorem 4.2). Every other plane ζ' which contains α and is orthogonal to ε, will intersect ε along a line β. From X draw the orthogonal XY' to β on plane ζ'. Line XY' will be also orthogonal to ε because besides line β it will also be orthogonal to a orthogonal to it contained in ε, because of the assumption that ζ' and ε intersect orthogonally. Consequently line XY' will coincide with the preceding XY and plane ζ' will contain α and XY, therefore it will be coincident with plane ζ.

Line β, defined in the preceding proposition, as the intersection of the unique plane ζ which contains the line α and intersects orthogonally the plane ε, is called **projection** of the line α to the plane ε.

Proposition 4.3. *The projections α', β' of two parallel lines α and β on the plane ε are either coincident or they are two points or they are two parallel lines of ε.*

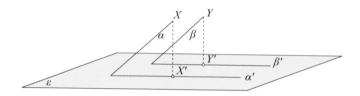

Fig. 4.7: Projection of parallels to parallels

Proof. If lines α, β are orthogonal to ε, then their projections are two points: the points at which they intersect plane ε. If plane ε is orthogonal to their plane, but not orthogonal to α, β then the lines project on the same line of ε. In the case where none of the above happens (See Figure 4.7), considering a point X on α and a point Y on β and projecting them onto ε, we define the planes orthogonal to ε which contain the projections α' and β' of the two lines. These two planes, as containing two pairs of parallel lines are parallel (Proposition 3.8) therefore lines α', β' by which they intersect plane ε will be parallel (Proposition 3.9).

Corollary 4.1. *The projection of a parallelogram of the plane ε onto another plane ζ, which is not orthogonal to ε, is a parallelogram.*

Corollary 4.2. *Two triangles $AB\Gamma$ and ΔEZ, contained in plane α, which are congruent and with parallel sides are projected onto plane β, not orthogonal to α, to two triangles $A'B'\Gamma'$ and $\Delta'E'Z'$, which are also congruent and have parallel sides.*

Proof. (See Figure 4.8) Follows from the preceding corollary, since the parallelogram, for example $B\Gamma ZE$, which is defined by two equal and parallel sides of the triangles on α, will be projected onto a parallelogram $B'\Gamma'Z'E'$, forming equal and parallel sides $B'\Gamma'$ and $E'Z'$.

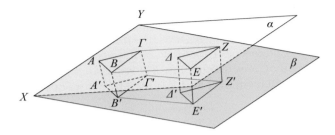

Fig. 4.8: Projection of congruent triangles onto congruent triangles

Corollary 4.3. *Two polygons $ABΓΔE...$, $ΠPΣT...$, contained in plane $α$, which are congruent and with parallel sides, are projected onto plane $β$, not orthogonal to $α$, as two polygons $A'B'Γ'Δ'E'...$, $Π'P'Σ'T'...$, which are also congruent and have parallel sides.*

Proof. Follows from the preceding corollary by dividing the two polygons into triangles and projecting.

Exercise 4.2. Show that the projection M' of the middle M of line segment AB on a plane $ε$ is the middle of the line segment $A'B'$, where A' and B' are the projections of A and B on that plane.

Exercise 4.3. If $A'B'Γ'$ is the projection of triangle $ABΓ$ on the plane $ε$, show that the projection of the centroid of $ABΓ$ is the centroid of $A'B'Γ'$.

Exercise 4.4. If line $ε$ is not orthogonal to plane $α$ and $ω$ is an angle $0 < ω < π/2$, show that there are two planes containing line $ε$ and making with $α$ dihedral angles equal to $ω$. Construct these planes and the bisector of the dihedral they form.

4.2 Trihedral angles

> If by chance ten or twenty centuries of history were eliminated, our knowledge relative to human nature wouldn't change a bit. The only irreversible loss would be if we lost the works of art these centuries bore.
>
> Claude Levi Strauss, *With an eye on things*

Three half lines OA, OB, $OΓ$, not contained in a plane and passing through a common point O, define pairwise, planes and create a solid shape, which divides space in two parts $\{Σ, Σ'\}$. From these two parts one, $Σ$ say, is convex (i.e. for every pair of points X, Y in $Σ$ the entire segment XY is contained

4.2. TRIHEDRAL ANGLES

in Σ), and the other non-convex. Selecting one of these parts, usually the convex part, defines a **trihedral angle** or simply **trihedral** (See Figure 4.9).

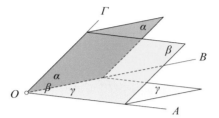

Fig. 4.9: Trihedral angle or trihedral

The part of space selected is called **interior** of the trihedral. The half lines which define the trihedral are called **edges** of the trihedral and the angles $\alpha = \widehat{BO\Gamma}$, $\beta = \widehat{\Gamma OA}$, $\gamma = \widehat{AOB}$, as well as the planes containing them, are called **faces** of the trihedral. Point O is called **vertex** of the trihedral.

The dihedral angles with edges the lines OA, OB, and $O\Gamma$, are called **dihedrals of the trihedral** and are often denoted below respectively by $\widehat{A}$, $\widehat{B}$, $\widehat{\Gamma}$. Often with the same letter we denote also the measure of the angle. This way, in the various propositions below α, β and γ may denote the faces or/and even the measures of the faces of the trihedral. Similarly $\widehat{A}$, $\widehat{B}$, $\widehat{\Gamma}$ may denote the dihedrals or/and their measures, which are defined by measuring the angles which these dihedrals excise on planes orthogonal to their edges. Two trihedrals $\widehat{OAB\Gamma}$ and $\widehat{O'A'B'\Gamma'}$ are called **congruent** when they have respective faces congruent and respective dihedrals also congruent.

Theorem 4.4. *In every trihedral $\widehat{OAB\Gamma}$ the sum of two of its faces is greater than the third.*

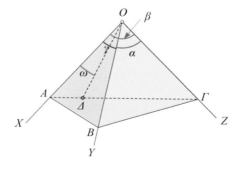

Fig. 4.10: Inequalities between faces

Proof. Let $\beta = \widehat{XOZ}$ be the greater of the three faces $\alpha = \widehat{YOZ}$, $\beta = \widehat{XOZ}$, $\gamma = \widehat{YOX}$. We'll show that $\beta < \alpha + \gamma$. In fact, consider two arbitrary points B and Γ on the edges OY and OZ respectively (See Figure 4.10). Then draw on the plane XOZ the triangle $\Delta O\Gamma$, so that it is congruent to $BO\Gamma$ and extend $\Delta\Gamma$ up to its intersection point A with the third edge. From the way the triangles $AB O$ and $A\Delta O$ were defined, they have AO common and BO and ΔO equal. Also from the triangle inequality we have

$$|A\Gamma| = |A\Delta| + |\Delta\Gamma| < |AB| + |B\Gamma|.$$

However $|\Delta\Gamma| = |B\Gamma| \Rightarrow |A\Delta| < |AB|$. From Corollary I-1.17 we have for the angle

$$\omega = \widehat{AO\Delta} < \widehat{AOB} \Leftrightarrow \widehat{AO\Gamma} - \widehat{BO\Gamma} < \widehat{AOB}.$$

Theorem 4.5. *In every trihedral $OAB\Gamma$ the sum of its faces is less than 2π.*

Proof. Consider arbitrary points A, B and Γ respectively on the edges OX, OY and OZ of the trihedral (See Figure 4.10). This forms three more trihedrals with vertices the points A, B and Γ. Applying the preceding proposition to these three trihedrals we have

$$\begin{aligned}\alpha + \beta + \gamma &= (\pi - \widehat{OB\Gamma} - \widehat{O\Gamma B}) + (\pi - \widehat{OA\Gamma} - \widehat{O\Gamma A}) + (\pi - \widehat{OAB} - \widehat{OBA}) \\ &= 3\pi - (\widehat{OB\Gamma} + \widehat{OBA}) - (\widehat{O\Gamma B} + \widehat{O\Gamma A}) - (\widehat{OA\Gamma} + \widehat{OAB}) \\ &< 3\pi - (\widehat{AB\Gamma} + \widehat{B\Gamma A} + \widehat{\Gamma AB}) \\ &= 3\pi - \pi = 2\pi.\end{aligned}$$

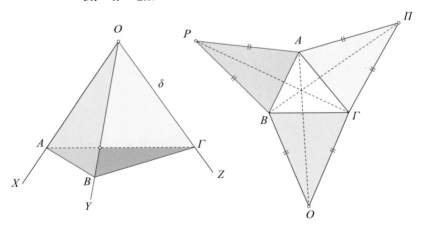

Fig. 4.11: Unfolding of trihedral

Considering equal line segments $|OA| = |OB| = |O\Gamma| = \delta$ respectively on the edges OX, OY, OZ, we define a triangle $AB\Gamma$, which, together with the length δ, describes the trihedral completely with the assistance of a plane figure

4.2. TRIHEDRAL ANGLES

which is called **unfolding of the trihedral**. This figure results by cutting the trihedral along its edges and rotating the isosceli triangles about their bases, which are the sides of the triangle $AB\Gamma$, until the planes of the isosceli coincide with that of $AB\Gamma$. The shape (See Figure 4.11), which results can be used also for the construction of the trihedral from paper. It suffices to rotate, inversely to the preceding direction, the isosceli triangles about their bases, until their three vertices coincide at the same point of space O, which defines then the vertex of the trihedral. Next two propositions show that the inequalities of the two preceding propositions are also sufficient conditions, which guarantee that such a process always leads to a trihedral.

Lemma 4.1. *Given three isosceli triangles with equal legs, whose apical angles, pairwise, have sum of angles greater than the third, we can construct a triangle, whose sides coincide with the bases of the isosceli.*

Proof. Suppose that the three triangles with equal legs are $OB\Gamma$, $\Pi\Gamma A$ and $PA'B'$. We place two of them, e.g. the first two, so that the angles at their apexes become adjacent. From the triangle inequality we have

$$|B\Gamma| + |\Gamma A| > |BA|.$$

Triangles $PA'B'$ and OAB are isosceli (See Figure 4.12) with the same length

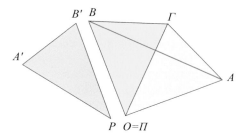

Fig. 4.12: Trihedral inequality

of legs and, by assumption, we have that their apical angles satisfy

$$\widehat{A'PB'} < \widehat{BOA}.$$

From Theorem I-1.5 follows

$$|A'B'| < |AB| < |B\Gamma| + |\Gamma A|,$$

consequently the triangle inequality is satisfied and a triangle with sides the bases of the isosceli can be constructed (Theorem I-2.6).

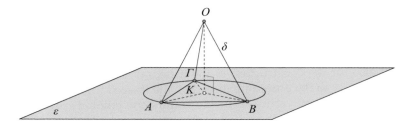

Fig. 4.13: Construction of trihedral

Theorem 4.6. *For every triple of angles α, β and γ with $\alpha + \beta + \gamma < 2\pi$ and such that the greatest of these is less than the sum of the two others, there exists a trihedral having these angles as faces.*

Proof. Construct isosceli triangles with apical angles α, β and γ and with their legs all equal to an arbitrary positive number δ (See Figure 4.13). The assumptions for the angles imply, according to the preceding lemma, that the bases of these isosceles make the sides of a triangle $AB\Gamma$. Consider the circumscribed circle of this triangle and draw the orthogonal to its center K. On this orthogonal consider then a line segment KO of length

$$\kappa = |KO| = \sqrt{\delta^2 - \rho^2},$$

where ρ is the radius of the circumscribed circle. From the pythagorean theorem it follows that $|OA| = |OB| = |O\Gamma| = \delta$ and the isosceli triangles $BO\Gamma$, ΓOA, AOB have equal legs and their bases are respectively equal to those of the initial isosceli triangles, therefore they are congruent to them and a trihedral with the given angles (faces) is formed at O. The fact that point O is outside the plane ε of the triangle $AB\Gamma$ is guaranteed by the assumption $\alpha + \beta + \gamma < 2\pi$, from which follows $\delta > \rho$. Indeed if point O were coincident with K, that is if $\delta = \rho$, then we would have $\alpha + \beta + \gamma = \widehat{AKB} + \widehat{BK\Gamma} + \widehat{\Gamma KA} = 2\pi$, contrary to the assumption. Also if it were $\delta < \rho$, then angle $\widehat{AOB}$ would be on plane ε of triangle $AB\Gamma$ and greater than $\widehat{AKB}$, because the isosceles AOB would be inside the isosceles AKB (Proposition I-1.8). Similar things would hold also for the other angles $\widehat{BO\Gamma}$ and $\widehat{\Gamma OA}$. Consequently $\alpha + \beta + \gamma = \widehat{BO\Gamma} + \widehat{\Gamma OB} + \widehat{AO\Gamma} > \widehat{BK\Gamma} + \widehat{\Gamma KA} + \widehat{AKB} = 2\pi$, again contrary to the assumption.

Remark 4.1. Last theorem combined with the next one implies that a trihedral angle is determined completely from its faces.

Before to proceed to the proof of the basic congruence theorem for trihedrals (alternatively: congruence of spherical triangles, remark 4.14), I'll note

4.2. TRIHEDRAL ANGLES

that congruence among trihedrals, as we defined it, is not coincident with coincidence in space, in the sense of moving the one until it coincides with the other. If the preceding action can be done, then, of course, the trihedrals

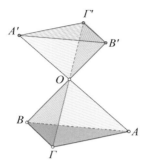

Fig. 4.14: Vertical trihedrals

are congruent. Our definition of congruence, however, includes also another case in which coincidence is not possible. This is the case of the **vertical trihedrals**. Here (See Figure 4.14) the two trihedrals $\widehat{OAB\Gamma}$ and $\widehat{OA'B'\Gamma'}$ have respective faces congruent and respective dihedrals also congruent. However it is not possible to place one onto the other so that they are coincident. This because their *orientations* are different. The two trihedrals result from three isosceli triangles using the preceding proposition but are placed in such a way so that the base triangles have different orientation (See Figure 4.15).

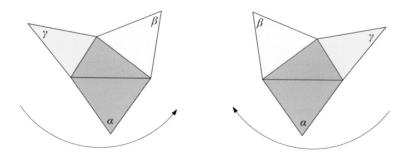

Fig. 4.15: Opposite orientation of trihedrals

Theorem 4.7. *Two trihedrals with respective equal faces are congruent, i.e. they have equal also the corresponding dihedrals lying opposite to equal faces.*

Proof. (See Figure 4.16) Replacing, if necessary, a trihedral with its vertical, we may suppose that the trihedrals have the same orientation. Adopting this assumption, we show that the dihedral $\widehat{A}$ is equal to $\widehat{A'}$. Analogously is proved the congruence of the other dihedrals. For the proof we define on

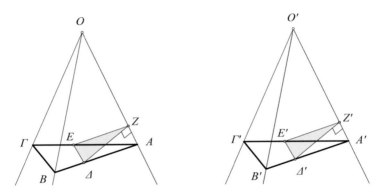

Fig. 4.16: Congruence of trihedrals with equal faces

the edges equal segments

$$OA = OB = O\Gamma = O'A' = O'B' = O'\Gamma',$$

by which are defined corresponding equal isosceli triangles

$$OAB = O'A'B', \quad OB\Gamma = O'B'\Gamma', \quad O\Gamma A = O'\Gamma'A'.$$

We take also equal segments $OZ = O'Z'$ respectively on $\{OA, O'A'\}$ and draw at points $\{Z, Z'\}$ the orthogonal planes to these edges. There are then defined the triangles $\{\Delta EZ, \Delta'E'Z'\}$ easily seen to be congruent. From the congruence of these triangles follows that their angles $\{\widehat{\Delta ZE}, \widehat{\Delta'Z'E'}\}$, which coincide with the corresponding dihedrals $\{\widehat{A}, \widehat{A'}\}$ are equal, as requested.

The congruence of triangles $\{\Delta EZ, \Delta'E'Z'\}$ results by comparing the right triangles $\{AZ\Delta, A'Z'\Delta'\}$ and showing them congruent. Similarly the triangles $\{AZE, A'Z'E'\}$ are congruent, and also triangles $\{AB\Gamma, A'B'\Gamma'\}$ are congruent and $\{A\Delta E, A'\Delta'E'\}$ are congruent. The proof of these congruences is trivial.

Remark 4.2. In § 4.15 (see also exercise 4.9) we'll see that the knowledge of the dihedrals of a trihedral also determines completely the trihedral. There is therefore an equivalence between faces and dihedrals of a trihedral angle: The former determine completely the latter and vice versa. Next theorem corresponds to the SAS congruence criterion for triangles (Proposition I-1.5).

Theorem 4.8. *Two trihedrals having one dihedral equal and the adjacent to it corresponding faces equal are congruent.*

4.2. TRIHEDRAL ANGLES

Proof. We define again on the two trihedrals equal segments on the edges

$$OA = OB = O\Gamma = O'A' = O'B' = O'\Gamma',$$

which in turn define isosceli triangles, like in the preceding theorem (See Figure 4.17). We suppose further that the dihedrals $\{\widehat{A}, \widehat{A'}\}$ are equal and the

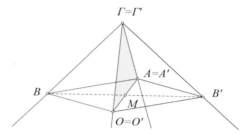

Fig. 4.17: Congruence of trihedrals from a dihedral and adjacent faces

adjacent to them faces $\{\widehat{OAB}, \widehat{OA\Gamma}\}$ are correspondingly equal to $\widehat{O'A'B'}$ and $\widehat{O'A'\Gamma'}$. Then we place the trihedrals so that two corresponding equal faces coincide e.g. the $\{OA\Gamma, O'A'\Gamma'\}$. If the trihedrals have the same orientation, then, because of the equality of the angles, the third corresponding edges $\{OB, O'B'\}$ will coincide too. If not, then the $\{\widehat{OB}, \widehat{O'B'}\}$ will lie on different sides of the plane, as in the figure. But then, considering the congruent to $O'A'B'\Gamma'$ vertical trihedral, we lead the case to the preceding one. An alternative direct proof, without the reduction to the vertical trihedral gives the next exercise.

Exercise 4.5. Show, without the use of the vertical trihedral, that two inversely oriented trihedrals, which have one dihedral equal and the adjacent to it corresponding faces equal, are congruent.

Hint: With the preparation of the preceding theorem (See Figure 4.17), the edges $\{OB, OB'\}$ lie on different sides of the plane $OA\Gamma$. We then show that the plane $OA\Gamma$ is the medial plane of the segment BB' and the third faces $\{OB\Gamma, O'B'\Gamma'\}$ are equal, thus reducing the proof to theorem 4.7. Attention is due to the fact that in figure 4.17 the quadrilateral $OBAB'$ is in general skew (BB' and OA may not intersect).

Exercise 4.6. A trihedral, which has two faces equal has the opposite to these dihedrals also equal.

Hint: Take equal segments on the edges $OA = OB = O\Gamma$, assuming that the faces $\{\widehat{AOB}, \widehat{AO\Gamma}\}$ are equal (See Figure 4.18-I). If M is the middle of $B\Gamma$, show (Theorem 4.7) that the trihedrals $\{OABM, OA\Gamma M\}$ are congruent.

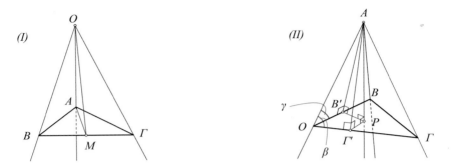

Fig. 4.18: Relative measure of dihedrals - faces

Exercise 4.7. In every trihedral opposite to the bigger face lies a bigger dihedral and opposite to a bigger dihedral lies a bigger face.

Hint: For the first claim suppose that the face $\beta = \widehat{AO\Gamma} \geq \gamma = \widehat{AOB}$ (See Figure 4.18-II) and project point A on face $\widehat{BO\Gamma}$ to point P. Then project point P onto the edges $\{OB, O\Gamma\}$ at the points $\{B', \Gamma'\}$. The triangles $\{OAB', OA\Gamma'\}$ are right with common hypotenuse OA and acute angles $\beta \geq \gamma$, hence the opposite orthogonal sides $|A\Gamma'| \geq |AB'|$. It follows that the right triangles $\{APB', AP\Gamma'\}$ with common the orthogonal side AP have corresponding hypotenuses $|A\Gamma'| \geq |AB'| \Rightarrow \widehat{A\Gamma'P} \leq \widehat{AB'P}$. But the last angles are precisely the dihedrals opposite to the faces $\{\gamma, \beta\}$.

The second claim follows from the first and preceding propositions. In fact, if we had $\widehat{B} > \widehat{\Gamma}$ but $\beta \leq \gamma$ then we would get a contradiction from exercise 4.6 or from the first claim.

Exercise 4.8. Show that, if a trihedral has two dihedrals equal, then the opposite to them faces are also equal.

Exercise 4.9. Show that two trihedrals which have corresponding dihedrals equal are congruent.

Hint: Use a figure similar to the 4.18-II, with points $\{A, A'\}$, such that $|AP| = |A'P'|$.

Proposition 4.4. *For every trihedral, the three planes which pass through an edge and are orthogonal to the opposite face intersect along a line (See Figure 4.19).*

Proof. ([69, p.69]) Consider a plane orthogonal to one edge of the trihedral, for example OA, which intersects the trihedral along a triangle $AB\Gamma$. Consider also the altitudes $A\Delta$, BE and ΓZ of the triangle $AB\Gamma$. These altitudes intersect at the orthocenter H of the triangle and the line OH coincides with the aforementioned intersection of the three planes. Indeed, line BE is orthogonal to $A\Gamma$ and skew orthogonal to OA, therefore also orthogonal to

4.2. TRIHEDRAL ANGLES

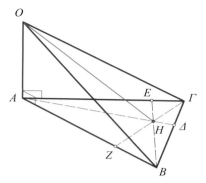

Fig. 4.19: Altitudes of trihedral

their plane, which is face $OA\Gamma$ (Proposition 3.18). Therefore plane OBE is the one passing through edge OB and orthogonal to the opposite face $OA\Gamma$. Similarly, the plane $O\Gamma Z$ is the one passing through the edge $O\Gamma$ and orthogonal to the opposite face OAB. Finally, line $B\Gamma$ is orthogonal to $A\Delta$ and skew orthogonal to OA, therefore it is orthogonal to their plane OAH and consequently plane $B\Gamma O$ which passes through $B\Gamma$ is also orthogonal to OAH (Theorem 4.2).

Exercise 4.10. In every face of a trihedral we construct the line through the vertex which is orthogonal to the opposite edge. Show that the three resulting lines are coplanar.

Exercise 4.11. Show that the three bisecting planes of the dihedrals of a trihedral intersect along a line ε, the points of which are equidistant from the faces of the trihedral.

Theorem 4.9 (Pythagoras 3D). *Given is a triply orthogonal trihedral $\widehat{OXYZ}$ (all three of its faces are right) and points A, B, Γ on its edges OX, OY and OZ respectively. Then, for the areas of the triangles holds (See Figure 4.20-I)*

$$\varepsilon(OA\Gamma)^2 + \varepsilon(OBA)^2 + \varepsilon(O\Gamma B)^2 = \varepsilon(AB\Gamma)^2.$$

Proof. ([68, p.102], [67, p.967]) Using the labels of figure 4.20-I:

$$\begin{aligned}
4\varepsilon(AB\Gamma)^2 &= a^2 \cdot v_A^2 \\
&= a^2 \cdot (x^2 + |O\Delta|^2) \\
&= a^2 \cdot |O\Delta|^2 + (y^2 + z^2) \cdot x^2 \\
&= 4\varepsilon(OB\Gamma)^2 + (y^2 \cdot x^2) + (z^2 \cdot x^2) \\
&= 4\varepsilon(OB\Gamma) + 4\varepsilon(OAB)^2 + 4\varepsilon(OA\Gamma)^2.
\end{aligned}$$

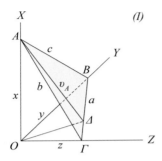 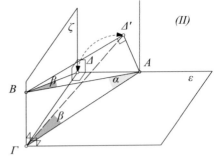

Fig. 4.20: Pythagoras 3D   Orthogonal planes

The second equality results from the fact that $O\Delta$ is orthogonal to $B\Gamma$ (altitude of triangle $OB\Gamma$), because $B\Gamma$ is orthogonal to the altitude $A\Delta$ of triangle $AB\Gamma$ and skew orthogonal to OA, hence (skew) orthogonal to every line of the plane $OA\Delta$.

Exercise 4.12. The points $\{A, B\}$ lie on corresponding planes $\{\varepsilon, \zeta\}$ which intersect orthogonally. Show that the corresponding angles $\{\alpha, \beta\}$ of the line AB with the planes have $\alpha + \beta < \pi/2$.

Hint: Let $\{\Delta, \Gamma\}$ be the projections of $\{A, B\}$ on the corresponding planes $\{\zeta, \varepsilon\}$ (See Figure 4.20-II). The aforementioned angles are $\alpha = \widehat{BA\Gamma}$, and $\beta = \widehat{AB\Delta}$. We rotate the plane $BA\Delta$ about BA till to coincide with plane $B\Gamma A$. Then the right triangle $B\Delta A$ will fall upon triangle $B\Delta' A$ of the plane $B\Gamma A$. The quadrilateral $B\Gamma\Delta\Delta'$ has two opposite angles right, hence is inscriptible and $\beta = \widehat{AB\Delta'} = \widehat{A\Gamma\Delta'}$. Since $|A\Delta'| = |A\Delta|$ the triangle $A\Gamma\Delta$ right and $|A\Delta'| = |A\Delta| < |A\Gamma|$. Hence the opposite angles will be correspondingly unequal: $\beta < \frac{\pi}{2} - \alpha$.

Exercise 4.13. One dihedral of a trihedral is right and the adjacent faces are acute of measures α and β. Calculate the third face of the trihedral.

Exercise 4.14. Show that if a trihedral has two faces right then it also has two dihedrals right and conversely, if it has two dihedrals right, then it will have also two faces right.

Exercise 4.15. Given is a trihedral $\widehat{OAB\Gamma}$. Is there a plane in space which intersects none of (the line carriers of) its edges?

Exercise 4.16. Construct a triply orthogonal trihedral of which we are given some triangular section.

Exercise 4.17. Construct a trihedral of which are given the bisectors of its faces.

4.3 Pyramids, polyhedral angles

> Hence we thought it fitting to guide our actions (under the impulse of our actual ideas [of what is to be done]) in such a way as never to forget, even in ordinary affairs, to strive for a noble and disciplined disposition, but to devote most of our time to intellectual matters, in order to teach theories, which are so many and beautiful, and especially those to which the epithet "mathematical" is particularly applied.
>
> *Ptolemy, Almagest, preface book I*

Given a polygon $AB\Gamma\Delta$... on a plane ε and a point O not contained in this plane, we call **pyramid** the solid shape created by joining point O with the polygon's vertices. The sides of the polygon as well as the line segments OA, OB, ... are called **edges** of the pyramid. Edges OA, OB, ... are called **lateral edges** of the pyramid (See Figure 4.21). The polygon, is called **base** of the pyramid. When the polygon is convex the pyramid is also called **convex**. Often with the term *base of pyramid* we mean also the plane ε which contains the polygon. The distance of O from the base is called **altitude** of the pyramid. In the following we'll deal exclusively with convex pyramids. Point O

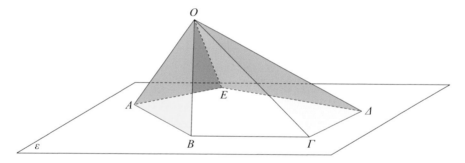

Fig. 4.21: Pyramid

is call **apex** of the pyramid, the angles at the apex, as well as the triangles AOB, $BO\Gamma$, $\Gamma O\Delta$, ... are called **lateral faces** of the pyramid. In the following the distinction between angles and triangles will be made clear from context. The trihedral angles which are formed at the vertices of the base polygon are called **trihedrals of the pyramid**.

At point O (apex of the pyramid) is defined something similar to the trihedral angle, which is called **polyhedral angle** or **solid angle**. Angles $\widehat{AOB}$, $\widehat{BO\Gamma}$, ... (and the planes containing them) are called **faces of the polyhedral angle** and the half lines which define them are called **edges** of the polyhedral angle.

The polyhedral angle is called **convex** when the corresponding polygon which results from a plane intersecting all its edges is convex.

Two polyhedral angles $\widehat{OAB\Gamma}...$, $\widehat{O'A'B'\Gamma'}...$ are called **congruent**, when their corresponding faces are congruent and their corresponding dihedrals are also congruent.

Pyramids are created from a polyhedral angle and a plane ε intersecting all its edges. Fixing the polyhedral angle and varying the plane ε, we create infinitely many pyramids which have the same polyhedral angle and their bases are the polygons defined as the intersections of the polyhedral angle by the plane ε. Two pyramids $OAB\Gamma...$ and $O'A'B'\Gamma'...$ are called **congruent**, when the corresponding triangles-faces are congruent and their corresponding dihedrals are also congruent.

Theorem 4.10. *In every convex polyhedral angle the sum of the measures of its faces is less than 2π.*

Proof. The proof is essentially the same with that of Theorem 4.5. Suppose that the convex polyhedral angle has v faces. We consider a plane intersecting the polyhedral and defining a pyramid $OAB\Gamma\Delta...$. We write the angles (of the faces) at the apex of the pyramid O as functions of the other angles:

$$\widehat{AOB} = \pi - \widehat{BAO} - \widehat{ABO},$$
$$\widehat{BO\Gamma} = \pi - \widehat{\Gamma BO} - \widehat{B\Gamma O},$$
$$\widehat{\Gamma O\Delta} = \pi - \widehat{\Delta\Gamma O} - \widehat{\Gamma\Delta O},$$
$$... \quad ... \Rightarrow$$

$$\widehat{AOB} + \widehat{BO\Gamma} + \widehat{\Gamma O\Delta} + ... = (\pi - \widehat{BAO} - \widehat{ABO}) + (\pi - \widehat{\Gamma BO} - \widehat{B\Gamma O}) + ...,$$
$$= (\pi - \widehat{ABO} - \widehat{\Gamma BO}) + (\pi - \widehat{B\Gamma O} - \widehat{\Delta\Gamma O}) + ...$$
$$< (\pi - \widehat{AB\Gamma}) + (\pi - \widehat{B\Gamma\Delta}) + ...$$
$$= v \cdot \pi - (\widehat{AB\Gamma} + \widehat{B\Gamma\Delta} + \widehat{\Gamma\Delta E} + ...)$$
$$= 2\pi.$$

The inequality follows by applying Theorem 4.4 to the trihedrals at the base of the pyramid. The last equality follows from the fact that the sum of the measures of the angles of a convex polygon with v sides is $(v-2)\pi$ (Proposition I-2.10).

The figure which results from a pyramid by intersecting it with a plane parallel to its base which has the base and the apex on different sides, is called **truncated** pyramid or **pyramid frustum** (See Figure 4.22). The two polygons contained in the parallel planes are called **bases** of the truncated pyramid.

Proposition 4.5. *In every truncated pyramid the polygons defined by its parallel faces are similar.*

Proof. The proof is a direct consequence of the proposition of Thales in space (Corollary 3.11).

4.3. PYRAMIDS, POLYHEDRAL ANGLES

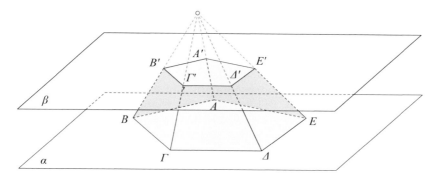

Fig. 4.22: Truncated pyramid

Exercise 4.18. Show that two similar polygons $AB\Gamma\Delta...$, $A'B'\Gamma'\Delta'...$ contained in two parallel planes define lines AA', BB', $\Gamma\Gamma'$, ... which pass through a common point O and consequently define also a truncated pyramid which has these polygons as bases.

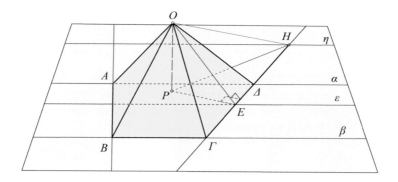

Fig. 4.23: Construction of pyramid $OAB\Gamma\Delta$

Theorem 4.11. *There is a pyramid $OAB\Gamma\Delta$ whose base $AB\Gamma\Delta$ is a trapezium, with given face OAB, given altitude from O, given direction (slope) of the parallel sides $\{A\Delta, B\Gamma\}$ relative to AB and with given angles of the triangular face $O\Gamma\Delta$ (See Figure 4.23).*

Proof. ([32, p.272]) Let us begin with *analysis*, assuming that we have constructed such a pyramid. Triangle OAB, the plane of the base $AB\Gamma\Delta$, its position relative to OAB and the lines α, β of the parallel sides of the trapezium are given. Also given is the shape of the triangle $O\Gamma\Delta$ (its angles), however not its exact position, in other words not the exact position of Γ, Δ respectively on β and α. The position of these points determines the pyramid completely. It suffices therefore to find the position of these points.

For this we observe, that if we draw the altitude OP of the pyramid and we project point P onto $\Gamma\Delta$ at point E, line OE is the altitude of the triangle $O\Gamma\Delta$, which is known by similarity. Therefore the ratio $\frac{E\Gamma}{EA}$ will be known and point E will be contained in a known line ε parallel to α, β. Similarly, if we consider point H on $\Gamma\Delta$, such that $|EH| = |EO|$, this point's ratio $\frac{E\Gamma}{EH}$ will be known and consequently point H will be contained in a known line η parallel to α, β. From the right triangle OPE we'll then have $|HE|^2 - |PE|^2 = |OE|^2 - |PE|^2 = |OP|^2$. However point P and $|OP|$ is known and as such the construction of the pyramid is reduced to the plane construction of the right triangle PEH whose vertex P is a given point, the other vertices E, H are contained in known lines ε and η respectively, and the difference of squares $|EH|^2 - |EP|^2$ is known. Such a construction of a right triangle is done in Exercise I-3.47 and with this the construction of the pyramid is easily completed.

4.4 Tetrahedra

> The author, as a rule, does not usually read below an exercise, before he, himself finds the solution or some other pertinent properties.
>
> J. Steiner, Werke v.I, p. 17

Tetrahedron is called a pyramid with triangular base. A tetrahedron in space is the analogue of the triangle on the plane. Every quadruple of points in general position (i.e. not all contained in a plane) defines a tetrahedron, which has these points as **vertices** (See Figure 4.24).

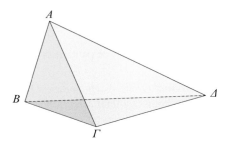

Fig. 4.24: Tetrahedron

Theorem 4.12. *Let $\{A', B', \Gamma'\}$ be points on the plane ε of the triangle $AB\Gamma$ satisfying the relations (See Figure 4.25)*

$$|A'B| = |B\Gamma'|, \quad |B'\Gamma| = |\Gamma A'|, \quad |\Gamma'A| = |AB'|.$$

4.4. TETRAHEDRA

Then the orthogonals from A', B', Γ' respectively to the sides $B\Gamma$, ΓA, AB pass through a common point O. If point O is internal to the three circles $A(|AB'|)$, $B(|B\Gamma'|)$, $\Gamma(|\Gamma A'|)$ and if also the triangles $A'B\Gamma$, $AB'\Gamma$, $AB\Gamma'$ are rotated respectively about $B\Gamma$, ΓA, AB until their vertices A', B', Γ' take a position on the orthogonal line η to the plane ε at point O, then points A', B', Γ' will coincide with a point Δ of η and this will define a tetrahedron $\Delta AB\Gamma$.

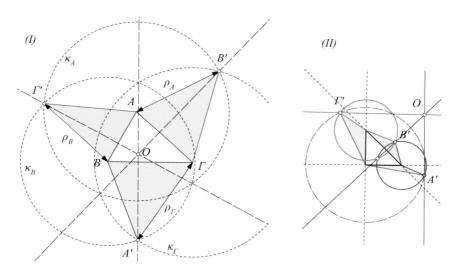

Fig. 4.25: Tetrahedron construction   Impossible construction

Proof. In figure 4.25-I plane ε is that of the paper and line η is the orthogonal to ε at O. The fact that the aforementioned lines pass through a common point O is a consequence of Exercise I-3.34. Another proof results also from the fact that these three lines coincide with the radical axes of the pairs of the circles κ_A, κ_B, κ_Γ with centers the vertices of the triangle and radii the sides of the other triangles which pass through them. Point O is the radical center of these circles (Theorem I-4.5). The property of the radical center, to have equal powers relative to the three circles, is used for the determination of Δ on line η. Let us denote with ρ_A, ρ_B, ρ_Γ the radii of these circles. Then the property of the radical center implies

$$\rho_A^2 - |OA|^2 = \rho_B^2 - |OB|^2 = \rho_\Gamma^2 - |O\Gamma|^2.$$

This means that the right triangles with hypotenuse and one orthogonal respectively $(\rho_A, |OA|)$, $(\rho_B, |OB|)$, $(\rho_\Gamma, |O\Gamma|)$, have the other orthogonals equal, therefore after the rotation of the external triangles about the sides of $AB\Gamma$, their vertices will coincide with point Δ of η. Figure 4.25-II shows one case where the construction is impossible.

Remark 4.3. The tetrahedron $AB\Gamma\Delta$, constructed according to the preceding proposition, will have point O as the projection of its vertex Δ on the plane ε of face $AB\Gamma$. Consequently $|O\Delta|$ will be the altitude of the tetrahedron from Δ and this will be expressed with the power of O relative to one, anyone, of the three circles ($|O\Delta|^2 = \rho_A^2 - |OA|^2$). The plane figure 4.25-I, which corresponds to the tetrahedron $AB\Gamma\Delta$ is called **unfolding of the tetrahedron**.

Remark 4.4. The converse of Theorem 4.12 also holds. If a tetrahedron $AB\Gamma\Delta$ is given and we project vertex Δ onto plane ε of $AB\Gamma$, the projection O will have the properties mentioned in the proposition. This way the orthogonals from O towards the sides of $AB\Gamma$ will contain the vertices A', B', Γ', which will result when we cut the adjacent faces along the edges ΔA, ΔB, $\Delta \Gamma$ and unwind the triangles onto plane ε, by rotating them respectively about the sides of $AB\Gamma$. This will create a plane figure similar to 4.25-I, the unfolding of the tetrahedron. It is obvious that this process can be applied also to any pyramid and will lead to a similar plane figure called **unfolding of the pyramid**. Using a process similar to that of Theorem 4.12 we can, conversely, construct the pyramid from its unfolding.

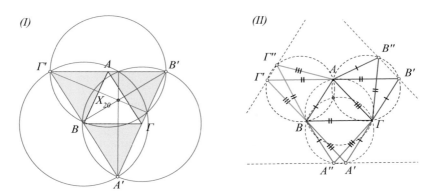

Fig. 4.26: Tetrahedrons with congruent faces

Figure 4.26-I shows one case of application of theorem 4.12 which results from an acute triangle $AB\Gamma$ and its anticomplementary $A'B'\Gamma'$. In this case the point corresponding to O of the theorem is the orthocenter of the anticomplementary, which is denoted with X_{20} (called *De Longchamps point of the triangle* one of the relatively recently studied *triangle centers* [72]). Triangle $A'B'\Gamma'$ is similar to $AB\Gamma$, therefore acute, and its orthocenter is contained in all three circles which the theorem requires. This figure is, therefore, the unfolding of a tetrahedron which has all its faces congruent to the triangle $AB\Gamma$.

Exercise 4.19. Show that, if in a tetrahedron all its faces are congruent to the triangle $AB\Gamma$, then $AB\Gamma$ is acute and the unfolding of the tetrahedron has the form of figure 4.26-I.

4.4. TETRAHEDRA

Hint: The unfolding of the tetrahedron must have the vertices A', B', Γ' exclusively in the suggested order (See Figure 4.26-II).

Exercise 4.20. Show that if a dihedral of the tetrahedron is obtuse, then the unfolding of the tetrahedron with respect to one of the triangular faces of this dihedral defines a point O as in the theorem 4.12 lying outside this triangle. Conclude that the tetrahedron, whose all faces are acute triangles, has also all its dihedrals acute (for the converse see exercise 4.118).

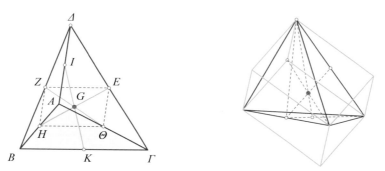

Fig. 4.27: Joining middles of opposite edges, Circumscribed parallelepiped

Exercise 4.21. Show that the three line segments, which join the middles of opposite edges in a tetrahedron pass through a common point G, which bisects them (See Figure 4.27-I).

In the next exercise, running ahead until the definition of the parallelepiped (§ 4.7), we define the **circumscribed parallelepiped** of a tetrahedron, which plays an important role in the study of the properties of the tetrahedron.

Exercise 4.22. Show that, in a tetrahedron, drawing from each edge the parallel plane to its opposite edge, defines a parallelepiped which has the edges of the tetrahedron as diagonals of its faces. Show that the intersection point of the diagonals of this parallelepiped coincides with point G of the preceding exercise (See Figure 4.27-II).

Exercise 4.23. Show that every tetrahedron possesses a trihedral whose faces are all acute angles.

Hint: ([75, p.38]) The sum of all angles of triangular faces is 4π. Hence there is a trihedral for which the sum of the adjacent angles is less or equal to π. At this vertex the adjacent angles must be acute, otherwise the obtuse would be greater than the sum of the two others (Theorem 4.4).

Exercise 4.24. Show that in a tetrahedron all its faces are congruent to the triangle $AB\Gamma$, if and only if its opposite edges are equal.

Remark 4.5. It is proved that the equality of the faces of a tetrahedron can be expressed equivalently using other different ways (Arnold [59, p.188], [78]), among which is that of the equality of their areas (Exercise 4.117).

4.5 Regular pyramids

> It [the Pyramids] seems to have been erected only in compliance with that hunger of imagination which preys incessantly upon life, and must be always appeased by some employment...I consider this mighty structure as a monument of the insufficiency of human enjoyments.
>
> S. Johnson, "Rasselas"(1759) ch. 32

We call **regular** a pyramid whose, (i) its base is a regular polygon and (ii) the line KO, which joins the center K of the regular polygon with the apex of the pyramid is orthogonal to the plane ε of the base.

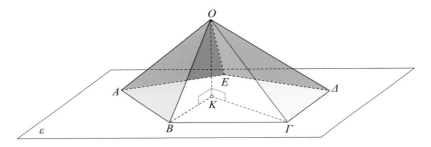

Fig. 4.28: Regular pyramid

Proposition 4.6. *In every regular pyramid the lateral faces are congruent isosceli and the lateral dihedrals are congruent. And conversely, if the lateral faces are congruent isosceli and the lateral dihedrals are also congruent, then the pyramid is regular.*

Proof. To begin with, the right triangles which are formed from the orthogonal OK on plane ε of the base and the radii KA, KB, $K\Gamma$, ... towards the vertices of the regular polygon (See Figure 4.28) are congruent, having the orthogonal OK common and the radii KA, KB, $K\Gamma$, ... equal. From this follows also that triangles OAB, $OB\Gamma$, $O\Gamma\Delta$, ... are congruent isosceli. From the congruence of the isosceli triangles follows the congruence of the faces at O. For the congruence of the lateral dihedrals we need the angles which are formed from planes orthogonal to the lateral edges. One such plane orthogonal to the edge $O\Gamma$, for example, results by drawing the altitudes of the triangles $BO\Gamma$ and $\Gamma O\Delta$ from B and Δ respectively (See Figure 4.29). Because of the congruence of these triangles, the altitudes pass through the same point Z of $O\Gamma$ and define a plane orthogonal to the edge $O\Gamma$ (§ 3.5). From the way they are constructed, though, all these triangles, like $BZ\Delta$, which are constructed in a similar way, also for the other edges $O\Delta$, OE, ... are congruent, therefore also the angles which correspond to $\widehat{BZ\Delta}$ are equal. These angles however are exactly the angles of the dihedrals, which therefore are congruent.

4.5. REGULAR PYRAMIDS

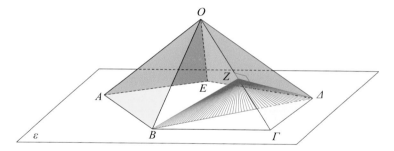

Fig. 4.29: Lateral dihedrals of a regular pyramid

For the converse consider the same figure. From the congruence of faces follows the existence of the isosceles triangle $B\Delta Z$. From the congruence of the dihedrals follows the congruence of the isosceli triangles like $B\Delta Z$, which are constructed similarly on the lateral edges. From the last congruence fol-

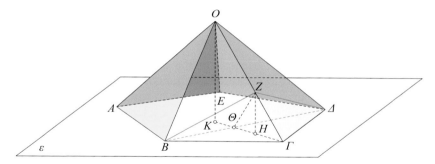

Fig. 4.30: Characterization of a regular pyramid

lows the congruence of triangles $\Gamma Z\Theta$ (Θ is the middle of $B\Delta$ in figure 4.30) and of its analogues on the other edges and from this last congruence follows that $AB\Gamma\Delta$... is regular and point K is the center of the polygon (alternatively apply theorem 4.7).

The polyhedral angles of regular pyramids are called **regular polyhedral angles**. Theorem 4.10 sets some restrictions on how big the angles (faces) can be at the apex of a regular polyhedral angle with κ number of faces. Next proposition gives all the cases where the lateral faces are congruent with a regular polygon with ν sides.

Theorem 4.13. *If a regular polyhedral angle has κ faces, congruent with the angle of a regular ν-gon, then the only possible values for κ and ν are those of the next table:*

$$\begin{array}{c|ccc} \kappa & 3 & 4 & 5 \\ \hline v & 3 & 3 & 3 \\ & 4 & & \\ & 5 & & \end{array}$$

Proof. To begin with, the angles of a regular v-gon are all equal to (Proposition I-2.10)
$$\alpha_v = \frac{v-2}{v}\pi.$$
If we have κ such equal angles at the apex of the polyhedral angle, then according to Theorem 4.10
$$\kappa \cdot \frac{v-2}{v}\pi < 2\pi \quad \Leftrightarrow \quad v < \frac{2\kappa}{\kappa-2}.$$
By assumption, κ and v are integers greater than or equal to 3. We easily see that the sequence $\beta_\kappa = \frac{2\kappa}{\kappa-2}$ is decreasing and its first few values are given by the following table, as:

κ	3	4	5	6
$\frac{2\kappa}{\kappa-2}$	6	4	3.333...	3

The values for $\kappa=7, 8, \ldots$ etc., because of the decreasing nature of the sequence, will be less than 3 and because simultaneously must hold $v \geq 3$ and $v < \frac{2\kappa}{\kappa-2}$ in these cases there cannot exist v which satisfies them. κ then can only have the values 3, 4, and 5 and for each one of these cases the allowed v with $v < \frac{2\kappa}{\kappa-2}$ are the written values in the column below the corresponding value of κ.

Remark 4.6. This simple proposition is the reason of the dramatic limitation in the number of the so called *regular polyhedra* or *Platonic solids*, which are examined in the next section. The proposition says, that if one wants to construct a polyhedral angle whose faces are equal regular polygons, then there are only 5 possibilities:

1. Trihedral angles ($\kappa = 3$): There exist three only possibilities of regular trihedral angles. The ones which have their three faces equilateral triangles ($v = 3$), the ones which have their three faces squares ($v = 4$) and the ones which have their three faces regular pentagons ($v = 5$).
2. Tetrahedral angles ($\kappa = 4$): There exists one and only regular tetrahedral angle, which has faces equilateral triangles ($v = 3$).
3. Pentahedral angles ($\kappa = 5$): There exists one and only regular pentahedral angle, which has faces equilateral triangles ($v = 3$).

Exercise 4.25. Using the triangle $B\Delta Z$ of figure 4.30, express the cosine of the angle $\frac{\widehat{A}}{2} = \frac{\widehat{B Z \Delta}}{2}$ of the dihedral as a function of the angle $\omega = \widehat{AOB}$ of their face and the number q of these angles, which surround its vertex.

4.5. REGULAR PYRAMIDS

Hint: If $a = |OA|$ is the length of the edge, then the following formulas hold:

$$|BZ| = a\sin(\omega), \quad |AB| = 2a\sin(\omega/2), \quad |K\Gamma| = \frac{a\sin(\omega/2)}{\sin(\pi/q)}$$

$$|B\Delta|^2 = 2|AB|^2\left(1 - \cos\left(\frac{(q-2)\pi}{q}\right)\right), \quad \cos(\widehat{A}) = \frac{2|BZ|^2 - |B\Delta|^2}{2|BZ|^2},$$

which, using simple calculations lead to the formula

$$\sin\left(\frac{\widehat{A}}{2}\right) = \frac{\cos\left(\frac{\pi}{q}\right)}{\cos\left(\frac{\omega}{2}\right)}.$$

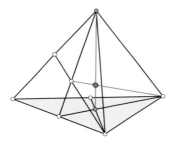

Fig. 4.31: Centroid of the tetrahedron

Exercise 4.26. Show that in every tetrahedron the lines which join its vertices with the centroids of its opposite faces pass through a common point G, which divides them into ratio 1:3 (See Figure 4.31).

Hint: Apply Exercise I-5.113.

The lines of the preceding exercise are called **medians** of the tetrahedron. Their intersection point is called **centroid** of the tetrahedron.

Exercise 4.27. Show that the lines which join the middles of opposite edges of the tetrahedron pass through its centroid.

Exercise 4.28. Show that the medial planes of the edges of a regular pyramid pass all trhough a common point O, which is equidistant from all vertices of the pyramid.

4.6 Polyhedra, Platonic solids

> I shall never get you put together entirely,
> Pieced, glued, and properly jointed.
>
> *Sylvia Plath, The Colossus*

We call **polyhedron** a solid of finite extent, which is enclosed by polygons called **faces** (See Figure 4.32). Often we use this term also for the entire plane carrying the face and it is not excluded the case in which two faces are on the same plane. Two different faces either have no common points or they have in common one (or more than one) entire polygon side of the two polygons-faces, called **edge** of the polyhedron. The common points between different edges, if any, are called **vertices** of the polyhedron. The dihedral angles

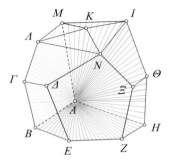

Fig. 4.32: Polyhedron

formed at the edges are called **dihedrals** of the polyhedron. The polyhedral angles which are formed from three or more faces at a vertex are called **polyhedral angles** of the polyhedron.

A polyhedron is called **convex** when, for each of its faces, it is contained entirely in one of the two half spaces defined by the plane of the face.

Two polyhedra $AB\Gamma\Delta...$ and $A'B'\Gamma'\Delta'...$ are called **congruent** if they have respective polygon-faces congruent and respective dihedrals also congruent. The polyhedra are called **similar** if they have respective polygon-faces similar, all with the same similarity ratio, and respective dihedrals also congruent.

In these lessons we deal exclusively with convex polyhedra, which are the analogues in space of the convex polygons on the plane, their characteristic property being that they lie entirely on one side of each of their plane-face. An important difference from polygons lies in the many different kinds of angles we meet in space: dihedral, trihedral and generally polyhedral angles. The pyramid and the truncated pyramid we met already in the preceding sections are examples of the simpler kinds of convex polyhedra. Other kinds of polyhedra are: prisms, parallelepipeds, tetrahedrons, cubes,

4.6. POLYHEDRA, PLATONIC SOLIDS

etc., which we'll examine next. A special category of polyhedra are the ana-

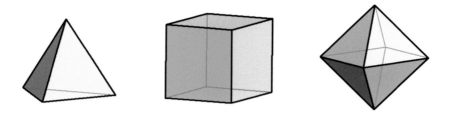

Fig. 4.33: Regular tetrahedron, cube and regular octahedron

logues in space of the regular polygons of the plane called **regular polyhedra**. These are defined as the polyhedra which have all their faces congruent and also congruent with a regular polygon and all their dihedral angles congruent. While on the plane there exist infinitely many regular polygons, in space, their analogues, which are the regular polyhedra, there are only five. These are the five **Platonic bodies** or **Platonic solids** : The **regular tetrahedron**, the **cube**, the **regular octahedron** (See Figure 4.33), and the **regular dodecahedron** and **regular icosahedron** (See Figure 4.34).

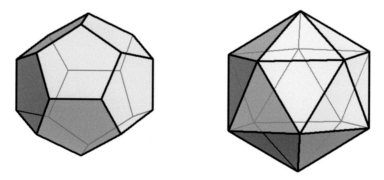

Fig. 4.34: Regular dodecahedron and icosahedron

Next figures (4.35, 4.36 and 4.37) present templates for the construction of these polyhedra from paper. After the figures are drawn in the requested size they must be folded along the edges which are common to two of the polygons. The edges which are free must be glued, each with a corresponding one which has an auxiliary flap, which is where the glue goes. To make things easier the auxiliary flaps which get the glue are also designed.

The limitation of the regular polyhedra to 5 is due to the corresponding limitation for regular polyhedral angles (Theorem 4.13). That these five platonic solids exist is relatively easy to show. The regular tetrahedron, the cube and the regular octahedron are constructed easily, in the sense that given a specific length δ, the corresponding solid whose edges have all length δ is easily constructed. For the dodecahedron and the icosahedron the corresponding construction is more involved, it can however be done using the suggested models (exercises 4.108, 4.111). The proof that there are no more

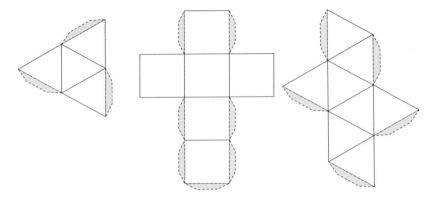

Fig. 4.35: Models of regular tetrahedron, cube and octahedron

than these 5 is also simple, using the aforementioned Theorem 4.13. The proof could be sketched as follows. In the beginning we construct the five solids with edge length δ. Assuming then that there also exists some other solid Σ' with edge length δ, then one polyhedral angle B of Σ' must coincide

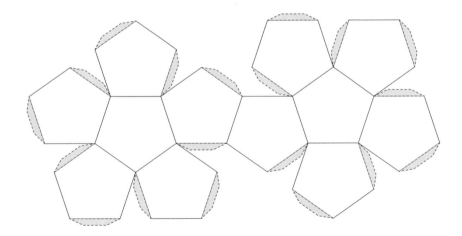

Fig. 4.36: Model of regular dodecahedron

with one of the 5 possibilities of the aforementioned proposition. Suppose that it coincides with the polyhedral angle A of one of the 5 specific solids, call it Σ. Then we put Σ' onto Σ, so that the polyhedral angle B coincides

4.6. POLYHEDRA, PLATONIC SOLIDS

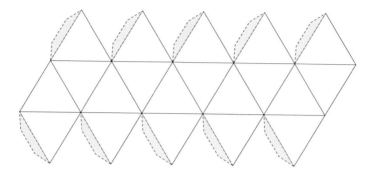

Fig. 4.37: Model of regular icosahedron

with A. Next will coincide also the non-adjacent dihedrals of $\widehat{A}$ and $\widehat{B}$ because of the congruence of the dihedral angles (Exercise 4.25). We proceed then inductively relative to the number of the faces of Σ' for which we have proved their coincidence with the corresponding ones of Σ, until we show coincidence for all the edges of Σ' with the corresponding ones of Σ.

Even though the sketch of the preceding proof can be completed into a full proof, I do not consider necessary, in the context of this book, to go deeper into the various details, which are required and deal with the congruence of polyhedral angles and the similarity between polyhedra.

In his 13th book of *Elements* Euclid proves the existence of the five platonic solids and the fact that they can be inscribed in a sphere. He also finds the relation which connects the edge of the solid with the radius of its circumscribed sphere ([71, pp.520-534]). A short sketch for the construction of the dodecahedron is given in exercise 4.108. Also exercise 4.111 sketches an alternate construction to that of Euclid, of the icosahedron. Complementary material for the platonic solids are contained in the tables of exercises 4.79 and 5.44.

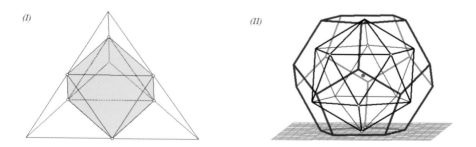

Fig. 4.38: Octahedron from tetrahedron , Icosahedron from dodecahedron

Exercise 4.29. Show that the middles of the edges of a regular tetrahedron define the vertices of a regular octahedron (See Figure 4.38-I). Similarly the centers of the square faces of a cube define the vertices of an octahedron.

The second part of the exercise is generalized in any platonic solid. Joining the centers of the faces of a given platonic solid we get a second platonic solid, we would say *inscribed* in the first, which is traditionally called **dual** of the first. However the first part of the exercise, that is the joining with line segments, of the middles of the edges of a platonic solid does not always lead to a new platonic solid. This way, for example, this process applied to the cube leads to the so called *cubeoctahedron* and applied to the dodecahedron leads to the so called *icosidodecahedron* § 7.8.

Exercise 4.30. Show that by joining the centers of the faces of a platonic solid results in a convex polyhedron with faces congruent regular polygons and congruent dihedrals, as well as polyhedral angles. Conclude that the polyhedron coincides with a platonic solid. Also conclude that the existence of the icosahedron is warranted by the existence of the dodecahedron (See Figure 4.38-II).

4.7 Prisms

> The pencil and the eraser are more useful in thought than an entire staff of assistants.
>
> *Theodor Adorno, Minima Moralia p. 216*

The surface which extends to infinity, consisting of the strips between parallel lines (AA', BB'), $(BB', \Gamma\Gamma')$, ... is called **prismatic surface** (See Figure ??).

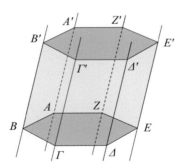

Fig. 4.39: Prismatic surface and Prism

The lines AA', BB', $\Gamma\Gamma'$, ... are called **edges** or **generators** of the prismatic surface. The dihedrals formed by two adjacent strips are called **dihedrals** of the

4.7. PRISMS

prismatic surface. Every plane, not parallel to the edges of a prismatic surface, intersects it along a polygon. When the plane is orthogonal to the edges then the corresponding polygon, which is excised from the prismatic surface, is called a **orthogonal section** of the prismatic surface. **Prism** is called the solid which results by intersecting a prismatic surface with two parallel planes, which are not parallel to an edge. The distance between these two planes is called **altitude of prism**. The polygons $AB\Gamma\Delta...$ and $A'B'\Gamma'\Delta'...$, which are defined by the intersection of the prismatic surface with the parallel planes are congruent (Corollary 3.7) and are called **bases of the prism**. The prism is called **convex** when the corresponding polygon of the base $AB\Gamma\Delta...$ is convex. The equal (Proposition 3.11) line segments AA', BB', $\Gamma\Gamma'$, ... are called **lateral edges** of the prism. Each strip of the prismatic surface intersected by the parallel planes of the bases by parallelogram is called **face** of the prism. The prism is called **right** when its lateral edges are orthogonal to its bases. The prism is called **regular** when it is right and the polygon $AB\Gamma\Delta...$ is regular. Two prisms are called **congruent** when they have respective faces congruent and respective dihedrals also congruent.

Exercise 4.31. Show that all the intersections of a prismatic surface with parallel planes, not parallel to its edges, define on the intersecting planes congruent polygons.

One of the simplest prisms is the **parallelepiped**, whose bases are parallelograms (See Figure 4.40). Two vertices, which have no common faces are called **opposite** vertices of the parallelepiped. The line segments which join two opposite vertices are called **diagonals** of the parallelepiped.

Fig. 4.40: parallelepiped and right or rectangular parallelepiped

Exercise 4.32. Show that the four diagonals of the parallelepiped pass through the same point which bisects them. Also show that the diagonals of the right parallelepiped are equal.

Exercise 4.33. Show that, if in a prism with square bases the four diagonals pass through the same point, then they are bisected by it and the prism is a parallelepiped.

A right parallelepiped on which additionally the bases are rectangles is called **rectangular parallelepiped** (See Figure 4.40). The **cube** is a rectangular parallelepiped whose faces are squares (See Figure 4.41).

Exercise 4.34. Show that the squares which constitute the faces of the cube are congruent to each other.

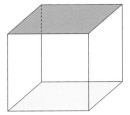

Fig. 4.41: Cube

The solid which results by intersecting a prismatic surface with non-parallel planes is called **truncated prism** (See Figure 4.42-I).

Exercise 4.35. Show that there exists an intersection of the cube with a plane, which defines on the intersecting plane a regular hexagon.

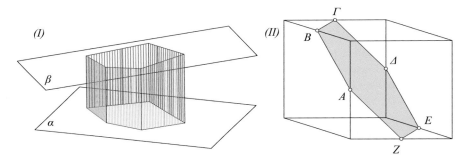

Fig. 4.42: Truncated prism   Hexagonal section of cube

Hint: The vertices of the hexagon are the middles of the edges of the cube (See Figure 4.42-II). This guarantees that all the sides of the hexagon are equal to half the diagonal of the square face of the cube. For the fact that the figure lies on a plane and also has equal angles it suffices to show that four successive points are coplanar and form a trapezium (like for example $AB\Gamma\Delta$ in the figure).

Exercise 4.36. Given is a triangular prismatic surface with orthogonal section an isosceles triangle $AB\Gamma$ (See Figure 4.43-I). Translating parallel the base of the triangle $z = |B\Gamma|$ in the containing face at distance h we define the section of the prismatic surface along the isosceles triangle $AB'\Gamma'$. To find the relation between the angles $\{\omega = \widehat{BA\Gamma},\ \omega' = \widehat{B'A\Gamma'}\}$ and to prove that, for variable h the angle ω' can obtain all possible measures in the interval $0 < \omega' < \omega$.

4.7. PRISMS

Hint: Compute that $\cos \omega' = \frac{h^2 + x^2 \cos(\omega)}{h^2 + x^2}$.

Exercise 4.37. Given is a prismatic surface with orthogonal section a parallelogram $AB\Gamma\Delta$ and the angle at A obtuse (See Figure 4.43-II). Apply the preceding exercise to show that there is a rectangular section $AB'\Gamma'\Delta'$ with a plane through A.

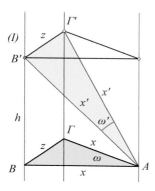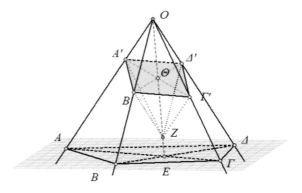

Fig. 4.43: Varying intersection angle   Rectangular section

Exercise 4.38. To intersect a tetrahedral angle $OAB\Gamma\Delta$ with a plane along a parallelogramic section $A'B'\Gamma'\Delta'$.

Hint: The intersection of the planes $OB\Delta$ and $OA\Gamma$ is a line which passes through the center of symmetry Θ of the parallelogram (See Figure 4.44).

Fig. 4.44: Intersection along a parallelogram

The direction of the diagonals $A'\Gamma'$ and $B'\Delta'$ is determined from Exercise I-2.50. Next theorem, of Simon Lhuilier (1750-1840), implies, that an arbitrary triangular prismatic surface can be intersected with an appropriate plane, so that the intersection is similar to a given triangle.

Theorem 4.14. *Given a prismatic surface with orthogonal section the triangle $AB\Gamma$ and given a triangle $A_1B_1\Gamma_1$, there exists a plane which intersects the prismatic surface along a triangle $AB_0\Gamma_0$ similar to $A_1B_1\Gamma_1$.*

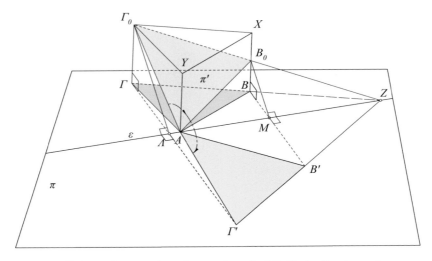

Fig. 4.45: Prismating section along triangle $AB_0\Gamma_0$ similar to a given one

Proof. There are at least two ways to see the problem. Both ways are interesting and reduce the problem to one of plane geometry.

According to the first way, the problem is reduced to the construction of a pyramid with base the trapezium $B\Gamma\Gamma_0B_0$ and vertex A (See Figure 4.45). In this way given is its face $AB\Gamma$, the direction of the parallels $\Gamma\Gamma_0$, BB_0 and the angles of the triangle $AB_0\Gamma_0$. The construction of the pyramid from these elements is done in Theorem 4.11.

A second way of proving the theorem is that of reducing it to two perspective triangles of the plane. Indeed, let us suppose that we found such a plane π', which intersects the prismatic surface along a triangle $AB_0\Gamma_0$, which is similar to the given triangle $A_1B_1\Gamma_1$. Suppose further that ε is the line-intersection of the planes $\pi = AB\Gamma$ and $\pi' = AB_0\Gamma_0$. We perform the so called **Unfolding** of plane π' onto plane π, that is we rotate π' around its intersection line ε with π, until it coincides with π. Then the triangle $AB_0\Gamma_0$ takes the position $AB'\Gamma'$ (See Figure 4.45).

The problem then transforms into the one of constructing the plane figure which consists of two triangles $AB\Gamma$ and $AB'\Gamma'$ and the line ε and to perform the opposite of unfolding, that is to rotate the constructed $AB'\Gamma'$ about ε until it takes the position $AB_0\Gamma_0$ with $\Gamma_0\Gamma$, B_0B orthogonal to the plane $\pi = AB\Gamma$.

4.7. PRISMS

This means that $A\Gamma_0'$ will be the hypotenuse of the right triangle $A\Gamma_0\Gamma$ and consequently greater than $\Gamma\Lambda$.

The construction of this plane figure is equivalent to finding line ε, passing through A and such that the triangle $AB'\Gamma'$ is similar to the given $A_1B_1\Gamma_1$, lines BB', $\Gamma\Gamma'$ are orthogonal to ε and $B'\Gamma'$, $B\Gamma$ intersect at point Z of ε. Also, so that the reverse process of unfolding can be performed, we must guarantee that $|A\Gamma'| > |A\Gamma|$. Such a construction is precisely guaranteed by Theorem I-5.41.

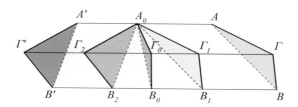

Fig. 4.46: Sections of prismatic surface similar to given triangles $AB\Gamma$, $A'B'\Gamma'$

Corollary 4.4. *Every triangular prismatic surface can be cut with appropriate planes along similar to given triangles $AB\Gamma$, $A'B'\Gamma'$ (See Figure 4.46).*

Proof. Application of the preceding theorem to an orthogonal section $A_0B_0\Gamma_0$ of the prismatic surface. According to the theorem, triangles $A_0B_1\Gamma_1$, $A_0B_2\Gamma_2$ similar to the given ones can be constructed. Every section of the prismatic surface parallel to these produces triangles similar to the given ones.

Exercise 4.39. Show that if a prismatic surface admits a section with a plane along a parallelogram $AB\Gamma\Delta$, then there exists a plane which intersects it along a square.

Exercise 4.40. Show that the opposite edges of a tetrahedron (i.e. edges with no common vertices) define two skew lines. Show also that the intersection of the tetrahedron with a plane parellel to two opposite edges is a parallelogram.

Exercise 4.41. Show that if a prismatic surface α, which has a parallelogramic orthogonal section, is intersected with a plane β non-parallel to its edges, then the intersection of $\{\alpha, \beta\}$ is also a parallelogram.

4.8 Cylinder

> In nature, everything results from three basic morphological shapes: the sphere, the cone and the cylinder. One must learn to draw these simplest of forms, in order to be able to do whatever he wants afterward.
>
> Paul Cézanne

Given a circle κ on the plane ε, the extended to infinity surface which is produced from the lines, which are orthogonal to ε and pass through the points of the circle is called **cylindrical surface**. The parallel lines XY which produce the surface are called **generators of the surface** and the parallel to them which passes through the center K of the circle is called **axis of the cylindrical surface**. To distinguish the preceding from other surfaces which are generated in similar ways we often use the more accurate term **right circular** cylindrical surface. The intersection of such a surface with two planes ε and ζ orthogonal to the generator defines a solid called (right circular)

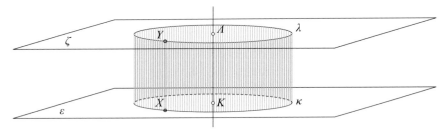

Fig. 4.47: Cylinder

cylinder (See Figure 4.47). The two congruent circles, which are excised by the planes ε and ζ from the cylindrical surface, are called **bases** of the cylinder. We often call bases the planes ε and ζ themselves. The distance between these planes is called **altitude of cylinder**. From the definition follows that the cylinder is the solid which results by rotating the rectangle $K\Lambda YX$ around its side $K\Lambda$. From the definition it also follows immediately that the intersections of the cylinder with planes parallel to its bases are circles congruent with κ. The radius ρ of the circle κ is called **radius** of the cylinder. Two (right) cylinders are called **congruent** when they have the same radius and the same altitude.

Cylinders are constructed easily from rectangles $XYY'X'$ which we bend in a circular manner and we join (glue) their opposite sides XY and $X'Y'$ (See Figure 4.48). It is proved with the help of *Differential Geometry* (theorem Hartmann-Nirenberg [61, p.77], see also references of section 4.9) that this process preserves the distances between points transferring the linear distance on the plane unfolding to the length of an arc of a *helicoidal* curve on the cylinder. This process is used in industry for the production of pipes. The

4.8. CYLINDER

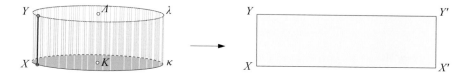

Fig. 4.48: Cylinder unfolding

reverse process, in which we cut a cylinder along one of its generators XY and we unwrap it to a plane rectangle $XYY'X'$ is called **cylinder unfolding**. The length of the side XX' of the unfolding is equal to the circumference of the circle κ. Consequently if ρ is the radius of the cylinder then $|XX'| = 2\pi\rho$.

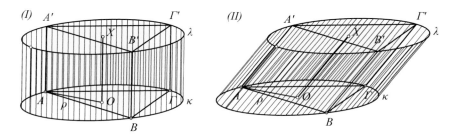

Fig. 4.49: Cylinder circumscribed to triangular prism

Exercise 4.42. Show that every right triangular prism can be inscribed in a cylinder, in the sense that there exists a right circular cylinder which contains the edges of the prism as generators.

Hint: Let $\kappa(O, \rho)$ be the circumcircle of the triangle $AB\Gamma$ of the base of the prism (See Figure 4.49-I). The cylinder with base κ and axis the orthogonal OX to its base at the point O is the wanted one.

Exercise 4.43. Show that every right regular prism can be inscribed in a cylinder.

Remark 4.7. The preceding two exercises, with small changes, can be generalized also for oblique prisms, whose bases are polygons inscriptible in a circle (See Figure 4.49-II). For these also it is proved that there exists an **oblique circular cylinder** which envelops them, that is a cylinder which is generated by parallel lines which join points of two *congruent* circles which are lying on parallel planes.

Exercise 4.44. Consider a plane ε parallel to the axis ζ of the cylindrical surface κ. Show that such a plane intersecting the surface, does it either (i) along two generators, or (ii) along a single generator.

In case (ii) of the preceding exercise we say ε is **tangent** to the cylinder (cylindrical surface) and the generator is called the **line of contact** of the plane with the cylinder.

Exercise 4.45. Given two, outside each other lying, cylindrical surfaces with parallel axes, show that there are exactly four planes simultaneously tangent to both cylinders.

4.9 Cone, conical surface

> Few things separate more profoundly the mode of life befitting an intellectual from that of the bourgeois than the fact that the former acknowledges no alternative between work and recreation.
>
> *Theodor Adorno, Minima Moralia, Timetable*

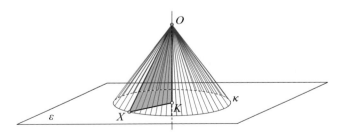

Fig. 4.50: Cone

Given a circle κ on the plane ε and a point O on the line orthogonal to plane ε which passes through the center of the circle K (O different than K), we call **cone** the solid which results from the lines which join the point O with all the points X of the circle (See Figure 4.50). Point O is called **vertex** or **apex** of the cone, the lines OX are called **generators** of the cone and the circle κ is called **base**. Often, to distinguish this kind of cone from other more general kinds, we call it **right circular cone**. Line OK is called **axis of cone**. The length $|OK|$ is called **altitude of cone** (See Figure 4.50). Two cones are called **congruent** when they have equal bases and equal altitudes.

From the definition follows that all triangles OKX, for points X on the circle κ, are right at K and congruent to each other. This way we can consider that the cone is the solid which is bounded by the plane ε and the surface generated by the hypotenuse OX of the right triangle OKX, by rotating the triangle about its orthogonal side OK. We call this surface **lateral surface** of the cone. This is part of the extended to infinity surface, which results

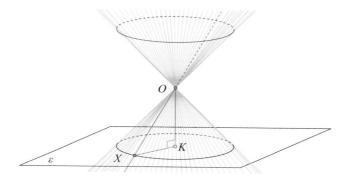

Fig. 4.51: Right conical surface

by rotating the line-carrier of the hypotenuse OX of the right triangle OKX (and not only its part OX). We call this surface a **conic surface** and more accurately **right circular conic surface**, in order to distinguish it from other surfaces which are generated in similar ways. The conic surface consists of two **nappes** with common apex point O (See Figure 4.51).

Proposition 4.7. *The right triangles OKX, which are formed from the points X of the base, the center K of the base and the apex O of the cone, are all congruent to each other.*

Proof. Obviously, because they have OK in common and equal all their other orthogonal sides as radii of the same circle.

The double of the angle $\alpha = \widehat{KOX}$, is called the **opening** or **angle of the cone**.

Proposition 4.8. *The intersection of a conical surface with a plane which is orthogonal to its axis and does not pass through its vertex is a circle with center on its axis.*

Proof. Suppose ε is a plane intersecting the axis orthogonally at K (See Figure 4.51). Suppose also X is a point of the intersection of one generator with ε. Triangles OKX are right at K and have fixed and independent of X $|OK|$ and the angle $\alpha = \widehat{XOK}$, therefore they are pairwise congruent and consequently $\rho = |XK|$ is also fixed. Therefore all points X are contained in the circle of center K and radius ρ. Conversely, every point of this circle defines a right triangle XOK, with the same angle $\alpha = \widehat{XOK}$, hence is contained in the cone.

Proposition 4.9. *The intersection of a conical surface with a plane ε, which passes through its vertex, is a pair of generators or exactly one generator or exactly one point, which coincides with its vertex.*

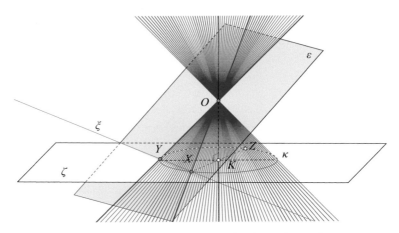

Fig. 4.52: Interesection of a cone with plane through its vertex

Proof. Suppose that the plane ζ is orthogonal to the axis of the conical surface and not passing through its vertex O, which, according to the preceding proposition, intersects it along a circle κ (See Figure 4.52). If the plane ε contains another point X of the generator (besides O), then it will contain the entire generator and consequently will intersect the circle κ, since the circle intersects every generator at exactly one point. We see then that the generators which are included in ε correspond exactly to the intersection points of the circle κ and the line ξ which is the intersection of the planes ε and ζ. The proposition follows from the fact that the line ξ and the circle κ can have two, one or no common points.

Corollary 4.5. *A plane, which contains the axis of a cone intersects it by two generators lying symmetrically relative to the axis.*

Proof. From the assumption follows that the line ξ, the intersection line of the planes ε and ζ of the (proof of the) preceding proposition, passes through the center of the circle κ and intersects it along two diametrically opposite points X and Z, hence the conclusion.

We call **tangent plane** of the cone a plane, which has exactly one generator in common with the cone.

Proposition 4.10. *A plane ε is tangent to a cone exactly when it contains one generator η of the cone and is orthogonal to the plane θ containing the generator η and the axis OK of the cone.*

4.9. CONE, CONICAL SURFACE

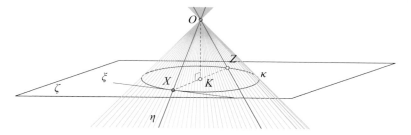

Fig. 4.53: Plane of lines $\{\xi, \eta\}$ tangent to cone

Proof. By definition the tangent plane ε of the cone contains exactly one generator of η (See Figure 4.53). We draw the plane ζ orthogonal to the axis of the cone which intersects it along the circle $\kappa(K)$. Suppose also that ξ is the line-intersection of the two planes ε and ζ. The generators of the cone, which are contained in ε, correspond exactly to the intersection points of line ξ with the circle κ. If then there exists only one intersection point X, then line ξ will be tangent to κ at X and ξ will be orthogonal to the radius KX of the circle κ and skew orthogonal to the axis of the cone, therefore ξ will be orthogonal also to the plane of the line $\eta = OX$ and the axis. Plane ε, containing ξ, will consequently also be orthogonal to the plane of η and OK. The argument can be reversed and this shows the truth of the proposition.

Theorem 4.15. *Given a conical surface Σ, every plane ε of the space intersects the surface Σ and as a matter of fact intersects either (i) all its generators or (ii) all except two or (iii) all except one.*

Proof. From the vertex O of the conical surface draw the plane ε' parallel to plane ε. There are three cases (Proposition 4.9): (i) plane ε' has in common with Σ only point O, (ii) plane ε' intersects Σ along two generators, (iii) plane ε' intersects Σ along one generator. In the first case, since ε' intersects all the generators, without containing any one of them, the same will happen for the parallel plane ε. In the second case plane ε' will intersect all generators without containing them, except two specific ones which it contains. Therefore the parallel plane ε will also intersect all the generators except the two which are contained in ε'. In the third case, plane ε' will contain exactly one generator (therefore it will be tangent to Σ) and its parallel will intersect all other generators except the one to which it will be parallel.

Remark 4.8. It is often useful to use an analogy which exists between the conical surface Σ and the plane shape which results by intersecting Σ with a plane θ, which passes through its axis. This shape (figure-4.54, in which θ coincides with the plane of the paper) consists of two generators forming an angle whose one bisector is the axis of Σ. This way, for example, to the last proposition corresponds the plane shape with line ε which describes the

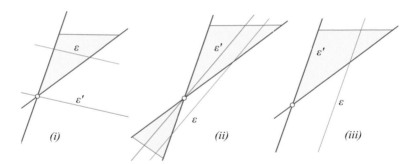

Fig. 4.54: Intersection of a conical surface with a plane

analogue of the three cases (i), (ii) and (iii), in which the intersecting plane ε of the conical surface meets (i) all (ii) all except two and (iii) all except one generators of Σ.

Corollary 4.6. *Given a conical surface Σ, if the plane ε does not pass through its vertex O but intersects all the generators or all except one, then the intersection of Σ and ε is contained in one of the two nappes of Σ. If plane ε intersects all except two generators of Σ, then it intersects both nappes of Σ.*

Proof. From the proof of Theorem 4.15 we saw that the parallel plane ε' of ε which passes through point O in cases (i) and (iii) does not contain a generator, respectively is tangent to one generator η. Consequently its parallel ε in the first case will intersect one of the two nappes (See Figure 4.54). In the second case plane ε will be parallel to the tangent plane ε' which contains each nappe in a different side. Therefore ε will intersect only one nappe. Finally in the case where ε' contains two generators α and β, these divide all the others which are contained in the same nappe into two groups and ε will intersect the generators of one group in the one nappe and the generators of the other group in the other nappe. Figure 4.54 shows the corresponding plane figure which results, according to the preceding remark. The shaded region corresponds to the nappe which intersects plane ε'.

A polyhedral angle $OXYZ...$ is called **inscribed in a (right) cone** or **can be inscribed in a (right) conical surface** when there exists a (right) conical surface, whose vertex coincides with point O and simultaneously the edges of the polyhedral angle are generators of the conical surface.

Proposition 4.11. *Every trihedral angle $OXYZ$ is inscribed in a right circular conical surface (See Figure 4.55).*

Proof. Consider the trihedral $OXYZ$ and the points A, B, Γ on its edges at equal distances from its vertex, so that $|AO| = |BO| = |\Gamma O| = \delta$. Consider next the medial planes of the line segments AB, $B\Gamma$ and ΓA. All these planes pass

4.9. CONE, CONICAL SURFACE

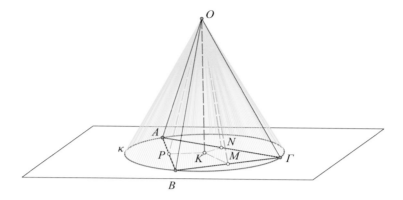

Fig. 4.55: Circumscribed cone of a trihedral angle

through line OK which is the line orthogonal to the plane $AB\Gamma$ at point O. This follows from the fact that the points of the medial plane, for example of segment AB, are equidistant from A and B. This way, by construction, point O belongs to all three medial planes. For the same reason the center K of the circumscribed circle κ of the triangle $AB\Gamma$ belongs also to all three medial planes. Consequently OK is also contained in all three medial planes. OK as (skew) orthogonal to two lines of the plane $AB\Gamma$ (for example AB and $A\Gamma$) is orthogonal to it (Proposition 3.18). It follows that the edges OX, OY and OZ of the trihedral angle are generators of the conical surface, which is defined by the circle κ and point O. This means exactly that the trihedral is inscribed in the conical surface.

The conical surface, whose existence is guaranteed by the preceding proposition, is called **circumscribed conical surface** of the trihedral angle. Analogously is also defined the **inscribed conical surface** of the trihedral, as the right circular conical surface which is tangent to the three faces of the trihedral. In what follows we'll see (Exercise 4.57) that every trihedral angle indeed has an inscribed conical surface.

Exercise 4.46. Show that every regular pyramid is inscribed in a right circular cone. That is, there exists a cone with vertex the pyramid's vertex which contains the edges of the pyramid as generators.

4.10 Truncated cone, cone unfolding

> Sure he that made us with such large discourse,
> Looking before and after, gave us not
> That capability and god-like reason
> To fust in us unused.
>
> *Shakespeare, Hamlet, act 4, sc. 4*

Given a cone with base the circle $\kappa(K)$ on the plane ε and vertex O, we call **truncated cone** or **cone frustum** the solid which is excised from the cone by a plane ζ parallel to its base ε and not going through its vertex O. The

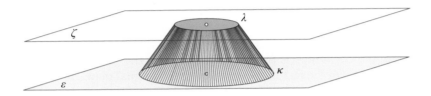

Fig. 4.56: Truncated cone

intersection of the plane ζ with the corresponding conical surface is a circle λ contained in the plane ζ with center on the axis of the cone (Exercise 4.8). The two circles κ and λ (as well as the planes ε and ζ), which define the truncated cone are called **bases** of the truncated cone (See Figure 4.56). The distance between the planes ε and ζ is called **altitude of the truncated cone**. Two truncated cones are called **congruent** when they have congruent bases and equal altitudes. The circle which results as the intersection of the cone

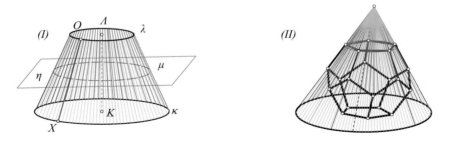

Fig. 4.57: Middle cut of truncated cone, Cone circumscribing dodecahedron

with the plane η, which is parallel to the base and passes through the middle of its altitude $K\Lambda$ is called **middle section** of the truncated cone (See Figure 4.57-I).

4.10. TRUNCATED CONE, CONE UNFOLDING

Exercise 4.47. To find the opening angle of the cone which circumscribes the dodecahedron, as well as, the length of the generator of the truncated cone defined by it and having for bases two opposite faces of the dodecahedron (See Figure 4.57-II).

Cones and truncated cones are used in many technical applications. They are constructed easily from circular sectors (§ 1.6) by gluing the initial and final bounding radius of these. Indeed, if we cut a cone along one of its gen-

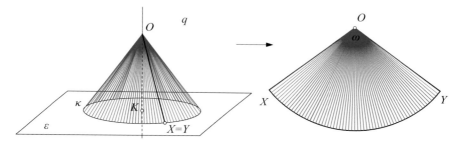

Fig. 4.58: Unfolding of cone

erators OX, we can next unfold it on the plane and produce a circular sector OXY, whose center corresponds to the vertex O of the cone and the arc of the circular sector corresponds to the base of the cone and has as length the length of the circle of its base (See Figure 4.58). This sector is called **unfolding of the cone**. As it is proved with the help of *differential geometry* (Theorem of Minding (1806-1885) [65, p.289], Theorema Eggregium of Gauss, [73, p.161], [76, p.206]), this process preserves the distances and areas, in the sense that two points of the cone define on it a shortest path connecting them and by the unfolding this path maps to a line segment. Also defined is the area of shapes on the cone and by the unfolding these shapes map to plane figures having the same area.

The elements of the unfolding are calculated easily from these of the cone. Thus, the radius of the sector is $\rho = |OX|$. The arc length of the sector is related to the length of the generator $|OX|$ and its opening. If we denote by $\omega = \widehat{XOY}$ the measure of the central angle of the cone in radians, then the length of the arc is $\sigma = |XY| = \rho \cdot \omega$. On the other hand the same length σ is the length of the circle κ of the base of the cone, whose radius is equal to

$$r = \rho \sin\left(\frac{\alpha}{2}\right),$$

where α is the opening of the cone. Thus, we have in totall

$$\sigma = |XY| = \rho \cdot \omega = 2\pi r = 2\pi \rho \sin\left(\frac{\alpha}{2}\right).$$

Simplifying the expression, we see that the angle ω of the unfolding depends exclusively on the opening α of the cone (all measurements in radians)

$$\omega = 2\pi \sin\left(\frac{\alpha}{2}\right).$$

In the reverse process, that of constructing the cone from the circular sector XOY, we wrap the sector joining (gluing) its radii OX and OY. The determination of the opening α of the cone from the angle ω of the sector leads to the solution of the last equation relative to α, and is done through the so called *inverse sine function*, a subject which is examined in *Calculus* [86, p.307].

The same process leads also to the unfolding of the truncated cone (See Figure 4.59). The cone is cut again across a generator XY and is unfolded on

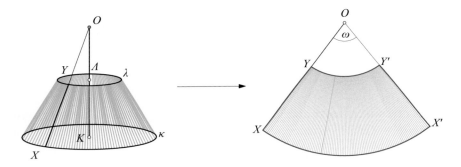

Fig. 4.59: Unfolding of truncated cone

the plane forming a circular annulus sector. The calculations of the relations between the truncated cone and the corresponding plane figure are similar to the preceding ones. Similar is also the reverse process of producing the truncated cone from the plane figure.

Exercise 4.48. To find the unfolding of the truncated cone of maximal area, which can be symmetrically inscribed in a rectangle τ, whose sides have lengths $\{a > b\}$ (See Figure 4.60).

Exercise 4.49. Continuing the preceding exercise, to find the relation of $\{a,b\}$, for which the corresponding unfolding of maximal area has a given fixed area $\varepsilon(AB\Gamma\Delta) = \varepsilon_0$.

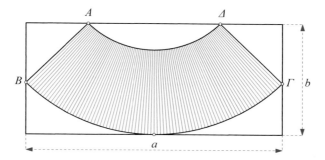

Fig. 4.60: Unfolding of maximal area

4.11 Sphere

> for the evil in the world far exceeds the good. The good we obtain hardly and with anxious endeavor, but the evil easily befalls us; for they say evils are round, linked together, and by a mutual dependence of causes follow one another, but the good lie scattered and disjoined, and with great difficulty are brought within the compass of our life.
>
> *Plutarch, Consolation to Apollonius 28*

Given a point O in space and a positive number ρ, we call **sphere** with center O and radius ρ the geometric locus of points X in space for which holds $|OX| = \rho$ (See Figure 4.61). We often denote such a sphere κ with $\kappa(O, \rho)$ or $O(\rho)$. The points Y of space, for which holds $|YO| < \rho$ constitute the **interior**

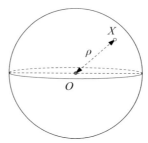

Fig. 4.61: Sphere

of the sphere and the points for which holds $|YO| > \rho$ build the **exterior** of the sphere.

Similarly, quite a few of the notions we meet in dealing with the circle are generalized to the sphere. This way we have the **chord of sphere**, which is a line segment whose endpoints are points on the sphere and the **diameter** of the sphere, which is a chord passing through the center of the sphere. Two points of the sphere which are endpoints of a diameter are called **diametrically opposite** or **diametral** points . Two spheres are called **congruent**, when they have the same radius.

Theorem 4.16. *The common points between a sphere σ and a plane ε, which intersects it, are either a circle of ε or a point of ε.*

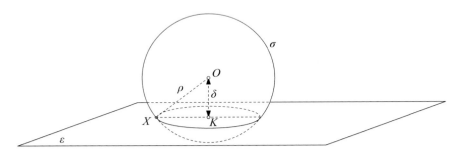

Fig. 4.62: Intersection of sphere and plane

Proof. The key to the theorem is the distance $\delta = |OK|$ between the center O of the sphere and the plane (See Figure 4.62). This is defined by drawing the orthogonal from point O to the plane and defining on it the line segment OK, where K is the projection of point O on ε. The following possibilities exist relative to the length δ and the radius ρ: (1) $\delta = 0$, (2) $\delta \neq 0$ and $\delta < \rho$, (3) $\delta = \rho$, (4) $\delta > \rho$.

(1) In this case ($\delta = 0$) the plane passes through the center O of the sphere ($K = O$) and the common points are coincident with the points of a circle of center O and radius ρ of the plane. Such a circle is called a **great circle** of the sphere. Each plane passing through the center of the sphere defines through its intersection such a circle.

(2) In this case ($0 < \delta < \rho$) there exist points of the plane X at distance from K: $|XK| = \sqrt{\rho^2 - \delta^2}$, for which, from the Pythagorean theorem, follows that $|XO| = \rho$, which means that they also lie on the sphere. Using the same argument we see easily that the circle of ε with center K and radius $r = \sqrt{\rho^2 - \delta^2}$ is coincident with the intersection of the plane and the sphere. Such circles are called **small circles** or **minor circles** or **parallel circles** of the sphere. Their characteristic is that they result from planes which intersect the sphere, but which do not pass through its center. The fact that $r < \rho$ explains their name, as well as the name of the circles of case (1).

(3) In this case ($\delta = \rho$) the distance from the plane is equal to the radius and K is a point of the sphere. For every other point X of the plane the triangle XKO will be right at K and its distance from O will be $|XO| = \sqrt{|XK|^2 + \rho^2} > \rho$. Therefore such a point is not contained in the sphere. Such

4.11. SPHERE

a plane is called **tangent plane** of the sphere. The only common point K of the tangent plane with the sphere is called **contact point** between the tangent plane and the sphere.

(4) In this case ($\delta > \rho$) the sphere and the plane have no common points, since for every other point X of the plane will hold $|XO| > |XK| > \rho$.

Corollary 4.7. *Every tangent plane to the sphere is orthogonal to its radius at the point of contact. Conversely, every plane orthogonal to the endpoint of the radius of the sphere is tangent to the sphere.*

Proof. We have shown in the preceding theorem, that when $|OK| = \rho$, that is when the distance between the center and the plane is equal to the radius of the sphere, then the plane is tangent to the sphere and has common with the sphere only the point of contact K. Conversely, if the plane ε is orthogonal to the endpoint K of a radius OK of the sphere, then, according to Pythagoras theorem, all the other points X of the plane will satisfy $|OX| > |OK| = \rho$, therefore point K will be the unique common point of ε with the sphere.

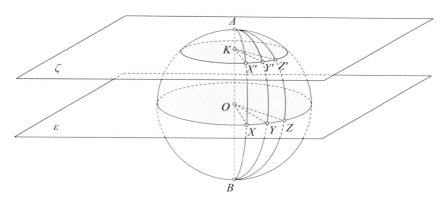

Fig. 4.63: Sphere by rotation of semicircle AXB

It is easy to see that the sphere is produced by a a semicircle AXB which we rotate about its diameter AB (See Figure 4.63). The middle X of the arc $\widehat{AB}$ describes upon rotation a great circle of the sphere. Every other point X' of the arc, different from A and B, describes a parallel circle. Points A and B define the axis of rotation and remain fixed, defining also a diameter of the produced sphere. The circles described by points X' of the arc $\widehat{AXB}$ are all contained in parallel planes, which are orthogonal to the diameter AB. Points A and B are the *poles* of these circles. Generally, for a circle contained in a sphere we call **poles of the circle** the endpoints of the diameter of the sphere, which is orthogonal to the plane of the circle.

Remark 4.9. It is proved with the help of *Differential Geometry* ([65, p.246]), that on the surface of the sphere the shortest path between two of its points A

and B is one of the arcs $\overset{\frown}{AB}$ defined on the great circle, which passes through A and B. When points A and B are not diametrically opposite, then the points A, B and O, where O is the center of the sphere, are not collinear. They define

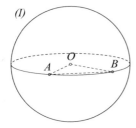
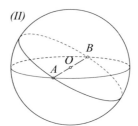

Fig. 4.64: Great circle from $\{A,B\}$ Great circles intersect at diametrals

consequently exactly one plane AOB and a corresponding great circle which contains them (See Figure 4.64-I). If we are also allowed to move off the surface of the sphere, then the shortest way between points A and B is the line segment AB. If, however, exiting the surface is not allowed, then the distance measured on the surface, coincides, as I mentioned, with the length of one of the arcs of the great circle, which is defined from A and B. This happens for example on Earth, which is approximately a sphere. The shortest sailings between two points A and B on the ocean lie along the great circle defined from A, B and the center O of the Earth. When points A and B are diametrically opposite, then all the planes which contain A and B pass through the center of the sphere (See Figure 4.64-II). Every such plane defines also a great circle and the shortest path between A and B, in this case, is not unique. Any semicircle of a great circle through A and B, gives the shortest distance between A and B.

If we exclude this case of diametrically opposite points, there is an analogy between great circles on a sphere and lines on the plane. As it is on the plane, so in the sphere, for every pair of (non diametrically opposite) points A and B there exists exactly one great circle which passes through them. Line segments on the plane correspond to arcs of great circles, which represent the shortest connection between points on the sphere.

Exercise 4.50. Show that the medial plane of a chord AB of the sphere passes through its center.

Hint: Obviously, since the medial plane is the geometric locus of the points which are equidistant from the endpoints A, B of the chord and the center O of the sphere has this property.

Exercise 4.51. Show that the middles of parallel chords of the sphere are contained in a plane which passes through the center of the sphere.

4.11. SPHERE

Hint: Consider one of the parallel chords, for example AB and its medial plane ε. This, according to the preceding exercise will pass through the center O of the sphere. Another parallel chord to AB, say $\Gamma\Delta$ will have medial line plane ε' also passing through point O but also parallel to ε (Theorem 3.15). Consequently the two planes will coincide.

Exercise 4.52. Show that, for three points A, B and Γ of the sphere, the line orthogonal to the plane of the triangle $AB\Gamma$ at the center of its circumcircle passes through the center of the sphere and the circumscribed circle of the triangle is contained in the sphere.

Exercise 4.53. Show that parallel planes intersecting the sphere $\sigma(O)$ define circles whose centers belong to a diameter of the sphere orthogonal to these parallel planes.

A line, which has exactly one common point with the sphere is called **tangent to sphere**. The common point of the tangent with the sphere is called **point of contact** of the tangent with the sphere.

Proposition 4.12. *A line ε is tangent to the sphere σ, if and only if it is orthogonal at the endpoint of a radius of the sphere.*

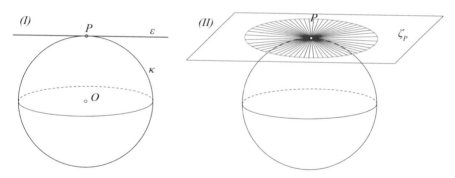

Fig. 4.65: Line tangent to sphere   Tangent plane at P

Proof. The line ε and the center O of the sphere define a plane η which intersects the sphere along a great circle κ (See Figure 4.65-I). If the sphere and the line ε have one common point, then also the circle κ and the line have a common point, therefore they are tangent and the line is orthogonal to the endpoint of a radius of the circle, which is also radius of the sphere (Theorem I-2.2). Conversely, if the line ε is orthogonal at the endpoint to a radius of the sphere, it will be orthogonal also to the endpoint of radius of κ, consequently it will have a common point with it, therefore also with the sphere.

Corollary 4.8. *A line ε is tangent to the sphere σ at its point P, if and only if it passes through P and is contained in the tangent plane of the sphere at P (See Figure 4.65-II).*

Theorem 4.17. *For every point A lying in the exterior of the sphere O(ρ), the tangents from A towards the sphere are all equal and form a cone which has common with the sphere a small circle of it, whose plane is orthogonal to the axis of the cone. The points of this circle coincide with the points of contact of the tangents from A with the sphere.*

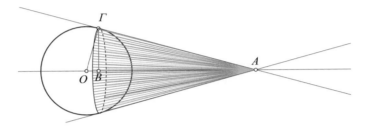

Fig. 4.66: Tangent cone to sphere from an outer point

Proof. Consider the plane ε, defined from the line OA and one tangent $A\Gamma$ from point A (See Figure 4.66). This plane intersects the sphere along a great circle which has a common point with $A\Gamma$, therefore it is a tangent to it. Triangle $O\Gamma A$ is consequently right, independent of the special position of ε, since it always has the same hypotenuse OA and orthogonal $O\Gamma$ with the same length $|O\Gamma| = \rho$. It follows, that the distance $\delta = |\Gamma B|$ of Γ from the axis, which is equal to the altitude of this right triangle, is fixed and independent of the plane ε. Consequently the points of contact, like Γ, are contained in the small circle with center B and radius δ. The radius δ of the small circle is calculated easily from the similar triangles $OB\Gamma$ and $O\Gamma A$:

$$\frac{\delta}{\rho} = \frac{|B\Gamma|}{|O\Gamma|} = \frac{|A\Gamma|}{|OA|} = \frac{\sqrt{x^2 - \rho^2}}{x} \Rightarrow \delta = \frac{\rho\sqrt{x^2 - \rho^2}}{x},$$

where $x = |OA|$.

The cone warranted by the preceding proposition is called **tangent cone** of the sphere from point A. When point A moves away to infinity, the tangents from it tend to become parallels and at the limit an (infinitely extended) a cylindrical surface is defined which is tangent to the sphere. The existence of this surface is guaranteed by the next proposition.

Exercise 4.54. *Through a given line, lying in the exterior of the sphere Σ, draw a tangent plane to Σ.*

4.11. SPHERE

Hint: Consider the plane ε through the center of the sphere and orthogonal to the given line. Let κ be the great circle along which ε intersects the sphere and P the intersection point of ε with the given line. Consider the tangents $\{\eta_1, \eta_2\}$ from P to κ and the two planes, each containing the given line and one of the $\{\eta_1, \eta_2\}$.

Proposition 4.13. *Given a sphere $O(\rho)$ and a line ζ, there exists a cylinder tangent to the sphere with generators parallel to ζ. The points of contact between the cylinder and the sphere coincide with a great circle, the plane ε of which is orthogonal to line ζ (See Figure 4.67).*

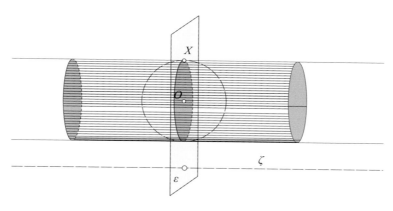

Fig. 4.67: Cylinder tangent to sphere

Proof. Consider the plane ε which is orthogonal to ζ and passes through the center O of the sphere. This plane intersects the sphere along a great circle κ. The lines parallel to ζ and passing through points X of κ are tangent to the sphere. This because they are orthogonal to the endpoint of the radius OY of the sphere. These lines therefore produce an (infinitely extended) cylindrical surface tangent to the sphere along the circle κ.

Exercise 4.55. Find the radius of the sphere inscribed in a cone of altitude v and angle ω.

Hint: (See Figure 4.68) The requested radius is calculated from the plane figure, which is excised by a plane passing through the axis of the cone. This intersects the cone along an isosceles triangle $AB\Gamma$ and the sphere along a great circle, which is simultaneously the inscribed circle of the triangle. The radius of the inscribed circle of the triangle is calculated as (Corollary I-3.4):

$$\rho = \frac{v \sin\left(\frac{\omega}{2}\right)}{1 + \sin\left(\frac{\omega}{2}\right)}.$$

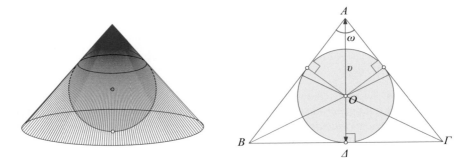

Fig. 4.68: Sphere inscribed in cone

Exercise 4.56. Show that every tetrahedron admits an inscribed sphere, that is, a sphere which is tangent to all four of its faces. Show that the position of the contact point of this sphere with one face $AB\Gamma$, is determined from the dihedral angles whose one face is $AB\Gamma$.

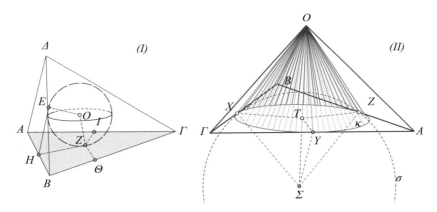

Fig. 4.69: Tetrahedron's inscribed sphere,   Cone inscribed in trihedral

Hint: Consider the three bisecting planes of the dihedral angles (of the base) with edges AB, $B\Gamma$ and ΓA. These three planes intersect at a point O, which has equal distances from the base $AB\Gamma$ as well as from all other faces of the tetrahedron (See Figure 4.69-I). Therefore the sphere with center O and radius $|OZ|$ where Z is the projection of O onto $AB\Gamma$, is tangent to all the faces of the tetrahedron. This proves the existence of the sphere. For the determination of Z observe that, if E is the projection of O on the face ΓAB,

4.11. SPHERE

then the plane ε which contains points O, Z and E intersects the edge AB orthogonally and defines the angle $\widehat{EHZ}$, whose measure is also the measure of the dihedral AB of the tetrahedron. This because plane ε contains two (non coplanar) orthogonals towards AB: OE and OZ. It follows that $\gamma = \widehat{EHZ}$ is the measure of the dihedral AB. Also, if ρ denotes the radius of the sphere, then

$$\tan\left(\frac{\gamma}{2}\right) = \frac{\rho}{|HZ|} \Rightarrow |HZ| = \rho \cdot \cot\left(\frac{\gamma}{2}\right).$$

It follows that the distances of the point of contact Z of the inscribed sphere with the face $AB\Gamma$, from the sides of the triangle $AB\Gamma$, satisfy

$$|ZH| = \rho \cdot \cot\left(\frac{\gamma}{2}\right), \quad |Z\Theta| = \rho \cdot \cot\left(\frac{\alpha}{2}\right), \quad |ZI| = \rho \cdot \cot\left(\frac{\beta}{2}\right),$$

where α, β, γ are respectively the measures of the dihedrals $B\Gamma$, ΓA and AB.

Exercise 4.57. Show that for every trihedral angle there exists a right circular cone tangent to its faces.

Hint: Construct a sphere $\sigma(\Sigma,\rho)$ inscribed to the surface of the trihedral, that is, tangent to all of its three faces (See Figure 4.69-II). The contact points X, Y, Z of the sphere with the faces of the trihedral angle define a plane ε, which intersects the sphere along a circle κ. The line segments ΣX, ΣY, ΣZ are radii of the sphere therefore equal. The triangles ΣXO, ΣYO, ΣZO, where O is the apex of the trihedral, are right and congruent, because they have the same hypotenuse ΣO and equal the orthogonal sides ΣX, ΣY, ΣZ. Therefore the altitudes of these triangles from X, Y, Z to the common hypotenuse ΣO will define the same trace T on ΣO. Therefore the plane ε of X, Y, Z will be orthogonal to ΣO, will intersect the sphere along a circle κ with radius $|TX| = |TY| = |TZ|$ and the trihedral angle along the triangle $AB\Gamma$, whose circle κ will be the inscribed one. The conical surface produced by the circle κ and the vertex O of the trihedral is the requested one.

Exercise 4.58. relying on the preceding figure 4.69-II, show that (i) OX, OY OZ are respectively orthogonal to $B\Gamma$, ΓA, AB, (ii) planes $OX\Sigma$, $OY\Sigma$, $OZ\Sigma$ are respectively orthogonal to $B\Gamma$, ΓA, AB and (iii) angles $\widehat{Y\Sigma Z}$, $\widehat{Z\Sigma X}$, $\widehat{X\Sigma Y}$ are respectively supplementary of the dihedrals OA, OB, $O\Gamma$.

Remark 4.10. The preceding exercise shows the way with which trihedral angles, which have an inscribed cone with given opening ω, are constructed. The opening of the cone is determined from the ratio λ of the radius ρ of the base to the altitude υ

$$\tan\left(\frac{\omega}{2}\right) = \lambda = \frac{\rho}{\upsilon}.$$

It suffices then to take a circle $\kappa(T,\rho)$ with arbitrary radius ρ and three points X, Y, Z on it (See Figure 4.70). At these points we draw respectively

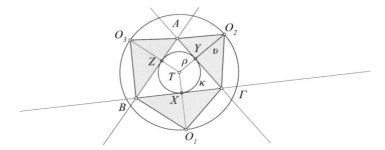

Fig. 4.70: Construction of trihedral

the tangents $B\Gamma$, ΓA, AB and we form the triangle $AB\Gamma$. Next we extend TX, TY, TZ by $\upsilon = \lambda\rho$ and we define respectively the points O_1, O_2, O_3. The trihedral results by rotating the triangles $B\Gamma O_1$, $\Gamma A O_2$, $AB O_3$ respectively about $B\Gamma$, ΓA, AB until points O_1, O_2, O_3 coincide at point O of the orthogonal on the plane $AB\Gamma$ at point T.

Exercise 4.59. Show that the intersection of two different spheres is either a circle or a point. Give a necessary and sufficient condition for the intersection of two spheres $O(\rho)$ and $O'(\rho')$ as a function of their radii ρ and ρ' and the distance of their centers $|OO'|$.

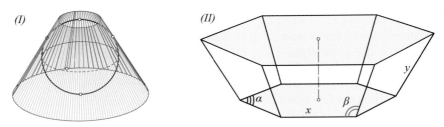

Fig. 4.71: Truncated circumscribing , Circumscribed truncated pyramid

Exercise 4.60. Which truncated cone has an inscribed sphere, which is tangent to the two bases and the lateral surface (See Figure 4.71-I)?

Exercise 4.61. Which truncated regular pyramid, with base a regular n-gon of side x and lateral edge of length y and lateral faces with angles β has an inscribed sphere, which is tangent to the two bases and the lateral surface (See Figure 4.71-II)?

4.12 Spherical and circumscribed polyhedra

> One does not play the piano with one's fingers, one plays the piano with one's mind.
>
> *Glenn Gould*

Not all polyhedra have a sphere passing through all of their vertices, or a sphere tangent to all their faces, or a sphere tangent to all their edges. Below we give some examples, which are exceptional and reminiscent of the incriptible/circumscriptibel quadrilaterals of the plane.

Theorem 4.18. *Four non coplanar points define uniquely a sphere containing them.*

Proof. The four points define a tetrahedron $AB\Gamma\Delta$. Consider the center K of the circumscribed circle of the face $AB\Gamma$ of the tetrahedron (See Figure 4.72). The points of the line KO, which is orthogonal to the plane of $AB\Gamma$ and passes through K, are equidistant from A, B and Γ. On the other hand the medial plane ε to one of the edges which pass through Δ, for example ΔA, intersects

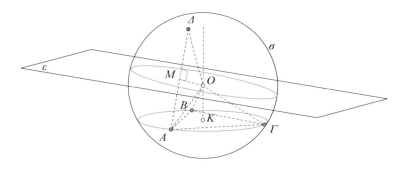

Fig. 4.72: Circumscribed sphere of tetrahedron

line KO. This because of the assumption that the four points A, B, Γ and Δ are non coplanar. If ε didn't intersect KO, then it would be parallel to it therefore itself also orthogonal to the plane of $AB\Gamma$, consequently $A\Delta$ would also be a line of the plane $AB\Gamma$, contrary to the assumption. The intersection point (O) of ε and KO has the property

$$|O\Delta| = |OA| = |OB| = |O\Gamma| = \rho,$$

therefore defines a sphere $\sigma(O, \rho)$ containing points $\{A, B, \Gamma, \Delta\}$.

Corollary 4.9. *The vertices of a tetrahedron are contained in a sphere.*

The sphere guaranteed by the preceding corollary is called **circumscribed sphere** of the tetrahedron. Polyhedra with more than four faces do not have in general a circumscribed sphere. When they do, then we call them **spherical polyhedra** and the sphere which passes through all of their vertices is called **circumscribed sphere**. We say then that the polyhedron is **inscriptible in a sphere**.

Exercise 4.62. Show that a parallelepiped is inscribed in a sphere, if and only if it is rectangular.

Exercise 4.63. Show that a pyramid is inscribed in a sphere, if and only if its base is a polygon inscriptible in a circle κ and its appex projects to the center of κ.

Exercise 4.64. Show that a right prism is inscribed in a sphere, if and only if its base is a polygon inscriptible in a circle.

Exercise 4.65. Construct a sphere of given radius, which passes through three given points. Examine when the problem has a solution.

Exercise 4.66. Construct a sphere of given radius, which passes through two given points and is tangent to a given plane. Examine when the problem has a solution.

Exercise 4.67. Construct a sphere of given radius, which passes through a given point and is tangent to two given planes. Examine when the problem has a solution.

A polyhedron Π is called **circumscribed** of the sphere $\Sigma(r)$ when all its faces are contained in planes tangent to the sphere and each of its faces has the polyhedron and the sphere on the same side. The sphere Σ is called then **inscribed** sphere of the polyhedron. The polyhedron is called **escribed** to the sphere, when all its faces are contained in planes tangent to the sphere and there exists at least one such plane which leaves the polyhedron and the sphere at different sides. The sphere then is called **escribed** to the polyhedron. Exercise 4.56 shows that every tetrahedron is a polyhedron *circumscribed* to a sphere. It is easily proved that every tetrahedron also admits four *escribed* spheres, contained in the four trihedral angles of the tetrahedron. The Platonic bodies are another example of polyhedra which admit an inscribed sphere.

Exercise 4.68. Suppose that the Platonic solid of edge length a, with faces regular p-gons, is inscribed in a sphere of radius R (See Figure 4.73) and show the relation of these data with the radius r of its inscribed sphere:

$$r^2 + \frac{a^2}{4\sin(2\pi/p)^2} = R^2.$$

4.13. SPHERICAL LUNE, ANGLE OF GREAT CIRCLES

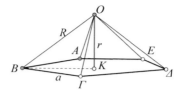

Fig. 4.73: Radius r of inscribed sphere of Platonic solid

On the occasion of this exercise let us note that there exists also a third sphere related to the Platonic solids. It is the one which is tangent to all the edges of a Platonic body (at their middles) and is called **medial** or **middle** sphere (Exercise 4.79). This sphere has the same center with the circumscribed one and radius ρ, which is related to the edge a and the radius of its circumscribed circle through the relation $\rho^2 + (a/2)^2 = R^2$.

Exercise 4.69. Show that, if a polyhedron has a middle sphere, then each face of it has an inscribed circle and the center of the sphere lies on the line orthogonal to the face at the center of the inscribed circle.

Exercise 4.70. Construct a tetrahedron with two congruent faces, which does not possesses a middle sphere.

4.13 Spherical lune, angle of great circles

> If a lunatic scribbles a jumble of mathematical symbols it does not follow that the writing means anything merely because to the inexpert eye it is indistinguishable from higher mathematics.
>
> E. Bell, *Men of Mathematics*, p. 232

Two planes α and β, passing through the center O of a sphere σ, intersect along a diameter AB. These planes also excise from the sphere two great circles respectively κ and λ. They define also four dihedral angles with edge AB. The angle of one of the dihedrals is defined through the plane angle formed on plane ζ, which passes through O and is orthogonal to the diameter AB (See Figure 4.74). This plane is parallel to the tangent planes at A and B (Corollary 4.7). This implies that the angle of the dihedral $\omega = \widehat{XOY}$ is congruent to the corresponding angle, formed by the tangents of the circles κ and λ at points A and B. Generally, two great circles κ and λ are defined respectively from planes which pass through the center of the sphere, therefore intersect along one diameter AB. The angles formed by the tangents of these circles at A and B are called the **angles of great circles** (there exist two supplementary to each other).

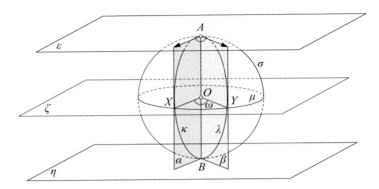

Fig. 4.74: Sphere and dihedral angle

We call **spherical lune** the part of the sphere, which is included between the faces of a dihedral angle, whose edge is a diameter of the sphere. The angle of the dihedral is called **angle of the spherical lune**. Two spherical lunes are called **congruent**, when the radii of the spheres containing them are equal and their angles are also equal. Next proposition follows directly from the definition of the dihedral angle (§ 4.1).

Proposition 4.14. *The angle of a dihedral, formed by the planes α and β, is congruent to the angle of the spherical lune, which it excises from a sphere having its center at the edge AB of the dihedral. Moreover this angle is congruent to one of the angles between the great circles which are defined by the two planes (See Figure 4.74).*

Exercise 4.71. How many spherical lunes are defined by three planes, pairwise not coincident, and passing through the center of the sphere?

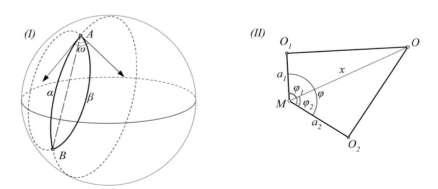

Fig. 4.75: Sphere from two intersecting circles in space

4.14. SPHERICAL TRIANGLES

Exercise 4.72. Show, that two circles in space, which intersect at the endpoints of a common chord AB, by a dihedral of measure ϕ, define uniquely a sphere which contains them (See Figure 4.75-I). Calculate the radius and the center of the sphere as a function of the radii r_1, r_2 of the two circles and of ϕ. Also calculate the angle ω, which is formed by the tangents of the two circles at the point A.

Hint: If there is such a sphere, then, in the middle plane of the chord AB will be formed the inscriptible quadrilateral O_1MO_2O, where O_1, O_2 are the centers of the circles, O is the center of the sphere and M the middle of AB (See Figure 4.75-II). The length x of the diagonal OM satisfies

$$\frac{O_1O_2}{\sin(\phi)} = x \Rightarrow x^2 = \frac{a_1^2 + a_2^2 - 2\cos(\phi)a_1a_2}{\sin^2(\phi)}, \quad a_i^2 = MO_i^2 = r_i^2 - \frac{AB^2}{4} \quad (i=1,2)$$

and the radius r of the sphere satisfies $r^2 = x^2 + \mu^2$, where $2\mu = |AB|$.

4.14 Spherical triangles

> This faith of his was relying on the theory, that everything in the world is alive, nothing is dead, and moreover, all these we consider inorganic, are not but the constituents of a giant organism which we are unable to determine.
>
> *Leo Tolstoi, Resurrection*

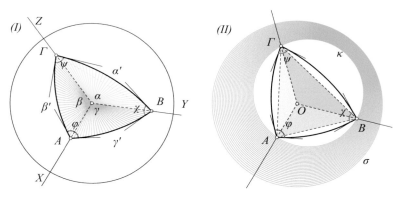

Fig. 4.76: Sphere and trihedral angle

Placing the apex of a trihedral angle at the center O of a sphere, we get on the sphere, through the intersection with the faces of the trihedral, a shape reminiscent of the plane triangle, called **spherical triangle** (See Figure 4.76). The figure consists of arcs of great circles, which the faces of the trihedral excise on the sphere. The length of these arcs is connected with the measure of the faces of the trihedral. If the faces of the trihedral have measures respectively α, β and γ (in radians), then the lengths of these arcs are, respectively

$$\alpha' = \alpha\rho, \ \beta' = \beta\rho, \ \gamma' = \gamma\rho,$$

where ρ is the radius of the sphere. These arcs are called **sides of the spherical triangle**. Their lengths are called **lengths of the sides of the spherical triangle**. Their endpoints A, B and Γ are the common points of the sphere and the edges of the trihedral and are called **vertices of the spherical triangle**. From the preceding section, we know that the dihedrals ϕ, χ and ψ of the trihedral are formed respectively at the vertices A, B and Γ through the tangents of the great circles, which are defined by the faces of the trihedral. These angles are called **angles of the spherical triangle**. Two spherical triangles $AB\Gamma$ and $A'B'\Gamma'$ are called **congruent** when they have equal respective sides and equal angles.

The analogy with triangles of the plane is obvious. Many propositions of plane geometry, which concern properties of triangles, are transferred also to spherical triangles. There exist however differences also, some of which we'll discuss in the next sections. The preceding correspondence between trihedrals and spherical triangles shows that every property of the trihedrals is translated into a property of the spherical triangles and vice versa, every property of the spherical triangles can be translated into a property of the trihedrals. An example of this analogy is given by the next proposition.

Theorem 4.19. *Every spherical triangle of the sphere σ is inscribed in a circle κ contained in σ (See Figure 4.77).*

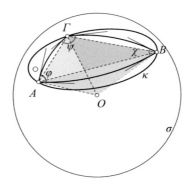

Fig. 4.77: Circumscribed circle of spherical triangle

4.14. SPHERICAL TRIANGLES

Proof. Let $OAB\Gamma$ be a spherical triangle and let us denote it using the same letters the corresponding trihedral angle. In Proposition 4.11 the circumscribed cone of the trihedral angle $\widehat{OAB\Gamma}$ is constructed using the circumscribed circle κ of triangle $AB\Gamma$. The points of this circle are at the same distance from O as are A, B and Γ, therefore they are on the sphere σ.

Exercise 4.73. Show that the intersection of a sphere and right circular cone with vertex at the center of the sphere is a parallel circle of the sphere.

Remark 4.11. Polyhedral angles (§ 4.3), similarly to trihedrals, by placing their apex at the center O of the sphere, define, through the intersections of their faces with the sphere, a **spherical polygon**, that is, a shape on the sphere which is bounded by successive arcs of great circles. This is the analogue on the surface of the sphere of plane polygons. Like with trihedrals so also with polyhedral angles we can use this correspondence to study properties of spherical polygons reducing them to corresponding properties of polyhedral angles. We'll see more about these interesting shapes on the sphere in § 5.3. Next proposition shows, how the dihedral angles of a trihedral depend on its faces. In the next section we'll see (Theorem 4.23) that, conversely, also the faces are completely determined from the dihedrals of the trihedral.

Theorem 4.20. *Let $\widehat{OAB\Gamma}$ be a trihedral with faces $\alpha = \widehat{BO\Gamma}$, $\beta = \widehat{\Gamma OA}$, $\gamma = \widehat{AOB}$ and dihedral angles $\phi(OA)$, $\chi(OB)$, $\psi(O\Gamma)$ (within parentheses the corresponding edge) (See Figure 4.78). Then it holds*

$$\cos(\alpha) = \cos(\beta)\cos(\gamma) + \sin(\beta)\sin(\gamma)\cos(\phi).$$

Proof. For the proof we place the trihedral with its apex O at the center of the sphere σ of radius ρ and we consider the intersection points A, B, Γ of its edges with the sphere. We also consider the tangent plane ε of the sphere at the vertex A. Suppose that OB and $O\Gamma$ intersect plane ε respectively at Δ and E (the simpler case where this does not happens will be examined below). We apply the cosine formula to triangles $O\Delta E$ and $A\Delta E$

$$|E\Delta|^2 = |OE|^2 + |OA|^2 - 2|OE||OA|\cos(\alpha)$$

$$|OE| = \frac{\rho}{\cos(\beta)}, \quad |OA| = \frac{\rho}{\cos(\gamma)}, \quad \Rightarrow$$

$$|E\Delta|^2 = \rho^2 \left(\frac{1}{\cos^2(\beta)} + \frac{1}{\cos^2(\gamma)} - 2\frac{\cos(\alpha)}{\cos(\beta)\cos(\gamma)} \right)$$

$$= \rho^2 \left((1 + \tan^2(\beta)) + (1 + \tan^2(\gamma)) - 2\frac{\cos(\alpha)}{\cos(\beta)\cos(\gamma)} \right).$$

We also have that

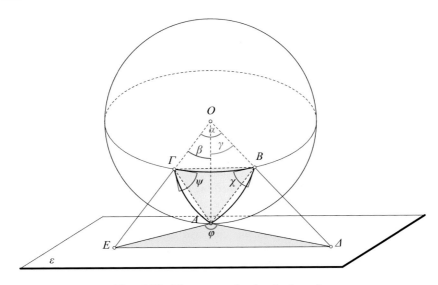

Fig. 4.78: Theorem of spherical cosine

$$|E\Delta|^2 = |AE|^2 + |A\Delta|^2 - 2|AE||A\Delta|\cos(\phi)$$
$$|AE| = \rho\tan(\beta), \ |A\Delta| = \rho\tan(\gamma), \ \Rightarrow$$
$$|E\Delta|^2 = \rho^2\left(\tan^2(\beta) + \tan^2(\gamma) - 2\tan(\beta)\tan(\gamma)\cos(\phi)\right).$$

Equating the two expressions for $|E\Delta|^2$ and simplifying, we find

$$1 - \frac{\cos(\alpha)}{\cos(\beta)\cos(\gamma)} = -\tan(\beta)\tan(\gamma)\cos(\phi) \ \Rightarrow$$
$$\cos(\beta)\cos(\gamma) - \cos(\alpha) = -\sin(\beta)\sin(\gamma)\cos(\phi),$$

which is equivalent to the requested relation. In the case where one of the two points Δ and E does not exist, for example Δ, this means, that the corresponding edge OB is parallel to the plane ε, consequently, angle γ is right. Then $\cos(\gamma) = 0$, $\sin(\gamma) = 1$ and the formula becomes

$$\cos(\alpha) = \sin(\beta)\cos(\phi) \Leftrightarrow |OE|\cos(\alpha) = |AE|\cos(\phi),$$

which is true, because $A\Delta$ and OB are parallel. When both points do not exist, then $O\Gamma$ and OB are respectively parallel to AE and $A\Delta$ and the angles β and γ are right. This implies that the angles α and ϕ are equal and the relation becomes $\cos(\alpha) = \cos(\phi)$. Consequently, the aforementioned formula holds in all cases.

Remark 4.12. This proposition can be expressed also as a proposition for spherical triangles using the lengths of the sides α', β', γ', which were de-

4.14. SPHERICAL TRIANGLES

fined in the beginning of this section (See Figure 4.79). We simply use the

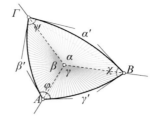

Fig. 4.79: Spherical triangle

proved formula, and make the substitution

$$\alpha = \frac{\alpha'}{\rho}, \quad \beta = \frac{\beta'}{\rho}, \quad \gamma = \frac{\gamma'}{\rho}.$$

The formula which results, is called **first formula for spherical cosine:**.

Theorem 4.21. *For every spherical triangle with sides of lengths α', β', γ' and corresponding opposite angles ϕ, χ and ψ holds:*

$$\cos\left(\frac{\alpha'}{\rho}\right) = \cos\left(\frac{\beta'}{\rho}\right)\cos\left(\frac{\gamma'}{\rho}\right) + \sin\left(\frac{\beta'}{\rho}\right)\sin\left(\frac{\gamma'}{\rho}\right)\cos(\phi).$$

Exercise 4.74. For every right spherical triangle with sides of lengths α', β', γ' and corresponding opposite angles $\phi = 90°$, χ and ψ holds:

$$\cos\left(\frac{\alpha'}{\rho}\right) = \cos\left(\frac{\beta'}{\rho}\right)\cos\left(\frac{\gamma'}{\rho}\right).$$

Exercise 4.75. Show that in a sphere there are infinitely many equilateral triangles, whose lengths of sides α' and the angles ϕ satisfy the equation:

$$\cos\left(\frac{\alpha'}{2\rho}\right) \cdot \sin\left(\frac{\phi}{2}\right) = \frac{1}{2}.$$

Exercise 4.76. Show that the three equal angles of the spherical equilateral triangles are greater than 60 degrees. (Exercise 5.43)

4.15 The supplementary trihedral

> The world embraces not only a Newton, but a Shakespeare; not only a Boyle, but a Raphael; not only a Kant, but a Beethoven; not only a Darwin, but a Carlyle. Not in each of these, but in all, is human nature whole. They are not opposed, but supplementary; not mutually exclusive, but reconcilable.
>
> J. Tyndall, Address, Belfast 1874

Given a trihedral $\widehat{OXYZ}$, the orthogonals on its faces define a new trihedral $\widehat{OX'Y'Z'}$ which is called **supplementary trihedral** of $\widehat{OXYZ}$. More precisely, the definition of this trihedral $\widehat{OX'Y'Z'}$ is the following. The half line OX' is the orthogonal to plane OYZ on the same side with OX, the half line OY'

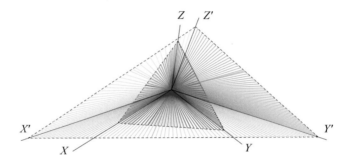

Fig. 4.80: Supplementary trihedral angle

is the orthogonal to OZX on the same side with OY and the half line OZ' is the orthogonal to OXY and on the same side with OZ. These three half lines define the edges and consequently the trihedral $\widehat{OX'Y'Z'}$ (See Figure 4.80).

Theorem 4.22. *The faces of the supplementary $\widehat{OX'Y'Z'}$ of the trihedral $\widehat{OXYZ}$ are the supplementary of the respective dihedral angles of $\widehat{OXYZ}$ (See Figure 4.81):*

$$\widehat{X'OY'} + \widehat{Z} = \widehat{Y'OZ'} + \widehat{X} = \widehat{Z'OX'} + \widehat{Y} = \pi.$$

Proof. In the plane ε, which is orthogonal to OZ at the point O are contained, first of all, angle $\widehat{AOB}$, which measures the dihedral $\widehat{Z}$ and second (by definition) OX' and OY'. Angle $\widehat{Z}$ is equal to $\phi = \widehat{AOB}$, where the line OA is the intersection of ε with the plane ZOX and OB is the intersection of ε with the plane ZOY. From the definition therefore, follows that the angles $\omega = \widehat{X'OY'}$ and $\phi = \widehat{AOB}$ have their sides orthogonal and equally oriented, therefore they are supplementary. Similarly we show also the analogous relation for the other angles of the trihedral $\widehat{OXYZ}$.

4.15. THE SUPPLEMENTARY TRIHEDRAL

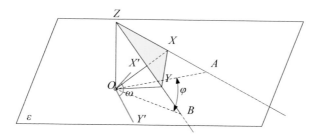

Fig. 4.81: Supplementary trihedral

Proposition 4.15. *The supplementary $\widehat{OX''Y''Z''}$ of the supplementary $\widehat{OX'Y'Z'}$ of the trihedral $\widehat{OXYZ}$ coincides with the initial trihedral $\widehat{OXYZ}$.*

Proof. According to the definition, the edge OZ'' of the supplementary of $\widehat{OX'Y'Z'}$ will be orthogonal to OX', OY' and will be on the same side as OZ'. However OZ has also this property.

Corollary 4.10. *If the trihedral angle $\widehat{OXYZ}$ has faces $\alpha = \widehat{YOZ}$, $\beta = \widehat{ZOX}$, $\gamma = \widehat{XOY}$ and dihedral angles $\phi(OX)$, $\chi(OY)$, $\psi(OZ)$ (inside parentheses the corresponding edge), then the supplementary of $\widehat{OX'Y'Z'}$ has respectively faces*

$$\alpha' = \pi - \phi, \ \beta' = \pi - \chi, \ \gamma' = \pi - \psi$$

and dihedrals

$$\phi' = \pi - \alpha, \ \chi' = \pi - \beta, \ \psi' = \pi - \gamma.$$

Proof. The first three equalities were proved in Theorem 4.22. The next three follow by applying the same proposition to the supplementary of the supplementary, which is (Proposition 4.15) the initial trihedral.

The existence of the supplementary trihedral and the relation of faces and dihedrals, which results from the preceding propositions, shows that the faces and the dihedrals of a trihedral are, in some sense, equivalent elements. The former determine the latter. This duality is responsible for the "second cosine theorem", which results from the first, applied to the supplementary trihedral.

Theorem 4.23. *Let $\widehat{OAB\Gamma}$ be a trihedral angle with faces $\alpha = \widehat{BO\Gamma}$, $\beta = \widehat{\Gamma OA}$, $\gamma = \widehat{AOB}$ and dihedral angles $\phi(OA)$, $\chi(OB)$, $\psi(O\Gamma)$ (inside parentheses the corresponding edge). Then holds*

$$\cos(\phi) = -\cos(\chi)\cos(\psi) + \sin(\chi)\sin(\psi)\cos(\alpha).$$

Proof. We apply the cosine formula (Theorem 4.20) to the supplementary trihedral, taking into account the relations of Corollary 4.10:

$$\cos(\alpha') = \cos(\beta')\cos(\gamma') + \sin(\beta')\sin(\gamma')\cos(\phi') \Leftrightarrow$$
$$\cos(\pi - \phi) = \cos(\pi - \chi)\cos(\pi - \psi) + \sin(\pi - \phi)\sin(\pi - \psi)\cos(\pi - \alpha) \Leftrightarrow$$
$$-\cos(\phi) = \cos(\chi)\cos(\psi) - \sin(\chi)\sin(\psi)\cos(\alpha).$$

This proposition also can be expressed, like the *first formula of spherical cosine* (Proposition 4.21), as a relation between elements of the spherical triangle, which is excised by the trihedral from a sphere. Adopting the notation of section 4.14, we translate the preceding proposition to the so called **second formula of spherical cosine**.

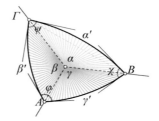

Fig. 4.82: Spherical triangle

Theorem 4.24. *For every spherical triangle with sides of length α', β', γ' and corresponding opposite angles ϕ, χ and ψ holds:*

$$\cos(\phi) = -\cos(\chi)\cos(\psi) + \sin(\chi)\sin(\psi)\cos\left(\frac{\alpha'}{\rho}\right),$$

where ρ is the radius of the sphere.

Corollary 4.11. *For every right at Γ ($\psi = 90°$) spherical triangle $AB\Gamma$, holds*

$$\cos(\phi) = \sin(\chi)\cos\left(\frac{\alpha'}{\rho}\right).$$

Remark 4.13. The formula for the second cosine rule, among other things, underlines also the fact that the spherical triangles are determined completely from their angles ϕ, χ and ψ (See Figure 4.82). Indeed, if these angles are given, then solving the preceding equation for α', we find the length of the side of the triangle and from the corresponding formulas for the other sides, we find in full all the sides of the triangle. We have here an essential difference from ordinary triangles of the plane. In the latter case, knowledge of the angles is not sufficient for the full determination of the triangle. There, from the angles, we can only conclude the triangle's shape, but not its size. There are many triangles, which have equal angles and are similar to each other. On the contrary, on the sphere there are no different triangles having equal angles. There is no similarity of shapes here.

4.15. THE SUPPLEMENTARY TRIHEDRAL

The analogy between spherical and euclidean triangles however still holds in many cases, like for example in the next analogue of the sine formula, which is mentioned as **formula of spherical sine**.

Theorem 4.25. *Let $\widehat{OAB\Gamma}$ be a trihedral angle with faces $\alpha = \widehat{BO\Gamma}$, $\beta = \widehat{\Gamma OA}$, $\gamma = \widehat{AOB}$ and dihedral angles $\phi(OA)$, $\chi(OB)$, $\psi(O\Gamma)$ (inside parantheses the corresponding edge) (See Figure 4.82). Then holds*

$$\frac{\sin(\phi)}{\sin(\alpha)} = \frac{\sin(\chi)}{\sin(\beta)} = \frac{\sin(\psi)}{\sin(\gamma)}.$$

Proof. A simple proof results by squaring the formula of the cosine (Theorem 4.20):

$$\cos(\alpha) = \cos(\beta)\cos(\gamma) + \sin(\beta)\sin(\gamma)\cos(\phi) \Rightarrow$$
$$(\cos(\alpha) - \cos(\beta)\cos(\gamma))^2 = (\sin(\beta)\sin(\gamma)\cos(\phi))^2 \Rightarrow$$
$$\cos^2(\alpha) + \cos^2(\beta)\cos^2(\gamma) - 2\cos(\alpha)\cos(\beta)\cos(\gamma) = \sin^2(\beta)\sin^2(\gamma)\cos^2(\phi)$$
$$= \sin^2(\beta)\sin^2(\gamma)(1 - \sin^2(\phi))$$
$$= 1 - \cos^2(\beta) - \cos^2(\gamma) + \cos^2(\beta)\cos^2(\gamma) - \sin^2(\beta)\sin^2(\gamma)\sin^2(\phi) \Rightarrow$$
$$\cos^2(\alpha) - 2\cos(\alpha)\cos(\beta)\cos(\gamma)$$
$$= 1 - \cos^2(\beta) - \cos^2(\gamma) - \sin^2(\beta)\sin^2(\gamma)\sin^2(\phi).$$

The last equation is written

$$\sin^2(\beta)\sin^2(\gamma)\sin^2(\phi)$$
$$= 1 - \cos^2(\alpha) - \cos^2(\beta) - \cos^2(\gamma) + 2\cos(\alpha)\cos(\beta)\cos(\gamma) \Leftrightarrow$$
$$\sin^2(\phi) = \frac{1 - \cos^2(\alpha) - \cos^2(\beta) - \cos^2(\gamma) + 2\cos(\alpha)\cos(\beta)\cos(\gamma)}{\sin^2(\beta)\sin^2(\gamma)} \Leftrightarrow$$
$$\frac{\sin^2(\phi)}{\sin^2(\alpha)} = \frac{1 - \cos^2(\alpha) - \cos^2(\beta) - \cos^2(\gamma) + 2\cos(\alpha)\cos(\beta)\cos(\gamma)}{\sin^2(\alpha)\sin^2(\beta)\sin^2(\gamma)}.$$

The last formula is symmetric relative to α, β and γ, consequently it will be the same for the other quotients $\frac{\sin^2(\chi)}{\sin^2(\beta)}$, $\frac{\sin^2(\psi)}{\sin^2(\gamma)}$.

The translation of the preceding formula for spherical triangles is given by the next proposition.

Proposition 4.16. *For every spherical triangle with sides of lengths α', β', γ' and corresponding opposite angles ϕ, χ and ψ holds (See Figure 4.82):*

$$\frac{\sin(\phi)}{\sin\left(\frac{\alpha'}{\rho}\right)} = \frac{\sin(\chi)}{\sin\left(\frac{\beta'}{\rho}\right)} = \frac{\sin(\psi)}{\sin\left(\frac{\gamma'}{\rho}\right)},$$

where ρ is the radius of the sphere.

Next figure 4.83 shows the supplementary trihedral $\widehat{OX'Y'Z'}$ of a trihedral $\widehat{OXYZ}$, whose apex has been placed at the center of the sphere. Each face of the trihedral $\widehat{OXYZ}$ defines a plane through O and this intersects the sphere along a great circle. This way, plane XOY intersects the sphere along the great circle z and point Z' is the pole of z on the same side as Z. Similarly also the other points X' and Y' are the poles of the great circles x, y defined by the faces of the trihedrals, taken on the side of the face where the entire initial trihedral is to be found. $X'Y'Z'$ is called **supplementary triangle** of XYZ.

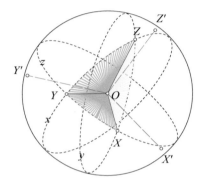

Fig. 4.83: The supplementary trihedral

Exercise 4.77. Show that the spherical equilateral triangle, whose all three angles are right, is identical to its supplementary.

Exercise 4.78. Show that the measure ω of the dihedral of a Platonic solid ($\{p,q\}$), in which in each vertex converge q edges and each face is a regular p-gon, satisfies the formula (See Figure 4.84)

$$\sin\left(\frac{\omega}{2}\right) = \frac{\cos\left(\frac{\pi}{q}\right)}{\sin\left(\frac{\pi}{p}\right)}.$$

Hint: Application of the corollary 4.11 to the spherical triangle $AB\Gamma$, which results on top of a small sphere which we place with its center at the vertex O of the Platonic solid. The sphere intersects the solid along a spherical regular polygon σ_q with q equal sides and center point A. Drawing the orthogonal $A\Gamma$ to one of its sides BB', we create a right spherical triangle with respective angles $\phi(A) = \pi/q$, $\chi(B) = \omega/2$ and $\psi(\Gamma) = \pi/2$. The measure of $\alpha = \frac{p-2}{p}\frac{\pi}{2}$ follows from the corresponding angle $\widehat{BOB'} = \frac{p-2}{p}\pi$ at O of the regular polygon with p sides. The formula follows from the corollary, by which $\cos(\phi) = \sin(\chi)\cos(\alpha)$ (see application in Exercise 4.79, alternatively see Exercise 4.25).

4.15. THE SUPPLEMENTARY TRIHEDRAL

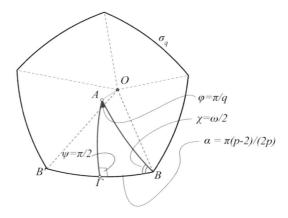

Fig. 4.84: The dihedral of the Platonic solid $\{p,q\}$

Remark 4.14. A direct consequence of the complete determination of the spherical triangle from its three angles or its three sides is also the fact, that a trihedral is determined completely from its three faces or its three dihedrals. One triple of angles determines fully the other triple. To see this, it suffices to place the trihedral with its vertex at the center of a sphere and to consider the spherical triangle, which is excised on the sphere. From the congruence of the spherical triangles follows the congruence of the trihedrals and vice versa.

Next table records the dihedral angles ω of the Platonic solids, through $\sin\left(\frac{\omega}{2}\right)$ and the ratio of the golden section $x = \frac{\sqrt{5}-1}{2}$ (Exercise 4.78), as well as the radii of the corresponding inscribed, middle (tangent to all edges) and circumscribed sphere, of the Platonic solid of edge of length a. For the rest of the elements see [13, III, p.492], where, using a different method than that of Euclid, is calculated, in a simple way, the radius of the circumscribed sphere of the icosahedron and next that of the dodecahedron. relying on these radii, the radii of other spheres can be subsequently calculated.

Exercise 4.79. Verify the truth of the first column and the first three rows of the preceding table.

Platonic solid	$\sin\left(\frac{\omega}{2}\right)$	Inscribed	Middle	Circumscribed
Tetrahedron	$\frac{1}{\sqrt{3}}$	$\frac{a}{12}\sqrt{6}$	$\frac{a}{4}\sqrt{2}$	$\frac{a}{4}\sqrt{6}$
Cube	$\frac{1}{\sqrt{2}}$	$\frac{a}{2}$	$\frac{a}{2}\sqrt{2}$	$\frac{a}{2}\sqrt{3}$
Octahedron	$\frac{2}{\sqrt{6}}$	$\frac{a}{6}\sqrt{6}$	$\frac{a}{2}$	$\frac{a}{2}\sqrt{2}$
Dodecahedron	$\sqrt{\frac{5+\sqrt{5}}{10}}$	$\frac{a}{20}\sqrt{250+110\sqrt{5}}$	$\frac{a}{4}(3+\sqrt{5})$	$\frac{a}{4}(\sqrt{15}+\sqrt{3})$
Icosahedron	$\sqrt{\frac{3+\sqrt{5}}{6}}$	$\frac{a}{12}(3\sqrt{3}+\sqrt{15})$	$\frac{a}{4}(1+\sqrt{5})$	$\frac{a}{4}\sqrt{10+2\sqrt{5}}$

4.16 Axonometric projection, affinities

> Man is a freak, colloidal combination of thirteen elements which happen to have a chemical affinity for each other, and is the strangest and one of the most amusing accidents of nature. The market value of the substance of the average man is about 98 cents.
>
> *J. S. Pickering, The Stars Are Yours, p.230*

In this and in the next section we return back to planes, their intersections, the dihedral angles and the prisms, to examine two processes (mappings) which are used extensively by architects, technical drawers, painters and photographers. These processes, among others, allow the representation of

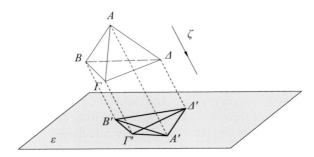

Fig. 4.85: Axonometric projection parallel to ζ

three dimensional objects on the plane (paper) and the preparation of designs, relying on which we can reconstruct the real object. Parallel line pencils in space give us the possibility to *project* a shape Σ of space into a shape Σ' of a fixed plane ε (See Figure 4.85). In this case there exists a fixed line ζ and to every point A of Σ we correspond the intersection A' of the parallel to ζ from A with ε. We call this process parallel or lateral **axonometric projection**, of Σ onto ε in the direction of ζ. Line ζ is called **direction of**

4.16. AXONOMETRIC PROJECTION, AFFINITIES

the projection and plane ε is called **plane of projection**. When ζ is orthogonal to ε then the process is called **orthogonal axonometric** projection. Next theorem gives the basic properties of the axonometric projection.

Theorem 4.26. *The axonometric projection, parallel to the line ζ, onto the plane ε, has the following properties (See Figure 4.86):*

1. *All the lines parallel to ζ, are mapped to their intersection point with plane ε.*
2. *All the lines non-parallel to ζ, are mapped to lines of plane ε.*
3. *Two parallel lines in space, are mapped respectively to either two points or to a line or to two parallel lines of the plane ε.*
4. *If the line ξ in space is mapped to the line ξ' of the plane, then two line segments AB, $\Gamma\Delta$ of ξ are mapped to two line segments $A'B'$, $\Gamma'\Delta'$ of ξ' and the ratios are preserved: $\frac{|AB|}{|\Gamma\Delta|} = \frac{|A'B'|}{|\Gamma'\Delta'|}$ (See Figure 4.87).*
5. *Two congruent angles in space, $\alpha = \beta$, with parallel respective sides, are mapped to two congruent angles $\alpha' = \beta'$ of the plane ε.*
6. *A parallelogram $AB\Gamma\Delta$, contained in a plane η non-parallel to ζ, is mapped to a parallelogram $A'B'\Gamma'\Delta'$ of plane ε.*
7. *Two congruent triangles $\sigma = \tau$, with respective sides parallel are mapped to two congruent triangles $\sigma' = \tau'$ of the plane ε.*

Proof. (1) is a direct consequence of the definition. If X is a point of line ξ parallel to ζ, the parallel from X to ζ will coincide exactly with ξ, therefore the intersection point of this parallel with plane ε will be the intersection point of ξ with ε.

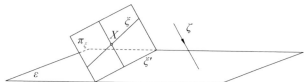

Fig. 4.86: Mapping of line ξ onto line ξ' through axonometry

(2) follows from the fact that a non-parallel line ξ to ζ in space, defines uniquely a plane π_ξ which contains all the parallels to ζ, which pass through points X of ξ. The line-intersection ξ', of the planes π_ξ and ε, is the line onto which ξ is mapped.

(3) follows from the preceding properties. If the lines in space ξ and ξ' are parallel to ζ, then they will be mapped, according to (1), to two points. If ξ, ξ' are not parallel to ζ, yet they are parallel themselves, then, according to (2) two planes π_ξ, $\pi_{\xi'}$ will be defined and what happens exactly will be determined by these planes. The two planes are coincident, if and only if line ζ is parallel to the plane which contains lines ξ and ξ'. If ζ is not parallel to the plane of ξ, ξ', then the two planes π_ξ, $\pi_{\xi'}$ are parallel and consequently they intersect two parallel lines at ε (Proposition 3.9).

(4) follows by applying the theorem of Thales on the plane π_ξ.

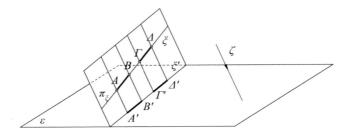

Fig. 4.87: Preservation of ratio of collinear segments by an axonometry

(5) is a consequence of (3), according to which the lines of the sides of the angles will be mapped respectively to parallel (or coincident) lines of ε.

(6) is also a direct consequence of (3) and (7) is also a direct consequence of (6) (see also Exercise 4.80).

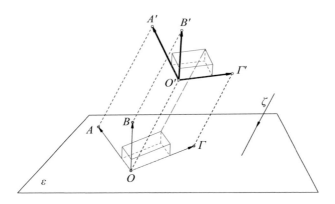

Fig. 4.88: Theorem of Pohlke

Theorem 4.27. *(of Pohlke (1810-1876)) Given four non collinear points O, A, B, Γ of the plane ε, there exists a system of three line segments in space $O'A'$, $O'B'$, $O'\Gamma'$, which are equal and pairwise orthogonal, as well as a line ζ, such that the axonometric projection relative to ζ corresponds O', A', B', Γ' to O, A, B, Γ.*

The theorem (See Figure 4.88), formulated by Pohlke, was proved by his student Schwarz in a somewhat more general form (Theorem 4.28). It is fundamental to the so called *Axonometry*, which is used for the representation of space shapes on paper. Figure 4.89 shows how the image of a house is transferred on paper by an axonometry. Usually there's also a shrinking (homothety with ratio $\lambda < 1$ called scale) present on ε, so that less paper

4.16. AXONOMETRIC PROJECTION, AFFINITIES

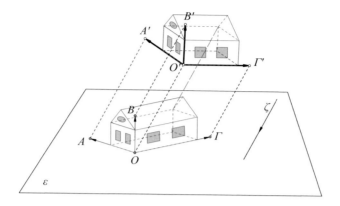

Fig. 4.89: Representation of a house on paper (plane ε)

is consumed. The properties of this representation, necessary for architects, engineers and designers of technical works, are studied by the so called *Descriptive Geometry* ([62], [70], [77]). Several of these properties concern the way with which an axonometry maps the points of a fixed plane α in space to the projection plane ε (See Figure 4.90). These properties are of interest,

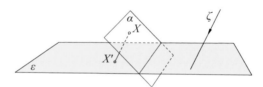

Fig. 4.90: Axonometry applied to the points of the plane α

because if we know them, we can, for example, design on paper (ε) the windows, doors, ornaments etc., which are seen on a wall (plane α) of a house, performing measurements only on the wall and forgetting the surrounding space. These properties follow from Theorem 4.26(4,6), according to which *the correspondence from plane α to plane ε is completely determined from three non collinear points A, B, Γ of α and their corresponding images A', B', Γ' on ε.* Indeed, given three non collinear points A, B, Γ on α and their projections on ε are A', B', Γ' on ε, then every other point X of the plane α defines a parallelogram $AYXZ$ and the projection X' of X onto ε defines a corresponding parallelogram $A'Y'X'Z'$. From the aforementioned properties follows that the following (signed) ratios are respectively equal

$$\frac{YA}{YB} = \frac{Y'A'}{Y'B'}, \qquad \frac{ZA}{Z\Gamma} = \frac{Z'A'}{Z'\Gamma'}.$$

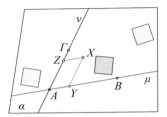 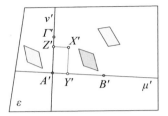

Fig. 4.91: Affinity between planes α and ε

This implies that, given point X, we can measure the ratios $\rho = \frac{YA}{YB}$, $\sigma = \frac{ZA}{Z\Gamma}$ and through them we can define the points Y', Z' on the lines $A'B'$, $A'\Gamma'$ respectively, such that $\frac{Y'A'}{Y'B'} = \rho$ and $\frac{Z'A'}{Z'\Gamma'} = \sigma$. Having points Y', Z', the other vertex X' of the parallelogram is determined directly and, consequently, we do not need to use the entire shape in space. Such a correspondence between two planes, like the one we just described, is called an **affinity**. I summarize again its main characteristics (See Figure 4.91):

1. An affinity is completely determined through three non collinear points A, B, Γ and their images A', B', Γ'.
2. An affinity corresponds lines (parallels) to lines (parallels).
3. An affinity preserves the ratios of collinear points and the ratios of areas of polygons.

Figure 4.91 shows also a peculiarity of affinity: *in general it does not respect congruence of two figures.* On plane α there are formed also three congruent squares. Two of them have their sides respectively parallel. These are mapped to respectively congruent parallelograms. The third square of α is congruent to the other two, it has however sides that are not parallel to them. We see that its image on ε is a parallelogram different from the other two.

Exercise 4.80. Show that an affinity of the plane α to the plane ε maps two polygons p, q of α, which are congruent and have respectively parallel sides, to two polygons p', q' of ε which are also congruent and have respective sides parallel (See Figure 4.92).

Hint: drawing the diagonals from a corresponding vertex of the polygons, we decompose them into triangles and we reduce the property to the corresponding one for triangles. The proof for triangles $AB\Gamma$, ΔEZ, which are congruent and have parallel respective sides follows from the properties of the affinity. According to these properties, parallelograms $ABE\Delta$, $A\Gamma Z\Delta$, $B\Gamma ZE$ will be mapped to corresponding parallelograms $A'B'E'\Delta'$, $A'\Gamma'Z'\Delta'$, $B'\Gamma'Z'E'$ of plane ε. From this follows the congruence of the triangles $A'B'\Gamma'$ and $\Delta'E'Z'$ (See also the corresponding properties for the orthogonal projection Proposition 4.3 and Corollary 4.2).

4.16. AXONOMETRIC PROJECTION, AFFINITIES

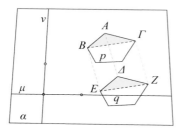
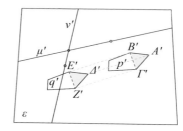

Fig. 4.92: Case of preservation of congruence between polygons from affinity

Exercise 4.81. Show that if f is a plane affinity between planes α and β and g is a plane affinity between planes β and γ, then the composition of the transformations $h = g \circ f$ is a plane affinity between planes α and γ.

Hint: Verify that h satisfies properties (1)-(3) of an affinity.

Exercise 4.82. Show that if an affinity f of the plane α onto the plane β, maps a triangle $AB\Gamma$ to a similar triangle $A'B'\Gamma'$, then f maps every polyon Π of α to a similar polygon Π' of β.

Hint: Use composition of affinities and the fact, that similarities also satisfy properties (1)-(3) of an affinity. Consequently they are also affinities and, if they coincide with an affinity on three non collinear points, because of (1), they will coincide everywhere.

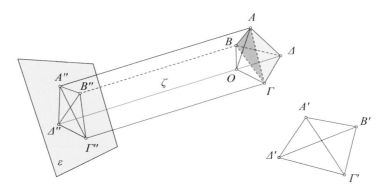

Fig. 4.93: Fundamental theorem of axonometry (Pohlke-Schwarz)

Next theorem (See Figure 4.93) is more general than that of Pohlke and was proved by Hermann Amandus Schwarz (1843-1921) [74] [66, p.305].

Theorem 4.28. *A given tetrahedron $OAB\Gamma$ in space can be projected onto a suitable plane with an inclined axonometry such that its image $O''A''B''\Gamma''$ is similar to a given quadrilateral $\Delta'A'B'\Gamma'$.*

Proof. To begin with, consider the affinity f between planes $\alpha = A'B'\Gamma'$ and $\beta = AB\Gamma$ which is defined through the correspondence between the vertices of the two triangles. Define Δ through $\Delta = f(\Delta')$ and the line $\zeta = \Delta O$. From the vertices of the triangle $AB\Gamma$ we draw parallels to ζ. This defines a triangular prism, which we intersect with a suitable plane ε along the triangle $A''B''\Gamma''$ similar to the given $A'B'\Gamma'$ (Theorem 4.14). This defines a parallel to ζ axonometric projection g on ε and the point $\Delta'' = g(\Delta)$. The composition $h = g \circ f$ is then an affinity and the triangles $A'B'\Gamma'$, $A''B''\Gamma''$ are by construction similar, therefore h is a similarity (Exercise 4.82) and the quadrilateral $A''B''\Gamma''\Delta'' = h(A'B'\Gamma'\Delta')$ is also similar to $A'B'\Gamma'\Delta'$.

4.17 Perspective projection

> It would please me if the painter were as learned as possible in all the liberal arts, but first of all I desire that he know geometry. I am pleased by the maxims of Pamfilos, the ancient and virtuous painter from whom the young nobles began to learn to paint. He thought that no painter could paint well who did not know much geometry.
>
> *Leon Batista Alberti, On Painting*

Pencils of lines, which pass through a fixed point O in space, give the possibility of a kind of *projection* of a shape Σ in space onto a shape Σ' on a

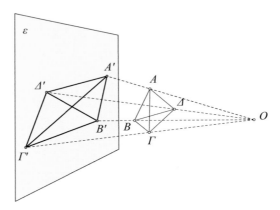

Fig. 4.94: Perspective projection from O onto the plane ε

fixed plane ε. This kind of projection corresponds to every point $A \neq O$ the intersection A' of OA with ε (See Figure 4.94). We call this process **central**

4.17. PERSPECTIVE PROJECTION

projection or **perspective** projection or **perspectivity** from O onto the plane ε. Point O is called **center** of the perspective (projection) and plane ε is called **projection plane** or **image plane**.

One of the basic differences between the axonometric and the perspective projections to plane ε is that the first preserves the parallelity of lines, while the second preserves this only for some special positions of the parallel lines but not in general. In what follows we examine some simple properties of the perspective projection from the point O onto the plane ε. In a perspective

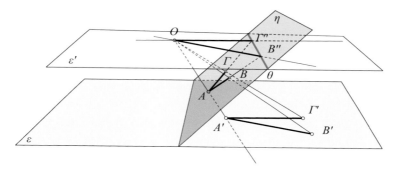

Fig. 4.95: Perspective projection

projection, the points of an arbitrary plane η, which does not pass through O and intersects plane ε, are mapped to points of the plane ε (See Figure 4.95). Every point A' of ε is the projection of exactly one point A of η. There do exist however also some points on η which do not have a corresponding projection on ε. These are the points of the so called **vanishing line** or **horizon** θ of the plane η. This line is the intersection of the plane η with the plane ε', which is parallel to ε and passes through point O. Points B'', Γ'', ... of θ define lines OB'', $O\Gamma''$, ... which are parallel to plane ε. Consequently these points do not have corresponding images on ε through the perspective projection from O. If ε is considered to represent a scenery and η represents the canvas of a painter placed at O, then the points at infinity of ε, which are defined from lines of ε (see remarks I-5.18 and I-5.19), are mapped on the canvas η to points of the vanishing line θ. In actual paintings, made with the rules of perspective projection, the vanishing line on the canvas corresponds to the remote points on the horizon of the corresponding scenery, whence the name.

Theorem 4.29. *The perspective projection has the following characteristic properties:*

1. *Every line θ contained in the plane ε', which passes through O and is parallel to ε, maps to the line at infinity of ε. Every line passing through O and not contained in ε' maps to a point. Finally, every line ζ not containing O and not contained in ε' maps to a line ζ' of ε.*

2. *If the line ζ of space maps to the line ζ' of plane ε, then the cross ratio $(AB; \Gamma\Delta)$ of four points A, B, Γ, Δ of ζ is equal to the cross ratio $(A'B'; \Gamma'\Delta')$ of their images A', B', Γ', Δ' on ζ'.*

Proof. The proof of (1) is a direct consequence of the definitions. The proof of (2) results from Theorem I-5.29.

Exercise 4.83. Show (using figure 4.95) that two parallel lines α' and β' of ε are the perspective projections from O of two lines α and β of the plane η, which intersect at a point of the vanishing line θ of η. And conversely two lines of η, which intersect at a point of θ, map to two parallel lines of ε.

Proposition 4.17. *The perspective projection $B'A'\Gamma'$, on plane ε from point O, of an angle $BA\Gamma$ of plane η, is congruent to the angle $B''O\Gamma''$ which is defined from the intersections B'', Γ'' of the escape line θ with the corresponding lines AB, $A\Gamma$.*

Proof. The proof follows immediately from the fact that the dihedral with edge OA, which is defined from the angle $\widehat{BA\Gamma}$ and whose intersection with ε is exactly $\widehat{B'A'\Gamma'}$ (See Figure 4.95) intersects also the parallel plane ε' of ε along an angle $\widehat{B''O\Gamma''}$ congruent to $\widehat{B'A'\Gamma'}$ (Corollary 3.11).

Proposition 4.18. *Given two intersecting planes ε and η and an angle $\widehat{BA\Gamma}$ on η, there exists a point O outside both planes, such that the perspective projection of η on ε maps angle $\widehat{BA\Gamma}$ to an angle $\widehat{B'A'\Gamma'}$ of ε with given measure $\phi = \widehat{B'A'\Gamma'}$.*

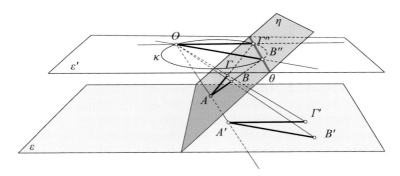

Fig. 4.96: Perspectivity of two angles of arbitrary measure

Proof. The proof is a simple application of the preceding proposition. Consider a plane ε' parallel to ε and the line θ of the intersection of η and ε' (See Figure 4.96). Let B'', Γ'' be respectively the intersection points of θ with AB and $A\Gamma$. Consider next the arc of circle κ of ε', whose points view $B''\Gamma''$ under an angle of measure ϕ. Any point O of this arc does the job. Indeed, if O is a point of this arc, then the angle $\widehat{B''O\Gamma''}$ has by construction measure ϕ and defines a dihedral angle with edge OA, which intersects plane ε along an angle $\widehat{B'A'\Gamma'}$ congruent to $\widehat{B''O\Gamma''}$.

4.17. PERSPECTIVE PROJECTION

Theorem 4.30. *Given two intersecting planes ε and η and a triangle $AB\Gamma$ on η, there exists a point O outside both planes, such that the perspective projection of η to ε maps the triangle $AB\Gamma$ to a triangle $A'B'\Gamma'$ of ε which is similar to a given triangle $A_0B_0\Gamma_0$ (See Figure 4.97).*

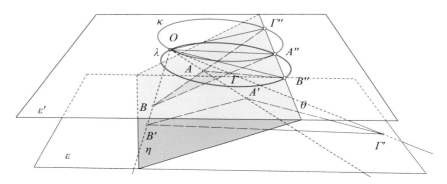

Fig. 4.97: Perspectivity of two triangles

Proof. For the proof we apply the preceding proposition. We consider a plane ε' parallel to ε and its intersection θ with the plane η of triangle $AB\Gamma$. Let A'', B'', Γ'' be the intersection points of θ with the side-lines of the triangle respectively $B\Gamma$, ΓA, AB. Let also α_0, β_0, γ_0 be the angles of the triangle $A_0B_0\Gamma_0$. We consider on plane ε' a circle arc κ the points of which view $A''\Gamma''$ under the angle β_0. We also consider a circle arc λ the points of which view $A''B''$ under the angle γ_0. Let O be the intersection point of the two arcs different from A''. The lines OA, OB, $O\Gamma$ intersect the plane ε at points A', B', Γ' which form a triangle $A'B'\Gamma'$. According to the preceding proposition angles $\widehat{\Gamma'B'A'}$ and $\widehat{A'\Gamma'B'}$ will have sides parallel to $\widehat{A''O\Gamma''}$ and $\widehat{B''OA''}$ respectively, therefore congruent to β_0 and γ_0 respectively.

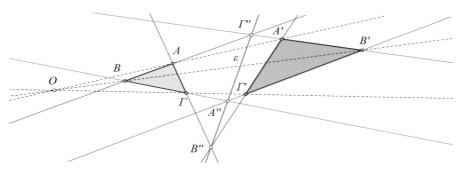

Fig. 4.98: Perspectivity by Desargues

Remark 4.15. From the theorem follows, that through a perspective projection the arbitrary triangle $AB\Gamma$ can be projected onto another $A'B'\Gamma'$, which may have any angles we want. Besides the method of the preceding proof, this can also be done through "folding" of the plane and with the help of the theorem of Desargues (I-§ 5.18). Indeed two triangles $AB\Gamma$ and $A'B'\Gamma'$ with arbitrary angles can, translated suitably (See Figure 4.98), be placed in such a way so that the lines AA', BB', $\Gamma\Gamma'$ pass through a point O (Exercise I-2.208). Then, by Desargues, the intersections of the lines $\Gamma'' = (AB, A'B')$, $A'' = (B\Gamma, B'\Gamma')$, $B'' = (\Gamma A, \Gamma'A')$ will be contained in line ε. Keeping one half plane with the first triangle fixed and rotating in space the other half plane of ε with the other triangle, forms two intersecting planes and a perspective projection of the one triangle to the other (see also Exercise 3.8).

4.18 Comments and exercises for the chapter

> One striking peculiarity of mathematics is its unlimited power of evolving examples and problems. A student may read through a book of Euclid, or a few chapters of Algebra; and within that limited range of knowledge it is possible to set him exercises as real and as interesting as the propositions themselves which he has studied ...
>
> I. Todhunter, *The Conflict of Studies*, ch.III, p.82

Exercise 4.84. Show that, if in a tetrahedron two pairs of opposite edges define skew orthogonal lines, then the same happens with the third pair of opposite sides.

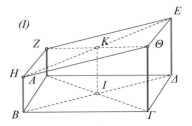
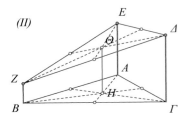

Fig. 4.99: Truncated prisms

Exercise 4.85. In the truncated prism $AB\Gamma\Delta EZH\Theta$ its base $AB\Gamma\Delta$ is a parallelogram. Show that the sum of two opposite edges is equal to the sum of two others: $|AZ| + |\Gamma\Theta| = |BH| + |\Delta E|$ (See Figure 4.99-I).

Exercise 4.86. In the truncated prism $AB\Gamma\Delta EZ$ its base $AB\Gamma$ is a triangle. Show that the sum of the lengths of its edges is equal to three times the length of the segment which joins the centroids of the two triangles: $|AE| + |BZ| + |\Gamma\Delta| = 3|H\Theta|$ (See Figure 4.99-II).

Exercise 4.87. Show that a plane parallel to two opposite sides of a skew quadrilateral divides the other sides in proportional parts.

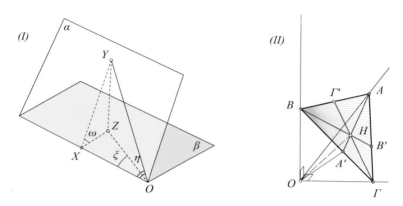

Fig. 4.100: Angle projection

Exercise 4.88. Let OX be the edge of a dihedral and ω the angle of its faces α and β. Draw a line OY on plane α and denote by OZ its projection to plane β (See Figure 4.100-I). Show that the angle $\eta = \widehat{XOY}$ and its projection on β: $\xi = \widehat{XOZ}$ satisfy the relation $\tan(\xi) = \tan(\eta)\cos(\omega)$.

Exercise 4.89. Show that if $AB\Gamma$ is the triangle formed from the intersection of a triply orthogonal trihedral (all its faces are right angles) with a plane (See Figure 4.100-II), then the traces $\{A', B', \Gamma'\}$ of the altitudes of the triangle coincide with the corresponding projections of the apex O of the trihedral on the sides of the triangle. Also the projection of O on the triangle's plane coincides with its orthocenter H.

Hint: Line $B\Gamma$ is (skew) orthogonal to lines $\{AA', OA\}$, hence is also orthogonal to every line of their plane, hence also to OA'.

Exercise 4.90. Show that a triply orthogonal trihedral can be intersected with a plane ε, so that the intersection triangle is equal with a given triangle $AB\Gamma$.

Hint: Use the preceding exercise. In the given triangle the traces $\{A', B', \Gamma'\}$ of the altitudes on its sides are known. Hence in the right triangle $OB\Gamma$ is known the hypotenuse $B\Gamma$ and the trace of the altitude A' from O, hence the triangle is constructible (See Figure 4.100-II) etc.

Exercise 4.91. Show that for tetrahedra, whose opposite edges are skew orthogonal, the altitudes, that is, the lines from a vertex which are orthogonal to the opposite face, all pass through a common point.

Hint: Use Proposition 4.4.

Exercise 4.92. Show that if every face of a polyhedron has an odd number of sides, then the polyhedron must have a pair number of faces.

A plane orthogonal to one edge of a tetrahedron and passing through the middle of its opposite edge is called a **Monge's plane** (1746-1818) of the tetrahedron. Next exercise ([64, p.196]) defines the **Monge's point** of the tetrahedron.

Exercise 4.93. Show that the six Monge's planes of a tetrahedron pass through a common point.

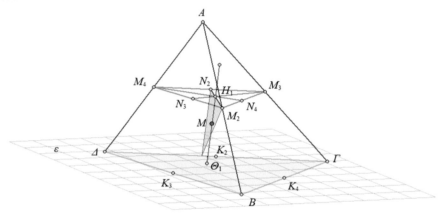

Fig. 4.101: Monge's point of the tetrahedron

Hint: Consider the middles M_2, M_3, M_4 of the edges, respectively, AB, $A\Gamma$, $A\Delta$ at the vertex A of the tetrahedron $AB\Gamma\Delta$ (See Figure 4.101). The Monge's plane μ_2 through M_2 is orthogonal to $\Delta\Gamma$, therefore also orthogonal to its parallel M_3M_4. Consequently, μ_2 intersects triangle $M_2M_3M_4$ along its altitude M_2N_2. Similarly Monge's plane μ_3 through M_3 intersects triangle $M_2M_3M_4$ along its altitude M_3N_3. Consequently, the intersection α of planes μ_2, μ_3 is a line $\alpha = H_1\Theta_1$ passing through the orthocenter H_1 of $M_2M_3M_4$. This line is simultaneously orthogonal to $\Delta\Gamma$, ΔB, therefore also to their plane $B\Gamma\Delta$. It follows that the three Monge's planes μ_2, μ_3, μ_4 of the edges around vertex A intersect along a line α. Similarly, the three Monge planes, of the edges around vertex B, intersect along a line β orthogonal to plane $\Gamma\Delta A$. The two lines α and β intersect at point M, because they are contained in plane μ_2 and they are not parallel. It follows that the Monge planes μ_2, μ_3, μ_4, as well as κ_3, κ_4 through the middles K_3, K_4, of $B\Delta$, $B\Gamma$ respectively, pass through the same point. Similarly also μ_2, μ_3, μ_4, κ_4, κ_2 will pass through the same point, therefore all six planes will pass through the same point.

Exercise 4.94. Show that the Monge point of the tetrahedron is the symmetric of the circumcenter of the tetrahedron relative to its centroid.

4.18. COMMENTS AND EXERCISES FOR THE CHAPTER

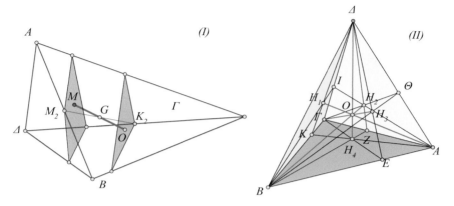

Fig. 4.102: Point M of Monge   Orthocentric tetrahedron

Hint: The circumcenter is the center O of the circumscribed sphere and is contained in the medial plane α of the edge $\Gamma\Delta$ (See Figure 4.102-I). The Monge's plane β, the one orthogonal to the edge $\Gamma\Delta$ is consequently parallel to α. The centroid G of the tetrahedron is the middle of M_2K_2, where M_2, K_2 are respectively the middles of the edges AB and $\Gamma\Delta$ (Exercise 4.27). It follows that the symmetric of O relative to G will be a point M contained in the Monge plane β. The argument shows that the symmetric of O relative to G is contained in each of the six Monge planes, therefore it coincides with their point of intersection.

Exercise 4.95. Show that if a vertex, e.g. A, of the tetrahedron $AB\Gamma\Delta$ is projected to the orthocenter of its opposite face, then the same happens also with the projections of the other vertices of the tetrahedron to their opposite faces. Conclude that, then, the altitudes of the tetrahedron intersect at a point, which coincides with the Monge point of the tetrahedron. Show also the converse, that is, if the altitudes of a tetrahedron intersect at a point, then each vertex is projected to the orthocenter of its opposite face and the intersection point of the altitudes coincides with the Monge point of the tetrahedron.

Tetrahedra like the one of the preceding exercise are called **orthocentric** (See Figure 4.102-II).

Exercise 4.96. Show that the tetrahedron $AB\Gamma\Delta$ is orthocentric, if and only if it admits two pairs of opposite edges which are skew orthogonal.

Hint: According to exercise 4.84, then all pairs of opposite edges will consist of skew orthogonal lines.

Exercise 4.97. Show that a **triply orthogonal tetrahedron**, that is a tetrahedron whose faces around a vertex are right angles, is orthocentric.

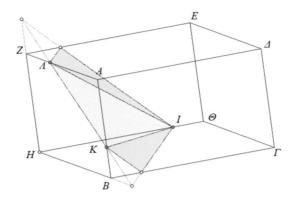

Fig. 4.103: Intersection of parallelepiped with plane

Exercise 4.98. Given are three points $\{I, K, \Lambda\}$ on three edges of parallelepiped $AB\Gamma\Delta EZH\Theta$. Find its intersection with the plane ε defined by these three points ([60, p.8]).

Hint: Besides the vertices of $IK\Lambda$, which are contained in ε, the other vertices of the polygon, formed by the intersection, must also be determined (See Figure 4.103). There are different cases depending on the position of these three points.

Exercise 4.99. Given is a plane ε, a line η, not contained in it and a point A outside both ε and η. Find a point B on η which is equidistant from A and the plane ε.

Hint: Consider the plane ζ orthogonal to η and passing through A. Let K be the intersection point of ζ with η. Consider also the plane κ, which contains line η and is orthogonal to ε. Let θ be the line-intersection of κ and ε (line θ is the projection of η to ε). The circle $\lambda(K, |KA|)$ and the plane κ intersect at two points Γ and Δ. The problem is reduced to the problem of the plane κ: To find point B on η equidistant from Γ (or Δ) and the line θ (Proposition I-4.7).

Exercise 4.100. Given are three non collinear points A, B, Γ and a plane ε not containing them. Find a sphere which passes through all three points and is tangent to the plane ε.

Hint: The problem reduces to the preceding one, by considering the circle λ which passes through all three points and the line η orthogonal to the plane $AB\Gamma$ at the center of λ.

Exercise 4.101. Show that all the Platonic solids are spherical polyhedra.

Hint: A simple way is to use the (congruent) circumscribed circles of the regular polygons-edges of the solids and Exercise 4.72. Several shapes and properties related to circles lead to corresponding shapes and properties of the sphere through its intersections with planes or/and rotating the circle around an axis passing through its center. Typical examples are these of the **power** of the sphere relative to a point and of the **radical plane** of two spheres, which are proved through reduction to the corresponding properties of the circles.

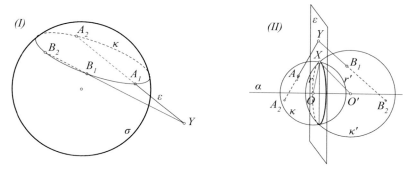

Fig. 4.104: Power of point relative to sphere , Radical plane of two spheres

Exercise 4.102. Show that the product $\delta = |YA_1||YA_2|$ of the segments, which result by intersecting a sphere σ with a line ε through a point Y, is independent of the direction of the line ε.

Hint: If YB_1 is a second line through Y, which intersects the sphere at points B_1, B_2 (See Figure 4.104-I), then the two intersecting lines define a plane which intersects the sphere along a circle κ. The property results from the corresponding property of the point relative to the circle κ.

By analogy to circles, the product $\delta = |YA_1||YA_2|$, of the preceding exercise, is called **power of point Y relative to sphere σ**.

Exercise 4.103. Show that the points Y of the plane ε, which contains the circle-intersection of two intersecting spheres $\kappa(O,r)$ and $\kappa'(O',r')$ have the property $|YA_1||YA_2| = |YB_1||YB_2|$ for every pair of lines which pass through point Y and intersect circle κ at points A_1, A_2 and circle κ' at points B_1, B_2 (See Figure 4.104-II).

Hint: Consider the circles which result by intersecting the two spheres with the plane defined from the line of the centers $\alpha = OO'$ and point Y (See Figure 4.104-II). Again, by analogy to the circles, the plane ε of the preceding exercise is called **radical plane** of two spheres. Next exercise shows the existence of the radical plane for every pair of non-concentric spheres.

Exercise 4.104. Show that for every pair of non-concentric spheres σ, σ' the geometric locus of the points in space, which have the same power relative to the two spheres is a plane ζ orthogonal to the center-line α of the two spheres.

Hint: Consider the intersection of the two spheres with a plane ε containing the center-line α, and the radical axis η of the two circles of ε which result. The requested plane ζ is that, which results by rotating the line η about the axis α.

The transfer and adaptation of properties of circles to corresponding properties of spheres offers an inexhaustible field of exercises. For example the **angle** of two intersecting spheres is defined by intersecting the spheres with a plane passing through their centers and considering the angle ω of the two intersecting great circles which result. From this follows trivially the definition of **orthogonality** for spheres ($\omega = 90°$).

Pencils of spheres are defined, similarly to those of circles: They are a set of spheres Σ, for which there exists a plane ε, such that, for every pair of spheres from Σ, their corresponding radical plane coincides with ε. The various sphere pencils result from *circle pencils*, which we rotate around their center-line. The rotation of the radical axis then produces the corresponding **radical plane** of the spheres. The intersection of a sphere pencil with a plane ε defines a circle pencil of ε. A subject related to the preceding one is also that of **inversion** relative to a sphere. Formally, the definition is the same with that of the transformation of an inversion relative to a circle. The properties also are similar to these of the plane inversion. Analogous is also the definition of the two **similarity centers** of two spheres, which, in the case of two spheres lying outside each other, defines two cones tangent to the two spheres.

Exercise 4.105. (**Radical center** of four spheres) Show that four spheres whose centers are in "general position", that is, they are non-collinear taken by three and not all four coplanar, define a point X, which has the same power relative to the four spheres.

Exercise 4.106. Show that the centers of similarity of three spheres, whose centers are in general position are contained by three in a line (see exercise I-5.110).

Exercise 4.107. (Sphere of Apollonius) Show that the geometric locus of the points X in space, for which the ratio of distances $\frac{|XA|}{|XB|} = \kappa$ from two fixed points A, B in space is fixed, is a sphere with center on line AB. Determine the elements of this sphere, in other words the position of its center O on line AB and its radius (see I-§ 4.4).

Next exercises describe the way Euclid used, in its "Elements" ([71, p.530], [13, III, p.493]), in order to construct a dedecahedron of edge-length a. The method starts from a cube of edge $a' = \phi \cdot a$, where $\phi = \frac{\sqrt{5}+1}{2}$ is the golden section (I-§ 4.2). Subsequently he constructs 6 "roofs" on top of the square

4.18. COMMENTS AND EXERCISES FOR THE CHAPTER 251

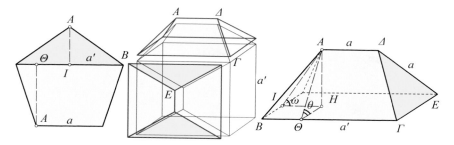

Fig. 4.105: Dedecahedron through addition of 6 roofs on the cube

faces of this cube, which in total give the requested dodecahedron (See Figure 4.105). The surfaces of the roof come from the two parts of a regular pentagon of edge a defined by a diagonal. The first exercise calculates the altitude of such a roof and the second shows that the two roofs, on top of two faces of the cube which have a common edge ($B\Gamma$), fit precisely together and give again a plane pentagon which has this edge as a diagonal and represents a face of the requested dodecahedron.

Exercise 4.108. On a square of edge $a' = \phi \cdot a$ construct a "roof" using the parts of a regular pentagon, which are defined by one diagonal of the pentagon.

Hint: The construction follows from the calculation of the altitude $|AH|$ of the roof and the position of H, which determines one of its vertices (See Figure 4.105). These elements come from the following, easily calculated, formulas:

$$|A\Theta| = \frac{a}{a'}\sqrt{a'^2 - (a/2)^2},$$
$$|AI| = \sqrt{a^2 - (a'/2)^2},$$
$$|\Theta H| = \frac{a'}{2},$$
$$|IH| = \frac{a' - a}{2},$$
$$|AH| = \sqrt{|AI|^2 - |IH|^2} = \frac{a}{2}.$$

Exercise 4.109. Show that two roofs, like these of the preceding exercise, fit precisely together on top of two faces of the cube, having a common edge $B\Gamma$ and give a plane regular pentagon of edge a (See Figure 4.105).

Hint: A precise fitting means that the trapezoid part of the one roof matches the triangular part of the other along $B\Gamma$, so that a *plane* pentagon results. We easily see that, in order for these two parts to be in a plane, it is necessary and sufficient for the angles ω and θ to be complementary, or equivalently to have inverse tangents. This however follows directly from the preceding formulas, which give:

$$\tan(\omega) = \frac{|AH|}{|IH|} = \frac{a}{a'-a}, \quad \tan(\theta) = \frac{|AH|}{|\Theta H|} = \frac{a}{a'} \quad \Rightarrow$$

$$\tan(\omega) \cdot \tan(\theta) - 1 = \frac{1 - \phi^2 + \phi}{\phi(\phi - 1)} = 0.$$

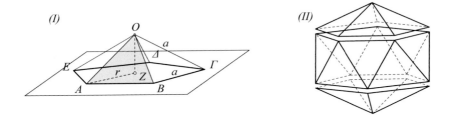

Fig. 4.106: Icosahedron cap with 5 equilaterals

Exercise 4.110. Construct a regular pyramid with base a regular pentagon of edge a and calculate the measures of its dihedrals.

Hint: For the construction of the regular pyramid with equilateral faces it suffices to calculate the radius $r = a\sqrt{\frac{5+\sqrt{5}}{10}}$ of the circumcircle of the pentagon and the altitude of the pyramid $|OZ| = a\sqrt{\frac{5-\sqrt{5}}{10}}$ (See Figure 4.106-I). The dihedrals are calculated using the formula for the spherical cosine.

The construction of the preceding exercise can be used for the construction of the icosahedron. We construct first one, symmetric relative to center, "drum" with bases two pentagons and adjacent surface consisting of ten equilateral triangles (See Figure 4.106-II). This kind of solid we'll name later (§ 7.8) *antiprism*. Subsequently on its two bases we attach two caps, like the one of the preceding exercise. This method is, in essence, the method Euclid follows, to inscribe an icosahedron in a given sphere ([13, III, p.486]). An alternate construction is described in the next exercise ([13, III, p.491]).

Exercise 4.111. Construct an icosahedron inscribed in a cube, beginning from a cube of edge a and determining two vertices of the icosahedron in each face of the cube.

Hint: In each pair of opposite faces of the cube we construct a segment ΔE (See Figure 4.107-I) with ratio $|\Delta E|/a = \phi$ (golden section). This forms totally three, pairwise orthogonal (See Figure 4.107-II), *golden* rectangles (I-§ 4.2). Their vertices are simultaneously vertices of an icosahedron.

4.18. COMMENTS AND EXERCISES FOR THE CHAPTER

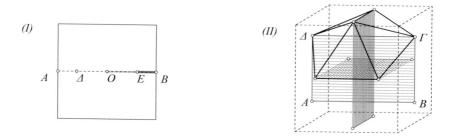

Fig. 4.107: Circumscribed cube of icosahedron

Exercise 4.112. Show that in a polyhedron with triangular faces, there exists an edge on which the adjacent angles of the two triangular faces are acute.

Hint: Consider the edge of maximum length of the polyhedron. Cones can

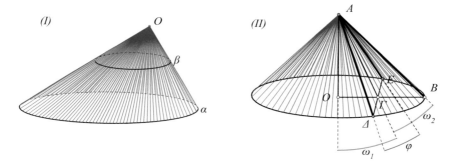

Fig. 4.108: Oblique cone   Angle between two generators

be defined more generally with the use of a curve α and a point O. The surface of the cone then results by connecting with lines point O with each point of the curve. Figure 4.108-I shows such a simple surface, which differs from the usual right cone on the fact that the apical vertex O does not project at the center of the circle of the base α. Such a cone is called **oblique** circular cone.

Exercise 4.113. Show that an intersection of an oblique circular cone with a plane parallel to its circular base α is also a circle β.

Exercise 4.114. A right cone with opening $\omega_1 + \omega_2$ is intersected with a plane $A\Delta E$ through the vertex A. If the inclination of the axis of the cone relative to the plane is $\widehat{OA\Gamma} = \omega_1$, find angle $\phi = \widehat{\Delta AE}$ formed by the two generators $A\Delta$, AE which are contained in the intersecting plane (See Figure 4.108-II).

Next exercise completes the propositions/exercises formulating the criteria of congruence of two trihedrals (§ 4.2). As remarked at the begining of § 4.14

for spherical triangles, these criteria correspond exactly to criteria of congruence of spherical triangles.

Exercise 4.115. Show that two trihedral angles, which have correspondingly two dihedrals equal and the contained face equal, are equal (congruent).

Hint: If the trihedrals $\{OAB\Gamma, O'A'B'\Gamma'\}$ have their dihedrals $\{\widehat{B} = \widehat{B'}, \widehat{\Gamma} = \widehat{\Gamma'}\}$ consider the points $\{A, A'\}$ on the edges, such that their projections $\{P, P'\}$ to the opposite faces $\{OB\Gamma, O'B'\Gamma'\}$ define equal segments $\{|AP| = |A'P'|\}$. Then project points $\{P, P'\}$ on the sides of the angles $\{\widehat{BO\Gamma}, \widehat{B'O'\Gamma'}\}$ etc. See the figure 4.18 of exercise 4.7.

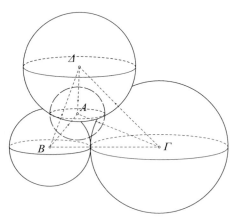

Fig. 4.109: Four tangent spheres

Exercise 4.116. Given is a tetrahedron $AB\Gamma\Delta$. When do there exist four spheres, pairwise tangent externally, with centers at the vertices of the tetrahedron (See Figure 4.109)?

Hint: See the analogue for the plane, Exercise I-2.13.

Exercise 4.117. Show that if the tetrahedron $AB\Gamma\Delta$ has faces of equal area then these faces are congruent triangles. Equivalently, its opposite edges are equal and the circumscribed parallelepiped is a rectangle.

Hint: ([75, p.226]) Consider the circumscribed parallelepiped of the tetrahedron (See Figure 4.110). Let α, β be its faces, respectively, which contain $B\Gamma$ and $A\Delta$. Project points B, Γ to B', Γ' on β and these to B'', Γ'' on $A\Delta$. Show that the right triangles $BB'B''$ and $\Gamma\Gamma'\Gamma''$ are congruent. Perform the corresponding process by projecting points A, Δ to α and conclude that $A\Delta$ and $B'\Gamma'$ are equal diagonals of a parallelogram.

Exercise 4.118. Show that if all the dihedral angles of a tetrahedron are acute, then all its faces are acute angled triangles (See Figure 4.111).

4.18. COMMENTS AND EXERCISES FOR THE CHAPTER

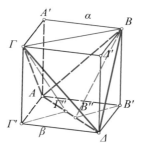

Fig. 4.110: Faces of equal area

Hint: Use contradiction, assuming that one face e.g. $\widehat{BA\Delta}$ is obtuse or right. Then, by hypothesis, the projection of $A\Gamma$ onto the plane $BA\Delta$ falls inside the angle $AB\Gamma$ (See Figure 4.111-I). From a point Γ_1 of this projection we draw the plane which is orthogonal to $A\Gamma$ and whose intersection with the dihedral $\widehat{A\Gamma}$ defines the measure $\widehat{B'\Gamma''\Delta'}$ of the dihedral. They result the following possibilities.

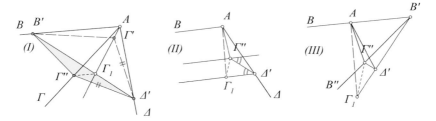

Fig. 4.111: Acute dihedrals ⇒ acute faces

(1) This plane, orthogonal to $A\Gamma$, intersects both other edges and defines a triangle $\Gamma''\Delta'B'$. Rotating this triangle about $B\Delta'$ we define the equal to it triangle $B'\Gamma'\Delta'$, with Γ' on line $A\Gamma_1$. Since the angle $\widehat{A\Gamma''\Delta'}$ is a right one, from the right triangle $A\Gamma''\Delta'$ results that $|\Gamma'\Delta'| = |\Gamma''\Delta'| < |A\Delta'|$. Analogoysly and $|B'\Gamma'| < |AB'|$. Hence, for the angles $\widehat{B'\Gamma'\Delta'} > \widehat{B'A\Delta'}$, which means that the dihedral $A\Gamma$ is obtuse, a contradiction.

(2) If the orthogonal to $A\Gamma$ plane at Γ'' is parallel to one side e.g. to AB of the angle $\widehat{BA\Delta}$ (See Figure 4.111-II), then its intersection with the plane $BA\Delta$ is a line $\Gamma_1\Delta'$ parallel to AB and its dihedral $A\Gamma$ is the supplementary of $\Gamma''\Delta'\Gamma_1$, which is smaller than the dihedral $A\Delta$ (Theorem 4.1), which is acute, a contradiction.

(3) If the orthogonal to $A\Gamma$ plane at Γ'' intersects AB externally at the point B', then the angle $B''\Gamma''\Delta'$, being external of the triangle $B'\Gamma''\Delta'$, is greater than the angle $\widehat{\Gamma''\Delta'B'}$ (See Figure 4.111-III), which is supplementary of $\widehat{\Gamma''\Delta'\Gamma_1}$, which, in turn is smaller from the dihedral of $A\Delta$ (Theorem 4.1), which is obtuse, a contradiction (see also exercise 4.20).

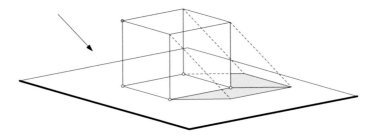

Fig. 4.112: The shadow of the cube

Exercise 4.119. Show that the shadow of a cube (See Figure 4.112), resulting by illuminating it with light beams parallel to a given direction, together with its base, makes a hexagon with the properties: (1) It is symmetric. (2) Has two opposite angles right and the adjacent to them sides equal to the side of the cube. (3) Its area is the sum of the area of the square face of the cube and the area of a parallelogram whose one side is the diagonal of the face of the cube. Determine the directions of illumination such that the hexagon has $\{2,3,\ldots\}$ times the area of the cube-face. Show that if the perimeter of the above hexagon remains constant, then its vertices, which are different from the cube-vertices are contained in circles with centers at the vertices of the cube and equal radii, their measure depending from the perimeter. How a point of one of the preceding circles determines the direction of the corresponding light beam? Find the area of the shadow, when the direction of the beam is parallel to a diagonal of the cube. If the perimeter of the hexagon remains constant, find the direction of the beam for which the shadow has maximal area.

Exercise 4.120. The right angled triangle $AB\Gamma$ of the space is projected orthogonally to the plane ε, onto the equilateral triangle $\Gamma\Delta E$ with side-length d. To find the relation satisfied by its orthogonal sides (See Figure 4.113-I).

Hint: If $\{x = |A\Gamma|, y = |AB|\}$ the orthogonal sides and $z = |B\Gamma|$ the hypotenuse, we can suppose, without affecting the generality that the vertex Γ lies on the plane ε and the vertices $\{A,B\}$ are projected correspondingly to $\{\Delta,E\}$ of the equilateral $\Gamma\Delta E$. We get the relations:

$$z^2 = |B\Gamma|^2 = x^2 + y^2, \ |BE|^2 = z^2 - d^2, \ |A\Delta|^2 = x^2 - d^2, \ y^2 = d^2 + (b-a)^2 \Rightarrow$$
$$4(x^2 - d^2)(y^2 - d^2) = d^4, \ b = |BE| = (2a^2 + d^2)/(2a), \ \text{where} \ a = |A\Delta|.$$

4.18. COMMENTS AND EXERCISES FOR THE CHAPTER

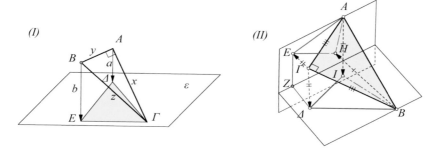

Fig. 4.113: Projection to equilateral   Projection to two equilaterals

Exercise 4.121. The right angled triangle $AB\Gamma$, with the hypotenuse-vertices $\{A,B\}$ lying correspondingly on two intersecting planes $\{\alpha,\beta\}$, is projected orthogonally to these planes and its projections are equilateral triangles of side-length d. To determine the dimensions of the right triangle.

Hint: If $\{I,H\}$ are the projections of $\{A,B\}$ on the planes (See Figure 4.113-II), then the right angled triangles $\{ABI,ABH\}$, by assumption will have in common the hypotenuse and their orthogonal sides $|IB| = |HA| = d$. Hence they will be congruent. Applying then the preceding exercise, we see that for $b = |AI| = |BH|$ and $a = |\Gamma\Delta| = |\Gamma E|$. Hence also the right angled triangles $\{\Gamma\Delta B, \Gamma EA\}$ are equal, hence the right angled triangle $AB\Gamma$ is isosceles. To compute $|AB|$ it suffices to apply the relation of the orthogonal sides of the preceding exercise for $x = y$.

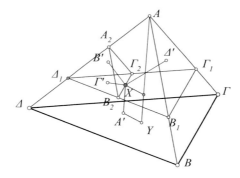

Fig. 4.114: Constant sum of distances of X from the faces

Exercise 4.122. Show that the sum of the distances of a point X of the interior of a regular tetrahedron $AB\Gamma\Delta$ from its faces is constant and equal to the altitude of the tetrahedron (See Figure 4.114).

Hint: If $\{XA', XB', X\Gamma', X\Delta'\}$ are the orthogonals from X to the faces, then the length of $|XA'|$ does not change when X moves on the plane $B_1\Gamma_1\Delta_1$ parallel to $B\Gamma\Delta$, and passing through X. Similarly the length $|X\Delta'|$ does not change when X moves on the plane $A_2B_2\Gamma_2$ parallel to $AB\Gamma$ and passing through X.

Exercise 4.123. Given is a dihedral of measure $\omega > 90°$. Find a plane intersecting the dihedral along a right angle (See Figure 4.115-I).

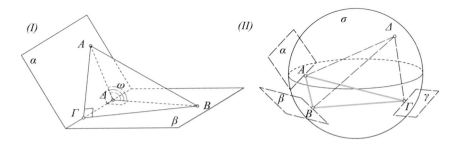

Fig. 4.115: Right from dihedral   Intersection of three planes

Exercise 4.124. The triangle $AB\Delta$ is isosceles and orthogonal to the intersection line $\Gamma\Delta$ of the planes $\{\alpha, \beta\}$, which define a dihedral of measure ω (See Figure 4.115-I). We consider also the sphere σ with diameter the base AB of the isosceles. Show that the sphere intersects the two planes along equal circles $\{\kappa_1, \kappa_2\}$ with a common chord contained in line $\Gamma\Delta$. Find the angle under which this chord is seen from the points of the circles $\{\kappa_1, \kappa_2\}$ in terms of ω and $d = |AB|$.

Exercise 4.125. Points $\{A, B, \Gamma, \Delta\}$ are on the sphere σ (See Figure 4.115-II). The planes $\{\alpha, \beta, \gamma\}$ are respectively orthogonal to the lines $\{\Delta A, \Delta B, \Delta \Gamma\}$ respectively at $\{A, B, \Gamma\}$. Locate the intersection point of the three planes. Locate also the intersection lines of the pairs of these planes.

Exercise 4.126. Continuing the preceding exercise, consider the three circles $\{\kappa_1, \kappa_2, \kappa_3\}$ intersected from the sphere σ respectively by the three planes $\{\alpha, \beta, \gamma\}$. Do these circles have a point in common? Do they pairwise intersect? Which are the intersection points?

Exercise 4.127. Two circles $\{\kappa_1(O_1, r_1), \kappa_2(O_2, r_2)\}$ in space intersect at a unique point A. Find a condition which guaranties the existence of a sphere σ containing both circles.

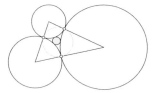

Fig. 4.116: Sphere tangent to three other spheres

Exercise 4.128. Show that three circles of the same sphere cannot be pairwise tangent at the same point of the sphere.

Exercise 4.129. Three spheres are pairwise externally tangent. Find the radius of the biggest possible sphere which passes through the hole formed by the three spheres (See Figure 4.116).

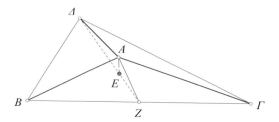

Fig. 4.117: Tetrahedron with right-angled faces at A

Exercise 4.130. The tetrahedron $AB\Gamma\Delta$ has its faces at A right angled and the projection of A on $B\Gamma\Delta$ falls onto the centroid E of this face. Show that $|AB| = |A\Gamma| = |A\Delta|$ and $B\Gamma\Delta$ is an equilateral triangle (See Figure 4.117).

Hint: By assumption AE and $A\Delta$ are (skew) orthogonal to $B\Gamma$. Hence $B\Gamma$ is orthogonal to their plane $A\Delta Z$, where Z is the middle of $B\Gamma$. Thus, the median AZ of $AB\Gamma$ is orthogonal to $B\Gamma$ and the right angled triangle $AB\Gamma$ is isosceles. Analogously $\{A\Gamma\Delta, A\Delta B\}$ are also right angled isosceli triangles.

Exercise 4.131. The rectangular parallelepiped $AB\Gamma\Delta A'B'\Gamma'\Delta'$ has the plane $A'B\Delta$ orthogonal to the diagonal $A\Gamma'$ (See Figure 4.118). Show that it is a cube.

Hint: Let Z be the intersection of the plane $A'B\Delta$ with $A\Gamma'$. The plane $A\Gamma\Gamma'$ intersects the plane $A'B\Delta$ along the line $A'E$, where E the center of the rectangle $AB\Gamma\Delta$. Triangles AZE and $\Gamma'ZA'$ are similar with ratio $|A'\Gamma'|/|EA| = 2$. Hence $|A'Z|/|ZE| = 2$ and Z is the centroid of $A'B\Delta$. Apply exercise 4.130 and deduce that $A'B\Delta$ is an equilateral triangle.

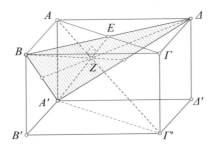

Fig. 4.118: Cube characterization

References

59. V. Arnold (2005) Arnold's Problems. Springer, Heidelberg
60. V. Arnlold (2004) Problems for children from 5 to 15. MCCME, Moscow
61. M. Berger (2002) A Panoramic View of Riemannian Geometry. Springer, Heidelberg
62. A. Church (1902) Elements of Descriptive Geometry.American Book Company, New York
63. N. Court (1980) College Geometry. Dover, New York
64. R. Crabbs, 2003. Gaspar Monge and the Monge Point of the Tetrahedron, *Mathematics Magazine*, 76:193-203
65. M. do Carmo (1976) Differential Geometry of Curves and Surfaces. Prentice-Hall, New Jersey
66. H. Doerrie (1965) 100 Great Problems of Elementary Mathematics. Dover,New York
67. F.G.M (1920) Exercises de Geometrie, 6e edition. Maison A. Mame et fils, Tours
68. E. Fourray (1900) Curiosites Geometriques. Vuibert, Paris
69. R. Glaser (1920) Stereometrie. Walther de Gruyter, Berlin
70. M. Hawk (1962) Theory and problems of descriptive geometry. McGraw-Hill, New York
71. J. Heiberg (1885) Euclidis Elementa. Teubner, Leipzig
72. K. Kimberling (2023) Encyclopedia of Triangle Centers. https://faculty.evansville.edu/ck6/encyclopedia/ETC.html
73. R. Millman, G. Parker (1977) Elements of differential geometry. Prentice-Hall, New Jersey
74. H. Schwarz, 1864. Elementarer Beweis des Pohlkeschen Fundamentalsatzes der Axonometrie, *Journal fuer die reine und angewandte Mathematik*, 63:309-314
75. I. Sharygin (1986) Problems in Solid Geometry. Mir, Moscow
76. I. Singer, J. Thorpe (1976) Lecture notes on Elementary Topology and Geometry. Springer, Berlin
77. E. Stiefel (1947) Lehrbuch der Darstellenden Geometrie. Springer, Basel
78. I. Vardi, 2000. Mekh-Mat Entrance Examination Problems, *Institut des Hautes Etudes Scientifiques*, 6:1-47

Chapter 5
Areas in space, volumes

5.1 Areas in space

> Accordingly, things arranged in a fixed order, like the successive demonstrations in geometry, are easy to remember (or recollect) while badly arranged subjects are remembered with difficulty.
>
> Aristotle, On Memory and Reminiscence

The area of polyhedral surfaces relies on the definitions and the properties of area of plane figures. Thus, the area of the surface of a rectangular parallelepiped with sides equal to α, β and γ is $2(\alpha\beta + \beta\gamma + \gamma\alpha)$, the area of the surface of a cube of size δ is $6\delta^2$ and similar calculations give the areas of any polyhedron, by adding the areas of their faces. The area of polyhedra, therefore, is not difficult to calculate and consequently possesses little theoretical interest (see however the exercises below).

Difficulties show up when trying to calculate areas of curved surfaces, like the cylinder, the cone, the sphere, as well as shapes contained in these surfaces. The definition of the area of these surfaces or/and their subsets in a general way, analogously to the definition of length for general curves, is the object of *Differential Geometry* (*area of surface region* [65, p.114], *volume element of manifold* [87, p.130]) and is beyond the scope of these lessons. It is proved, however, that the area of the cone, the cylinder and the sphere can be defined as a limit of areas of polyhedra inscribed in these surfaces. The process is, in essence, the same with that we used to define the perimeter and the area of the circle (§ 1.5). The **cone area** is defined as the limit of the areas of regular pyramids inscribed in the cone. Similarly the **cylinder area** is defined as the limit of the areas of regular prisms inscribed in the cylinder. In these two objects the area of the bases is that of the circle, consequently the interest is focused on calculating the area of the *lateral surface*.

As the **lateral area** of the cylinder we consider the limit of the lateral areas of regular prisms inscribed in the cylinder, while we increase the number

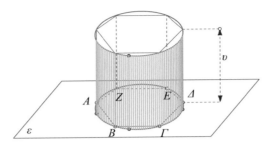

Fig. 5.1: Area of lateral surface of cylinder

of their sides (See Figure 5.1). Let us denote again with Π_1, Π_2, Π_3, ... the regular polygons which we used also in § 1.2. They begin with the square (Π_1) inscribed in the circle κ and each one is also inscribed in κ and has twice the number of sides of the preceding one. Denoting by p_ν the perimeter of Π_ν, the area of the lateral surface of the regular prism with base Π_ν is

$$\upsilon \cdot p_\nu,$$

where υ is the altitude of the prism. From Lemma 1.2 it follows that the sequence

$$\upsilon \cdot p_1, \quad \upsilon \cdot p_2, \quad \upsilon \cdot p_3, ...$$

is increasing. And from Lemma 1.3 follows that the sequence is bounded. We call the limit of this sequence the lateral **area of the cylinder**. From the preceding Corollary 1.1, follows then the next proposition.

Theorem 5.1. *The area of the lateral surface of the cylinder is equal to*

$$\varepsilon = 2\pi \cdot \rho \cdot \upsilon,$$

where ρ is the radius of the base of the cylinder and υ its altitude.

The calculation for the lateral area of the cone is done in a similar way. As **lateral area** of the cone we consider the limit of the lateral areas of the regular pyramids inscribed in the cone as we increase the number of their sides. Using the preceding notation, the area of the lateral surface of the regular cone with base Π_ν is

$$\frac{1}{2} \cdot \upsilon_\nu \cdot p_\nu,$$

where υ_ν is the altitude of the triangular face of the prism (not the altitude of the cone, neither its generator, but the length of the segment like OE in figure-5.2). We again verify easily that the sequence

$$\frac{1}{2} \cdot \upsilon_1 \cdot p_1, \quad \frac{1}{2} \cdot \upsilon_2 \cdot p_2, \quad \frac{1}{2} \cdot \upsilon_3 \cdot p_3, ...$$

5.1. AREAS IN SPACE

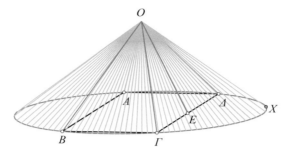

Fig. 5.2: Area of lateral surface of cone

is increasing and bounded by the number

$$\frac{1}{2} \cdot v \cdot (2\pi\rho),$$

where $v = |OX|$ is the length of the generator of the cone and ρ is the radius of the circle of its base. Again according to Axiom 1.1 the sequence will converge. Exactly this limit of this sequence we call **lateral area of the cone**.

Theorem 5.2. *The area of the lateral surface of the cone is equal to*

$$\varepsilon = \pi \cdot v \cdot \rho,$$

where ρ is the radius of the base of the cone and v the length of its generator.

Proof. I sketch the proof because I don't want to enter to details about limits. We then show first that the sequence

$$v_1, v_2, v_3, \ldots$$

is increasing and bounded, therefore converges, and in fact to $v = |OX|$. For the perimeters p_v, we know already that they converge to $2\pi\rho$. To complete the proof we need the rule for limits, according to which *the limit of a product of convergent sequences is the product of the limits*. According to this rule then the limit of the sequence $(v_v)(p_v)$ will be $v \cdot (2\pi\rho)$, from which the formula for the area results.

The calculation of the area of the lateral surface of a truncated cone is reduced to the corresponding calculation for cones. The lateral surface of the truncated cone can be seen as the difference of the lateral area of the cone relative to the larger base (circle κ in figure-5.3) minus the lateral area of the cone relative to the small base (circle λ in figure-5.3). Applying the formula, we consequently have that the area is given by the formula

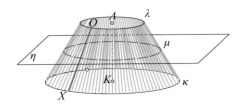

Fig. 5.3: Area of lateral surface of truncated cone

$$\varepsilon = \pi(v_2\rho_2 - v_1\rho_1),$$

where ρ_1, ρ_2, v_1, v_2 are the lengths of the radii and the generators of the cones whose difference is the truncated cone (See Figure 5.4). Setting $v = v_2 - v_1$, $\delta = \rho_2 - \rho_1$ and substituting in the formula, we have that

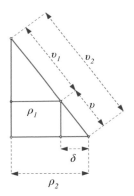

Fig. 5.4: Section of truncated cone

$$\begin{aligned} v_2\rho_2 - v_1\rho_1 &= (v_1 + v)(\rho_1 + \delta) - v_1\rho_1 \\ &= v\rho_1 + v_1\delta + v\delta \\ &= v\rho_1 + \left(v\frac{\rho_1}{\delta}\right)\delta + v\delta \\ &= v(\rho_1 + \rho_1 + \delta) \\ &= v(\rho_1 + \rho_2). \end{aligned}$$

The substitution in the middle equality follows from the similarity of the corresponding right triangles (See Figure 5.4). The half sum $\rho = \frac{1}{2}(\rho_1 + \rho_2)$ is the radius of the middle section of the truncated cone and as such we have the following proposition.

Proposition 5.1. *The area of the lateral surface of the truncated cone is given by the formula*

5.1. AREAS IN SPACE

$$\varepsilon = 2\pi \cdot \rho \cdot \upsilon,$$

where ρ is the radius of its middle section and υ is the length of its generator.

Corollary 5.1. *The area of the lateral surface of a cylinder, a cone and a truncated cone is given by the formula*

$$\varepsilon = 2\pi\rho\upsilon,$$

where υ is the length of the generator and ρ is the radius of the middle section of the solid.

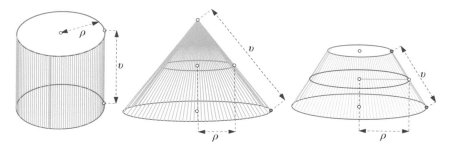

Fig. 5.5: Area of lateral surface, universal formula

Proof. The corollary simply unifies the specific results of the three preceding propositions, relying on the observation that, for the cone the middle section is the circle with radius half that of the base, while for the cylinder the middle section is a circle congruent to that of the base (See Figure 5.5).

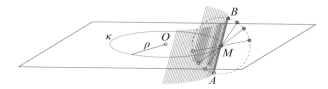

Fig. 5.6: Area of lateral independent of the inclination of AB

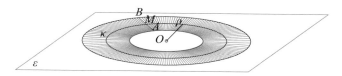

Fig. 5.7: Annulus when AB is contained in ε

Remark 5.1. Last formula shows that the line segment AB rotated about its middle M generates the lateral surface of a truncated cone with the same area, independent of its inclination to the plane of the rotation circle κ (Figure 5.6). When AB is orthogonal to the plane of κ, then the lateral surface is that of a cylinder. When AB is on the plane of κ and smaller than the radius of κ, then this generates a circular plane annulus (See Figure 5.7).

Corollary 5.2. *The area of the lateral surface of a cylinder, cone as well as of a truncated cone is given by the formula*

$$\varepsilon = 2\pi\sigma\eta,$$

where σ is the length of the vertical segment at the middle of the generator up to the axis and η is the altitude of the solid (See Figure 5.8).

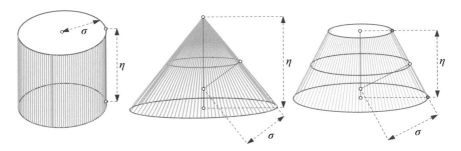

Fig. 5.8: Area of lateral surface, universal formula II

Proof. Proof by the figure. In the case of the cylinder the corollary is identical to the preceding one. In the cases of the cone and the truncated cone, from the similarity of the right triangles with sides (ρ, σ) and (η, υ) respectively (See Figure 5.9) follows

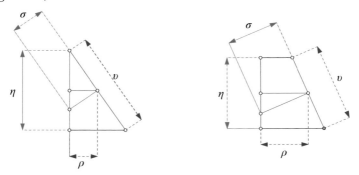

Fig. 5.9: Area of lateral surface, alternative universal formula

$$\frac{\rho}{\sigma} = \frac{\eta}{\upsilon} \quad \Rightarrow \quad \rho\upsilon = \sigma\eta.$$

Substituting $\rho\upsilon$ in the formula of the preceding corollary we prove the claim.

5.1. AREAS IN SPACE

Remark 5.2. If we use the "unfolding" terminology, mentioned in § 4.8, § 4.9, and specifically the fact that the process of unfolding preserves lengths and areas, the formulas, which were proved here with the usage of limits, are reduced to simple plane area calculations. Figure 5.10 shows the unfoldings of a cone, a cylinder and a truncated cone and their areas. In all three cases

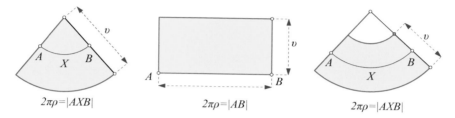

Fig. 5.10: Area through unfolding

v is the length of the generator and ρ is the radius of the middle section and is found from the corresponding equality already mentioned above. In it the right side is, in the cases of the cone and the truncated cone, the length of the arc AXB and in the case of the cylinder the length of the line segment AB.

Exercise 5.1. Show that from all parallepipeds inscribed in a given cylinder the one with square base has the maximum possible lateral area.

Hint: Use Exercise I-2.148.

Exercise 5.2. Find the cylinder with the lateral surface of maximal area which can be inscribed in a sphere.

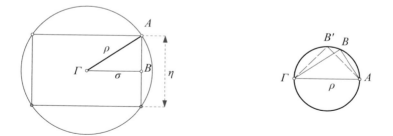

Fig. 5.11: Inscribed cylinder of maximal lateral area

Hint: The quantity $\sigma \cdot \frac{\eta}{2}$ is maximized when $\sigma = \frac{\eta}{2}$ (See Figure 5.11).

Exercise 5.3. Show that given two planes α and β and a triangle $AB\Gamma$ of plane α, such that its side $B\Gamma$ belongs to the intersection of α and β, the area of the projection $A'B\Gamma$ of $AB\Gamma$ on β is

$$\varepsilon(A'B\Gamma) = \cos(\omega)\varepsilon(AB\Gamma),$$

where ω is the angle of the planes α and β (See Figure 5.12-I).

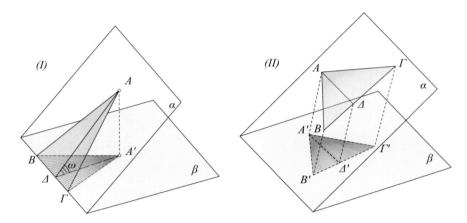

Fig. 5.12: Projection of a triangle to another plane

Exercise 5.4. Show that the preceding formula holds also when the triangle $AB\Gamma$ has its side $B\Gamma$ parallel to the intersection of α and β.

Hint: Apply Corollary 4.2 and transfer $AB\Gamma$ to the plane α in a parallel manner to itself, such that its edge $B\Gamma$ coincides with the intersection of α and β.

Exercise 5.5. Show that the preceding formula holds for every triangle $AB\Gamma$ of the plane α and its projection $A'B'\Gamma'$ on the plane β. That is

$$\varepsilon(A'B'\Gamma') = \cos(\omega)\varepsilon(AB\Gamma),$$

where ω is the angle of the planes α and β (See Figure 5.12-II).

Hint: From one vertex of $AB\Gamma$ for example A draw the parallel $A\Delta$ to the intersection of α and β. Apply the preceding exercise to the two triangles $AB\Delta$ and $A\Gamma\Delta$ and add or subtract, depending on whether Δ is inside or outside of $B\Gamma$.

Exercise 5.6. Show that for every convex polygon $AB\Gamma\Delta...$ of the plane α and its projection $A'B'\Gamma'\Delta'...$ on the plane β the following relation holds for their areas:
$$\varepsilon(A'B'\Gamma'\Delta'...) = \cos(\omega)\varepsilon(AB\Gamma\Delta...),$$
where ω is the angle of the planes α and β.

Hint: From one of the vertices of $AB\Gamma\Delta...$ draw all its diagonals and decompose it into triangles. Then, apply the preceding exercise and add.

5.2 Area of the sphere

> The first was to never accept something as real, unless I know it obviously as real.
>
> *Descartes, Discourse on the Method, b' part*

The area of the sphere is calculated through the areas of solids which result from inscribed regular polygons. Here we think that the sphere is generated through the rotation of a circle κ about one of its diameters AE. An approximation of the sphere is produced through the rotation of a regular polygon inscribed in the circle κ. Let us denote again with Π_1, Π_2, Π_3, ... the regular polygons we used also in § 1.2. They begin from the square (Π_1) inscribed in the circle κ and each polygon is inscribed in κ and has double the number

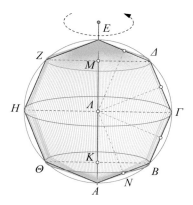

Fig. 5.13: Approximation of the surface of sphere

of sides of its predecessor, while all have the diagonal AE in common (See Figure 5.13), which is also a diameter of the circle κ. From the rotation of one such polygon about the diameter AE, is produced a surface consisting of cones and truncated cones.

The lateral surface of these cones and truncated cones was calculated in the preceding section and, according to the formulas which were proved there (Corollary 5.2), the total surface in the case of the octagon Π_2 is (See Figure 5.13):

$$\varepsilon_2 = (2\pi \cdot \eta_2) \cdot (|AK| + |K\Lambda| + |\Lambda M| + |ME|) = (2\pi \cdot \eta_2) \cdot |AE| = (2\pi \cdot \eta_2)(2\rho)$$

where $\eta_2 = |\Lambda N|$ is the distance of the side of the octagon (Π_2) from the center of the circle, which is equal to the radius of the inscribed circle of the regular polygon and ρ is the radius of the circle κ in which the polygons are inscribed. We admit then that the area of the surface of the sphere is the limit of the sequence which results in this way:

$$\varepsilon_1 = (4\pi\rho)\eta_1, \ \varepsilon_2 = (4\pi\rho)\eta_2, \ \varepsilon_3 = (4\pi\rho)\eta_3, \ldots.$$

The fact that this sequence converges follows from the fact that the sequence of radii
$$\eta_1, \eta_2, \eta_3, \ldots$$
of the circles inscribed to the polygons Π_ν (which in turn are inscribed in the circle with diameter $|AE| = 2\rho$) converges to ρ. Consequently the limit of these areas ε_ν is
$$\varepsilon = 4\pi\rho^2.$$

Thus, after several assumptions relative to what we call area of sphere, we arrive at the proposition.

Theorem 5.3. *The area of the surface of the sphere of radius ρ is $\varepsilon = 4\pi\rho^2$.*

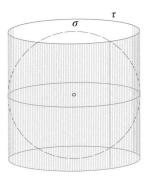

Fig. 5.14: Sphere and enveloping cylinder

Corollary 5.3. *A sphere σ has the same area with the lateral surface of its enveloping cylinder τ (See Figure 5.14).*

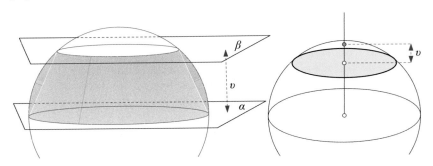

Fig. 5.15: Spherical zones and their area

We call **spherical zone/cap** the part of the sphere which is included between two parallel planes which intersect it (or are tangent to the sphere) (See Figure 5.15). The distance υ between the planes is called **height** of the spherical

5.2. AREA OF THE SPHERE

zone. The area of the spherical zone is defined, similarly to that of the sphere, as a limit of areas of solids inscribed in the spherical zone. We consider the spherical zone as being generated through the rotation of an arc $\widehat{AB}$ of the circle κ about a diameter XY of the circle, which does not intersect the arc (See Figure 5.16). The area of the surface produced is defined as the limit of the sum of lateral areas of the truncated cones which result through division of the arc into ν equal parts and considering the corresponding polygonal lines with vertices at the dividing points. As first polygonal line P_1 we take the line segment AB, which through the rotation generates a truncated cone. The second polygonal line results by considering the middle Γ of the arc and the polygonal line $A\Gamma B$. Continuing this way, we produce a sequence of polygonal lines $P_1, P_2, P_3, \ldots$ which approximate the arc $\widehat{AB}$. Each polygonal line has double the number of sides of its predecessor. As with the sphere, so also here, for each polygon P_ν is defined the corresponding radius η_ν of the inscribed circle. The lateral surface of the generated solid follows exactly as in the sphere

$$\varepsilon_\nu = 2(\pi \eta_\nu) \cdot |A'B'| = (2\pi \upsilon)\eta_\nu.$$

As we saw previously, the radii η_ν converge to the radius ρ of the circumscribed circle and, consequently, to that of the generated sphere, so we proved the following proposition.

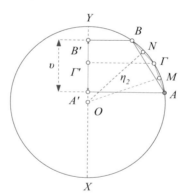

Fig. 5.16: Definition of area of spherical zone

Theorem 5.4. *The area of a spherical zone of height υ in a sphere of radius ρ is equal to*

$$\varepsilon = 2\pi \cdot \rho \cdot \upsilon,$$

consequently it is independent of the position of the zone on the sphere and depends only on its height.

Exercise 5.7. The sphere $\Sigma(r)$ is intersected with a sphere $\Sigma'(R)$, which passes through its center Σ. Show that the spherical zone defined on Σ' and contained in the interior of Σ has area independent of the radius R.

5.3 Area of spherical polygons

> Truth is so obscure in these times, and falsehood so well established, that, unless we loved the truth, we could not know it.
>
> *Pascal, Pensees*

Having the area of the sphere, we can take a quick look at the elementary shapes which are defined on its surface and their areas. Things are analogous to those of the plane. Besides, on a huge sphere (relative to our size), like Earth, it is difficult to distinguish the real line from the great circle of the sphere. As on the plane, so on the sphere the area is defined first for **spherical polygons**, in other words polygons on the sphere, whose sides are arcs of great circles (See Figure 5.17). Spherical polygons, in analogy to spherical triangles, are created on the sphere, by placing the apex of a polyhedral angle at the center of the sphere and considering the intersections of its faces with the sphere. These intersections define the **sides** of the spherical polygon. The dihedrals of the polyhedral angle define the **angles** of the spherical polygon. Two spherical polygons are called **congruent**, when they have respective sides equal and respective angles equal. For simplicity we limit ourselves to **convex spherical polygons**, in other words to polygons, whose sides (arcs of great circles), when extended, leave the entire polygon on one side (to one of the two hemispheres defined by the great circle-carrier of the side). The three first properties we require of the area, are the same

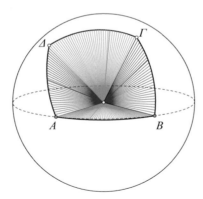

Fig. 5.17: Convex spherical quadrilateral $AB\Gamma\Delta$

with those of the area on the plane (I-§ 3.1). The big difference is the last property. No matter how big the sphere Σ, whose radius we suppose is ρ, contrary to the Euclidean plane, it has always finite area.

5.3. AREA OF SPHERICAL POLYGONS

Property 5.1 *Two congruent polygons have equal areas.*

Property 5.2 *A polygon Π consisting of other, finite in number, polygons Π', Π'', Π''', ..., which do not overlap, has area the sum of the areas of the polygons*

$$\varepsilon(\Pi) = \varepsilon(\Pi') + \varepsilon(\Pi'') + \ldots.$$

Property 5.3 *A polygon Π, contained in another Π', has area*

$$\varepsilon(\Pi) < \varepsilon(\Pi').$$

Property 5.4 *The area of the entire sphere is $4\pi\rho^2$.*

Remark 5.3. For these properties holds also the analogue of remark I-3.1, which we made for the area of plane polygons (I-§ 3.1). The results of this section will show, that, as in the Euclidean plane, properties 5.1 - 5.4, which we accept for the area, determine completely the area formula for spherical triangles (Theorem 5.6) and more generally the area of spherical polygons.

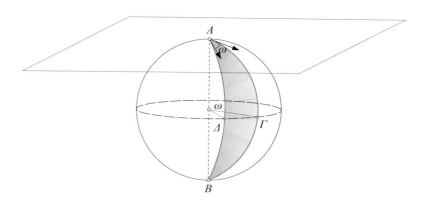

Fig. 5.18: Spherical lune of angle ω

Theorem 5.5. *The area of a spherical lune T of angle ω (in radians) is*

$$\varepsilon(T) = \omega(2\rho^2).$$

Proof. (See Figure 5.18) If the angle of the spherical lune $T = AB\Gamma\Delta$ is of the form $\omega = \frac{2\pi}{\nu}$, where ν is a positive integer, then the proposition holds. This follows from the additive Property 5.2 and Property 5.4 which we require of area. According to them, placing ν such lunes successively, we cover the entire sphere and we have

$$4\pi\rho^2 = \varepsilon(\Sigma) = \nu\varepsilon(T) \quad \Rightarrow \quad \varepsilon(T) = \frac{2\pi}{\nu}2\rho^2.$$

Similarly if the angle is a rational multiple of 2π the proposition holds. More specifically if
$$\omega = \frac{\mu}{\nu}(2\pi), \ \mu\varepsilon \ 0 < \frac{\mu}{\nu} < 1$$
and the quotient of the positive integers μ and ν is in lowest terms, then again holds
$$\varepsilon(T) = \left(\frac{\mu}{\nu}2\pi\right)(2\rho^2).$$
This, because the angle is written as a sum of μ angles of measure $\frac{2\pi}{\nu}$ and consequently the corresponding spherical lune T is written as a union of μ successive spherical lunes of area, according to the preceding step, $\frac{2\pi}{\nu}(2\rho^2)$. Consequently in this case also
$$\varepsilon(T) = \mu\frac{2\pi}{\nu}(2\rho^2) = \omega(2\rho^2).$$
For the general case, we use Lemma I-3.3 in the same way we used it for plane areas. Suppose then that the angle ω is of the form
$$\omega = \theta(2\pi), \ \text{with} \ 0 < \theta < 1.$$
According to the aforementioned lemma, for a given positive integer ν, there exists suitable integer μ such that
$$\frac{\mu}{\nu} < \theta < \frac{\mu+1}{\nu}.$$
The corresponding spherical lunes T_1, T, T_2 with angles
$$\omega_1 = \frac{\mu}{\nu}(2\pi), \ \omega = \theta(2\pi), \ \omega_2 = \frac{\mu+1}{\nu}(2\pi),$$
placing them in such a way so that their first side is coincident, will be contained one inside the other and we'll have
$$\varepsilon(T_1) = \frac{2\pi\mu}{\nu}(2\rho^2) < \varepsilon(T) < \varepsilon(T_2) = \frac{2\pi(\mu+1)}{\nu}(2\rho^2).$$
Let us then suppose that
$$\delta = \varepsilon(T) - (2\pi\theta)(2\rho^2),$$
is a positive number. Then the following inequality results

5.3. AREA OF SPHERICAL POLYGONS

$$\delta = \varepsilon(T) - (2\pi\theta)(2\rho^2) < \varepsilon(T) - \left(2\pi\frac{\mu}{\nu}\right)(2\rho^2)$$
$$< \varepsilon(T_2) - \left(2\pi\frac{\mu}{\nu}\right)(2\rho^2)$$
$$= \left(2\pi\frac{\mu+1}{\nu}\right)(2\rho^2) - \left(2\pi\frac{\mu}{\nu}\right)(2\rho^2)$$
$$= \frac{2\pi}{\nu}(2\rho^2),$$

which holds for every ν. Latter result however is contradictory, because the last term can become smaller than δ, if we take suitable ν. Similarly, we show that the assumption $\delta < 0$ leads to a contradiction, therefore $\delta = 0$. This completes the proof of the theorem.

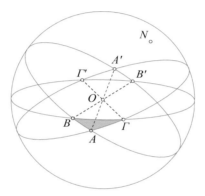

Fig. 5.19: Spherical triangle

Theorem 5.6. *The area $\varepsilon(T)$ of a spherical triangle T of the sphere $\Sigma(\rho)$ with (dihedral) angles α, β and γ is equal to*

$$\varepsilon(T) = (\alpha + \beta + \gamma - \pi)\rho^2.$$

Proof. The proof relies on the outstanding symmetry of the sphere. The spherical triangle $AB\Gamma$ defines through the diametrically opposite points of its vertices a congruent one and consequently having equal area (Property 5.1) triangle $A'B'\Gamma'$. The hemisphere Σ_N (with pole N in figure-5.19) which is defined from side (great circle) AB and contains the triangle I call occasionally *north* (and the other south), is covered completely from the spherical lunes of the angles of the triangle with some overlapping:
  1) The lune $BA\Gamma$ which contains the triangle.
  2) The lune $AB\Gamma$ which has common with the preceding lune the triangle $AB\Gamma$.
  3) The lune $A'B'\Gamma'$ which contains in the north hemisphere the triangle

$A'B'\Gamma'$ of equal area to $AB\Gamma$.
The cover of the north hemisphere then gives the equality of areas

$$\varepsilon(\Sigma_N) = 2\pi\rho^2 = [\alpha(2\rho)^2] + [\beta(2\rho^2) - \varepsilon(AB\Gamma)] + [\gamma(2\rho^2) - \varepsilon(AB\Gamma)].$$

Simplifying the equation we get the requested relation.

Corollary 5.4. *In every spherical triangle the sum of its angles is greater than π.*

The difference $(\alpha + \beta + \gamma - \pi)$, which according to the preceding discussion is positive, is called **spherical excess** of the triangle.

Corollary 5.5. *The area of a convex spherical polygon Π with angles $\alpha_1, \alpha_2, ..., \alpha_\nu$ is equal to*

$$\varepsilon(\Pi) = [(\alpha_1 + \alpha_2 + ... + \alpha_\nu) - (\nu - 2)\pi] \cdot (\rho^2).$$

Proof. If the polygon has ν vertices $A_1, A_2, ..., A_\nu$, choose one, for example vertex A_1, and omitting its adjacents $\{A_2, A_\nu\}$, join A_1 with $A_3, ..., A_{\nu-1}$ creating $(\nu - 2)$ triangles. The area of the polygon is the sum of the areas of the triangles and each one of them has area given by Theorem 5.6 and is of the form $(\alpha + \beta + \gamma - \pi)\rho^2$. Summing up all the angles of these various triangles we get the sum $(\alpha_1 + \alpha_2 + ... + \alpha_\nu)$, while from the $(\nu - 2)$ triangles we have the subtrahend $(\nu - 2)\pi$.

Remark 5.4. The congruence of spherical polygons defined on the surface of a sphere Σ of radius ρ includes the case where the two shapes are diametrically opposite, in other words they can be placed so that one can coincide with the diametrically opposite of the other. In this sense triangles $AB\Gamma$ and $A'B'\Gamma'$ are considered as congruent in the proof of Theorem 5.6.

5.4 Euler Characteristic

> I find that I never lose Bach. I don't know why I have always loved him so. Except that he is so pure, so relentless and incorruptible, like a principle of geometry.
>
> *Edna St Vincent Millay, letter 1920*

In this section we'll discuss an application of the formula of area of spherical polygons related to tilings of the sphere. A **tiling** of the sphere is a complete covering of it with tiles, that is, convex spherical polygons. As with real tiles, we require here also that there are no overlaps and further that the tiles have common vertices. Later means that if two tiles are adjacent, then they have common either an entire edge or some vertex. Euler's theorem says that in such a tiling the number of tiles, the total number of edges and the total number of vertices are numbers related to each other.

5.4. EULER CHARACTERISTIC

Theorem 5.7 (Euler's theorem). *In every tiling of the sphere with a number E of tiles, which have a total number A of edges and a total number K of vertices, holds*

$$K + E = A + 2.$$

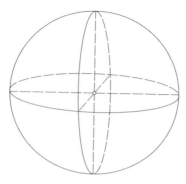

Fig. 5.20: Tiling of the sphere with eight equal tiles

Proof. The figure shows a simple case of tiling of the sphere with 8 congruent to each other triangular tiles, which are triply orthogonal (they have all three of their angles right). They correspond to a division of the sphere into 8 equal pieces, using three planes which are pairwise orthogonal and pass through the center. We have then $E = 8$, $K = 6$ and $A = 12$. Let us see now the general proof. We write the area of the sphere $4\pi\rho^2$ as a sum of the areas of the tiles

$$4\pi\rho^2 = \varepsilon_1 + \varepsilon_2 + ... + \varepsilon_E.$$

Each tile, i.e. spherical polygon, has area which is expressed with its angles and the number of its sides

$$\varepsilon_v = ((\alpha_1 + \alpha_2 + ...\alpha_\kappa) - (\kappa - 2)\pi)\rho^2.$$

And the total area will be a sum of such sums

$$4\pi\rho^2 = (\Sigma(\alpha) - \Sigma(\kappa\pi) + E(2\pi))\rho^2 \Leftrightarrow (2\pi)(2-E) = \Sigma(\alpha) - \pi\Sigma(\kappa).$$

The two terms to the right represent the sum of all the angles of all the tiles, the first ($\Sigma(\alpha)$) and the sum of all the numbers κ, which give the number of the sides of the tile, the second ($\Sigma(\kappa)$).

We consider the angles of all the tiles and we divide them into groups. In the same group we put all the angles from the different tiles which possess the same specific vertex. The angle-sum of such a group will be 2π, since all together cover completely the area around the specific vertex. We have K vertices, consequently the sum $\Sigma(\alpha)$ of all angles of all tiles will be $K(2\pi)$. The sum $\Sigma(\kappa)$ of all the numbers κ, which express the number of sides of each tile, will be $2A$, that is twice the number of edges. This, because each edge belongs exactly to two tiles, consequently it will be counted twice in the above sum. Consequently the last formula becomes

$$(2\pi)(2-E) = \Sigma(\alpha) - \pi\Sigma(\kappa) = (2\pi)K - \pi(2A) \Leftrightarrow 2 - E = K - A,$$

which is exactly the requested one.

The theorem of Euler is transferred to a similar property for convex polyhedra. The idea is to take a point O in the interior of a polyhedron and to project the polyhedron radially from O onto the surface of a great sphere which contains it completely. Then the edges of the polyhedron are projected to spherical polygons which in total give a tiling of the sphere. This way if E is the number of the faces of the polyhedron, A is the number of its edges and K the number of its vertices, this gives a tiling of the surrounding sphere with these elements as its tiles. Consequently applying the theorem of Euler we have

Theorem 5.8. *For every convex polyhedron with E faces, A edges and K vertices holds*
$$K + E = A + 2.$$

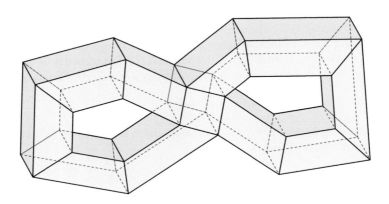

Fig. 5.21: Polyhedron with two holes

For every polyhedron Π, convex or not, the number $\chi(\Pi) = E - A + K$ is called **Euler characteristic** of the polyhedron ([81, p.236]). The preceding theorem implies, that for convex polyhedra the characteristic is 2. It is proved, that for every polyhedron this characteristic is of the form $\chi(\Pi) = 2 - 2\gamma$, where γ is the so called **genus** of the polyhedron ([81, p.256], [85, p.179]), which gives the number of its holes. This way in the polyhedral *eight* of figure 5.21, we have $E = 42$ faces, $A = 84$ edges and $K = 40$ vertices. The characteristic of this polyhedron then is $\chi(\Pi) = 42 - 84 + 40 = -2 = 2 - 2\gamma$, which verifies the fact that the polyhedron has $\gamma = 2$ holes.

5.5 Volumes

> The way of paradoxes is the way of truth. To test Reality we must see it on the tight-rope. When the Verities become acrobats we can judge them.
>
> *Oscar Wilde, The Picture of Dorian Gray*

The **volume** of the polyhedron Π is a number $o(\Pi)$ related to content, the part of the space which is enclosed by Π. By analogy to the plane areas, it is defined through four characteristic properties.

Property 5.5 *Two congruent polyhedra have equal volumes.*

Property 5.6 *A polyhedron Π composed of other, finite in number, polyhedra Π', Π'', ... which do not overlap has volume the sum of volumes of the polyhedra*

$$o(\Pi) = o(\Pi') + o(\Pi'') + \ldots.$$

Property 5.7 *A polyhedron Π contained inside another polyhedron Π' has volume*

$$o(\Pi) < o(\Pi').$$

Property 5.8 *The volume of the cube with side equal to the unit length is 1.*

From these properties we deduce immediately the formula for the volume of rectangular parallelepipeds. The process of proving this well known formula, relying on these properties, is exactly the same as that of the proof of the formula for the area of the rectangle (I-§ 3.2).

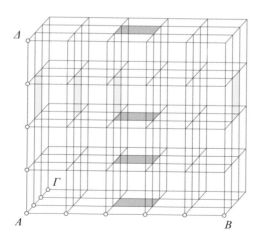

Fig. 5.22: Division of cube into $\mu \cdot \nu \cdot \xi$ parallepipeds

Lemma 5.1. *We divide the sides of the unit (side of length 1) cube AB, AΓ, AΔ respectively into μ, ν and ξ equal parts and we draw from the divison points parallel planes to the planes of the other sides. This creates $\mu \cdot \nu \cdot \xi$ parallepipeds, each having area $\frac{1}{\mu \cdot \nu \cdot \xi}$ (See Figure 5.22).*

Proof. Obviously there result $\mu\nu\xi$ congruent parallepipeds, which according to axiom 5.5, will have the same volume O. According to axiom 5.6, the volume of the cube which is 1 (Property 5.8) will be the sum of the volumes, or $1 = \mu \cdot \nu \cdot \xi \cdot O$.

Lemma 5.2. *The volume $o(\Sigma)$ of a rectangular parallelepiped Σ with sides AB, AΓ, AΔ, which have lengths rational numbers, is equal to the product of the side-lengths $o(\Sigma) = |AB||A\Gamma||A\Delta|$.*

Proof. We suppose that the rationals

$$|AB| = \frac{\alpha}{\beta}, \quad |A\Gamma| = \frac{\gamma}{\delta}, \quad |A\Delta| = \frac{\varepsilon}{\zeta}$$

are irreducible and we divide the unit cube, as in the preceding lemma, into $\beta \cdot \delta \cdot \zeta$ congruent rectangular parallelepipeds, each having volume $\frac{1}{\beta\delta\zeta}$. By assumption, side AB is divided into α equal segments of length $\frac{1}{\beta}$, side $A\Gamma$ into γ equal segments of length $\frac{1}{\delta}$ and $A\Delta$ into ε equal segments of length $\frac{1}{\zeta}$. By drawing from the divison points parallel planes to the planes of the other edges, we divide the parallelepiped into $(\alpha\gamma\varepsilon)$ congruent parallelepipeds, each having volume $\frac{1}{\beta\delta\zeta}$. The sum of the volumes of these parallelepipeds is equal to the volume of the original parallelepiped (Property 5.6) and is $\frac{\alpha\gamma\varepsilon}{\beta\delta\zeta}$.

Theorem 5.9. *The volume of the rectangular parallelpiped Π with edge lengths α, β and γ is (See Figure 5.23-I)*

$$o(\Pi) = \alpha \cdot \beta \cdot \gamma.$$

Proof. As we did with areas (I-§ 3.2), so also here we apply Lemma I-3.3 to the numbers α, β and γ and a fairly large positive integer ν. According to the lemma, there will be positive integers ρ, σ and τ so that it holds

$$\left|\alpha - \frac{\rho}{\nu}\right| < \frac{1}{\nu}, \quad \left|\beta - \frac{\sigma}{\nu}\right| < \frac{1}{\nu}, \quad \left|\gamma - \frac{\tau}{\nu}\right| < \frac{1}{\nu}.$$

From these follows that the initial rectangular parallelepiped Π contains in its interior Π_1, with side lengths $\frac{\rho}{\nu}, \frac{\sigma}{\nu}, \frac{\tau}{\nu}$, and is contained in the rectangular parallelepiped Π_2 with side lengths $\frac{\rho+1}{\nu}, \frac{\sigma+1}{\nu}, \frac{\tau+1}{\nu}$. It follows that, as much the volume $o(\Pi)$, as also the product $(\alpha\beta\gamma)$, will satisfy the same inequalities

5.5. VOLUMES

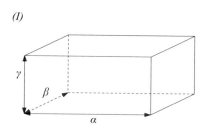 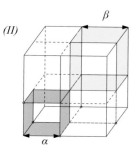

Fig. 5.23: Rectangular parallelepiped $o(\Pi) = \alpha\beta\gamma$ Identity for cubes

$$\frac{\rho}{\nu}\frac{\sigma}{\nu}\frac{\tau}{\nu} \leq o(\Pi) \leq \frac{\rho+1}{\nu}\frac{\sigma+1}{\nu}\frac{\tau+1}{\nu}, \quad \frac{\rho}{\nu}\frac{\sigma}{\nu}\frac{\tau}{\nu} \leq \alpha\beta\gamma \leq \frac{\rho+1}{\nu}\frac{\sigma+1}{\nu}\frac{\tau+1}{\nu}.$$

Consequently their difference will satisfy the inequality

$$\begin{aligned}|o(\Pi) - \alpha\beta\gamma| &\leq \frac{\rho+1}{\nu}\frac{\sigma+1}{\nu}\frac{\tau+1}{\nu} - \frac{\rho}{\nu}\frac{\sigma}{\nu}\frac{\tau}{\nu}\\ &= \frac{\rho\sigma+\sigma\tau+\tau\rho+\rho+\sigma+\tau+1}{\nu^3}\\ &= \frac{1}{\nu}\left(\frac{\rho}{\nu}\frac{\sigma}{\nu}+\frac{\sigma}{\nu}\frac{\tau}{\nu}+\frac{\tau}{\nu}\frac{\rho}{\nu}\right) + \frac{1}{\nu^2}\left(\frac{\rho}{\nu}+\frac{\sigma}{\nu}+\frac{\tau}{\nu}\right) + \frac{1}{\nu^3}\\ &\leq \frac{1}{\nu}(\alpha\beta+\beta\gamma+\gamma\alpha) + \frac{1}{\nu^2}(\alpha+\beta+\gamma) + \frac{1}{\nu^3}\\ &\leq \frac{(\alpha\beta+\beta\gamma+\gamma\alpha)+(\alpha+\beta+\gamma)+1}{\nu}.\end{aligned}$$

This inequality shows that the fixed quantity $|o(\Pi) - \alpha\beta\gamma|$ cannot be positive, because it is less than a number, which can become as small as we want, provided we choose a sufficiently large ν. Consequently this quantity will be zero.

Corollary 5.6. *The volume of a cube of edge α is equal to α^3.*

Exercise 5.8. Show with the help of a cube of edge $\alpha + \beta$ that the following identity holds (See Figure 5.23-II):

$$(\alpha+\beta)^3 = \alpha^3 + 3\alpha^2\beta + 3\alpha\beta^2 + \beta^3.$$

Exercise 5.9. Show with the help of the preceding figure and the identity $\alpha = (\alpha - \beta) + \beta$ that the following identity holds:

$$(\alpha-\beta)^3 = \alpha^3 - 3\alpha^2\beta + 3\alpha\beta^2 - \beta^3.$$

Exercise 5.10. Show that the middles of the sides of two opposite faces of a cube define a rectangular parallelepiped. Compute the volume of this, in dependenc of the side of the cube.

Exercise 5.11. Starting from a vertex of a cube, divide each edge starting at this vertex in parts proportional to the numbers $\{\kappa, \lambda, \mu\}$. Compute the volume of the parallelepipeds formed by drawing parallel planes to the faces from the division points.

Exercise 5.12. Given is a rectangular parallelepiped with side lengths a, b, and c. To compute as functions of $\{a, b, c\}$ the (minimal) distances of a diagonal and of each of the skew sides to this. To compute also the (minimal) distances of the diagonal and the skew to this diagonals of the faces.

Exercise 5.13. The rectangular parallelepiped of dimensions $\{a > b > c\}$ is divided in n equal parts along the big dimension and put the resulting parallelepipeds one on top of the other, so that pairwise they have in common a face of the same dimensions. To find the way to pile the parts, so that the resulting parallelepiped has maximal/minimal area.

5.6 Volume of prisms

> ... for he who seeks for methods without having a
> definite problem in mind, seeks for the most part in vain.
>
> David Hilbert, *Mathematical problems*

With the help of the volume of the rectangular parallelepiped and further use of the basic properties of volumes (§ 5.5), we can find the volume of right prisms and next also the volume of oblique prisms.

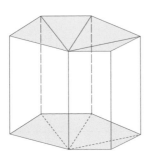 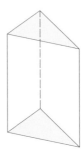 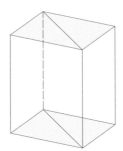

Fig. 5.24: Volume of right prism = area of base × altitude

Proposition 5.2. *The volume of a right prism Π is equal to the product of the area of its base times the altitude of the prism.*

5.6. VOLUME OF PRISMS

Proof. The proof for the general prism reduces to the triangular prism. By drawing the diagonals from one vertex of the base to the other vertices we divide the prism into triangular prisms (See Figure 5.24). It suffices then to show that for the triangular prism the volume is given by the product of the area of the triangle times the altitude. To show the latter it suffices to show that for every prism with base a parallelogram the corresponding property holds, since every triangular prism is half of such a prism. However the prism with base the parallelogram $ABΓΔ$ has the same volume and the same area of base as the other prism which is produced in the known way by drawing for example from $Δ$ the orthogonal $ΔE$ and from $Γ$ the orthogonal

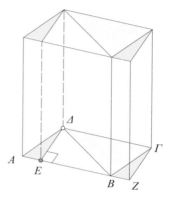

Fig. 5.25: Volume of triangular prism = area of base × altitude

$ΓZ$ and forming the rectangle $ΔEZΓ$ (See Figure 5.25). The volume of the prism on top of the triangle $AEΔ$ is transferred to that of the prism on top of the congruent triangle $BZΓ$, consequently the volume of the prism with base $ABΓΔ$ is the same with the volume of the prism with base the rectangle $ΔEZΓ$ and for this we already know that its volume is equal to the area of the base times the altitude.

A similar reduction of the calculation of volume can be applied also for oblique prisms. We can divide the polygonal base there again in triangles, double the triangles into parallelograms and reduce the parallelograms to rectangles. This reasoning shows, that to prove the formula for the volume of all oblique prisms, it suffices to prove it for the prisms with base a rectangle. For these however the following proposition holds.

Proposition 5.3. *The oblique prism with base the rectangle $ABΓΔ$ and the right with the same base and same altitude have the same volume.*

Proof. The proof uses the trick we use for areas. We compare the area of a rectangle and a parallelogram with the same base and same altitude using the congruence of triangles $AΔE$ and $BΓZ$ (See Figure 5.26), from which follows the equality of areas of the quadrilaterals $AHΓΔ$ and $BHEZ$. We do

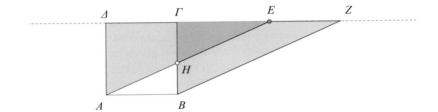

Fig. 5.26: Parallelograms with equal area

something similar for volumes. If $\Pi_1 = AB\Gamma\Delta - NO\Pi P$ is the right prism with base the rectangle $AB\Gamma\Delta$ and $\Pi_2 = AB\Gamma\Delta - IK\Lambda M$ is the oblique prism with the same base and same altitude, then parallel sides (PN, MI) and (NO, IK) define a sort of intermediate prism $\Pi_3 = AB\Gamma\Delta - EZH\Theta$ having

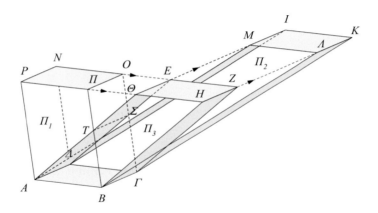

Fig. 5.27: Prisms of equal volume

also the same base (See Figure 5.27). Π_1 and Π_3 have the same volume, because the triangular prism with base $A\Theta P$ is congruent to the triangular prism with base $BH\Pi$, therefore if we subtract from the two congruent prisms their common part which is the triangular prism $T\Theta\Pi$ then the remaining prisms with bases the trapeziums ΠPAT and ΘTBH have the same volume. Adding to them also the triangular prism with base TBA we have the equality of volumes of Π_1 and Π_3. Similarly follows also the equality of volumes of Π_3 and Π_2 and in total the equality of volumes of Π_1 and Π_2.

Theorem 5.10. *The volume of a prism (right or oblique) is equal to the area of its base times its altitude.*

5.6. VOLUME OF PRISMS

Proof. Follows from the preceding propositions and subdivision in triangular prisms.

Proposition 5.4. *The volume of a prism (right or oblique) is equal to the area of its orthogonal section times the length of its generator.*

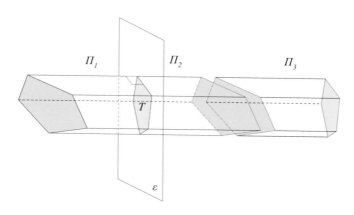

Fig. 5.28: Surgery of prisms

Proof. We use a sort of surgery on the prism (See Figure 5.28). We cut it with an orthogonal to its generator plane ε, and we divide it into two pieces Π_1 and Π_2. We next translate Π_1 to the same side as Π_2 and we adjust it so that the parallel faces of the initial prism coincide. This forms a right prism with base the orthogonal section T of the initial and length of generator the same with that of the initial prism. The proof is completed by applying Proposition 5.2.

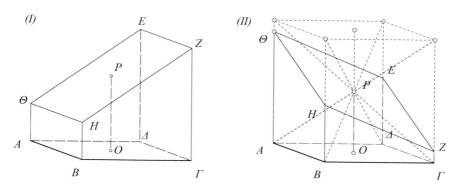

Fig. 5.29: Oblique section of a symmetric prism

Exercise 5.14. A prism whose base is a rectangle $AB\Gamma\Delta$ is intersected with a plane parallel to the edge AB (See Figure 5.29-I). Show that the intersection is a parallelogram ΘHZE and the volume of the solid Σ produced is equal to the product of the area of the base $\varepsilon(AB\Gamma\Delta)$ times the length $v = |OP|$, where O, P are respectively the intersection points of the diagonals of the two parallelograms $AB\Gamma\Delta$ and ΘHZE.

Hint: Take a copy Σ' of Σ, turn it upside down and attach it to Σ so that you have coincidence of $E' = \Theta$, $Z' = H$, $H' = Z$ and $\Theta' = E$.

Exercise 5.15. Examine if the volume formula of the preceding exercise is valid also for any plane section of the prism (See Figure 5.29-II). Examine finally, if the formula is valid for any right prism with a base $AB\Gamma\Delta...$, which is a convex point-symmetric polygon.

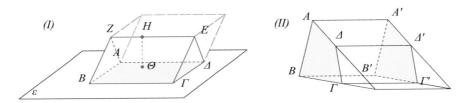

Fig. 5.30: Volume of triangular prism

Exercise 5.16. Show that the volume of a triangular prism is $\frac{1}{2}\varepsilon(AB\Gamma\Delta)|H\Theta|$, where $\varepsilon(AB\Gamma\Delta)$ is the area of one face of the prism and $|H\Theta|$ is the distance of this face from the opposite edge (See Figure 5.30-I).

Exercise 5.17. Find a formula similar to that of the preceding exercise for the volume of a prism which has an orthogonal section equal to a convex quadrilateral $AB\Gamma\Delta$ (See Figure 5.30-II).

Exercise 5.18. On a prismatic surface, whose orthogonal section is a square, we consider the points $\{X, Y\}$ of two opposite edges, of whose the distance is d. We consider also all the planes passing through these points and intersecting the prismatic surface along a parallelogram. To find the plane for which the corresponding parallelogram has the minimal area.

Exercise 5.19. Of all the right prisms with the same height and the same volume and polygonal base of n sides, to find the one which has the minimal lateral surface.

5.7 Volume of pyramids

> ... Geometry will draw the soul to the truth
> and will create the philosophical spirit...
>
> Plato, Republic VII-527

The simplest pyramid is the one with a triangular base. However, even in this case we have some difficulties trying to express its volume with the help of a formula. Next figure shows how we can use prisms (whose volume we know how to measure) to approximate the volume of a pyramid.

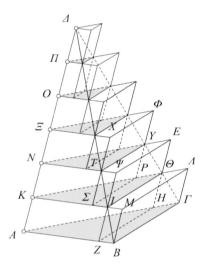

Fig. 5.31: Approximation of pyramid with prisms

Proposition 5.5. *Let Π be a triangular pyramid of altitude υ with base $AB\Gamma$ of area ε. Then for every positive number ν there are solids Π_1 and Π_2 which are unions of prisms and their volumes $o(\Pi_1)$, $o(\Pi_2)$ as well as the volume of the pyramid $o(\Pi)$ satisfy the inequalities*

$$o(\Pi_1) < o(\Pi) < o(\Pi_2), \quad \kappa\alpha\iota \quad o(\Pi_2) - o(\Pi_1) = \frac{\varepsilon \cdot \upsilon}{\nu}.$$

Proof. Divide one edge to the apex Δ of the pyramid, e.g. $A\Delta$ into ν equal parts and draw parallel planes to the base (See Figure 5.31). From the intersection points of these planes with the other edges draw parallels to the chosen edge $A\Delta$. These constructions form prisms with bases these parallel planes, from which, some are entirely in the interior of the pyramid (like $AZH\Theta IK$), while others include part of the pyramid in their interior (like

$AB\Gamma \Lambda KM$). The contained in the pyramid prisms I call *internal* and the others *external*. Π_1 is the union of the internal prisms and Π_2 is the union of the external prisms. Π_1 is contained completely in Π and consequently (Property 5.7) has

$$o(\Pi_1) < o(\Pi).$$

Π_2 contains the entire pyramid in its interior and therefore

$$o(\Pi) < o(\Pi_2).$$

The critical observation here is, that the two solids consist of congruent prisms except the bottom one. Indeed each prism, which is part of Π_1 has a congruent one which is part of Π_2. For example (See Figure 5.31), the interior $AZH\Theta KI$ is congruent to the exterior above it $NTY\Phi\Xi X$ etc. It follows that the difference of volumes between the two solids is the volume of the bottom prism with base $AB\Gamma$, therefore $o(\Pi_2) - o(\Pi_1) = \frac{\varepsilon v}{v}$.

Another critical observation concerns the ratio of volumes between two solids like Π_1 which are constructed for two different triangular pyramids with the same altitude. The ratio of their volumes translates into a ratio of areas between their bases.

Proposition 5.6. *Suppose that, for the given triangular pyramids $\Pi = AB\Gamma\Delta$ and $\Pi' = A'B'\Gamma'\Delta'$ with the same altitude the corresponding solids Π_1, Π_2 and Π'_1, Π'_2, are constructed, as in the preceding proposition. Then the ratios of volumes will be*

$$\frac{o(\Pi_1)}{o(\Pi'_1)} = \frac{o(\Pi_2)}{o(\Pi'_2)} = \frac{\varepsilon(AB\Gamma)}{\varepsilon(A'B'\Gamma')}.$$

Proof. Let us see the proof for the second ratio. The first is proved equally easily. The volume of Π_2 is the sum of the volumes of the v prisms which constitute it. The first at the bottom has volume $o_1 = \varepsilon \cdot \frac{v}{v}$, where $\varepsilon = \varepsilon(AB\Gamma)$ is the area of $AB\Gamma$ and v is the (common) altitude of the pyramids.

The critical observation is that the prism below has a base which is similar to $AB\Gamma$, suppose with ratio κ. The volume of the second prism therefore will be $o_2 = (\kappa^2 \varepsilon) \cdot \frac{v}{v} = \kappa^2 o_1$. Similarly $o_3 = \lambda^2 o_2 = \lambda^2 \kappa^2 o_1$ etc. Summing over the v volumes of prisms we have

$$o(\Pi_2) = o_1 + \ldots + o_v = o_1 + \kappa^2 o_1 + \kappa^2 \lambda^2 o_1 + \ldots = (1 + \kappa^2 + \kappa^2 \lambda^2 + \ldots)o_1 = \mu \cdot o_1.$$

Number μ, which represents the sum of the similarity factors, depends only on v and not on the base triangle. Indeed, from the similarity of the triangles (See Figure 5.31) ΔAB and ΔKI, follows that

$$\kappa = \frac{|KI|}{|AB|} = \frac{|\Delta K|}{|\Delta A|} = \frac{v-1}{v}.$$

Similarly follows that

5.7. VOLUME OF PYRAMIDS

$$\lambda = \frac{|NT|}{|KI|} = \frac{\nu-2}{\nu-1} \Rightarrow \lambda\kappa = \frac{\nu-2}{\nu}.$$

Thinking similarly we find then that

$$\mu = (1 + \kappa^2 + \kappa^2\lambda^2 + \ldots)$$
$$= 1 + \left(\frac{\nu-1}{\nu}\right)^2 + \left(\frac{\nu-2}{\nu}\right)^2 + \ldots + \left(\frac{1}{\nu}\right)^2$$
$$= \frac{1}{\nu^2}(1 + 2^2 + 3^2 + \ldots + \nu^2),$$

which shows that μ depends only on ν and not on the special triangle. By the way, even though we do not need it, it does not hurt to know that ([81, p.14])

$$1 + 2^2 + 3^2 + \ldots + \nu^2 = \frac{\nu(\nu+1)(2\nu+1)}{6} \Rightarrow \mu = \frac{(\nu+1)(2\nu+1)}{6\nu}.$$

Returning then to the proof, the ratios of volumes between the two solids will be

$$\frac{o(\Pi_2)}{o(\Pi'_2)} = \frac{\mu \cdot o_1}{\mu \cdot o'_1} = \frac{o_1}{o'_1} = \frac{\varepsilon \cdot \frac{\nu}{\nu}}{\varepsilon' \cdot \frac{\nu}{\nu}} = \frac{\varepsilon}{\varepsilon'}.$$

Corollary 5.7. *For two pyramids Π, Π' with bases of equal area and equal altitude the volume of the respective solids satisfies: $o(\Pi_1) = o(\Pi'_1)$ and $o(\Pi_2) = o(\Pi'_2)$.*

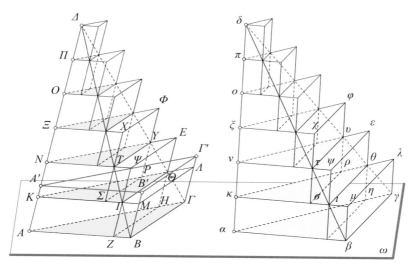

Fig. 5.32: Comparison of pyramids through approximations

Theorem 5.11. *Two triangular pyramids of the same altitude, which have bases with the same area, have the same volume.*

Proof. Place the two pyramids $\Pi = AB\Gamma\Delta$ and $\Pi' = \alpha\beta\gamma\delta$ with their base on the same plane ω (See Figure 5.32). Suppose that the pyramids have the same altitude v and their bases have the same area ε. Suppose that the first $AB\Gamma\Delta$ has greater volume and holds

$$o(AB\Gamma\Delta) > o(\alpha\beta\gamma\delta).$$

we'll show that this is contradictory. For this, construct the prism $\Pi_d = AB\Gamma\Gamma'B'A'$ with base $AB\Gamma$, which has volume equal to the difference $d = o(AB\Gamma\Delta) - o(\alpha\beta\gamma\delta)$. Next construct for the two pyramids Π, Π' the solids Π_1, Π_2, Π'_1, Π'_2 (of Proposition 5.5) for number v so that $\frac{v}{v}$ will be less than the altitude of the prism Π_d. Then also the respective volume of the base prism will be

$$o(AB\Gamma\Lambda KM) < o(\Pi_d).$$

On the other hand the difference between volumes of the two pyramids will be

$$o(\Pi_d) = o(AB\Gamma\Delta) - o(\alpha\beta\gamma\delta) < o(\Pi_2) - o(\Pi'_1) = o(\Pi_2) - o(\Pi_1) = o(AB\Gamma\Lambda KM).$$

An inequality which contradicts the preceding one. We get a similar contradiction also from the assumption that $o(AB\Gamma\Delta) < o(\alpha\beta\gamma\delta)$. Consequently it must hold $o(AB\Gamma\Delta) = o(\alpha\beta\gamma\delta)$.

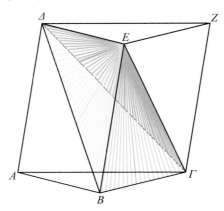

Fig. 5.33: Volume of pyramid: $o(AB\Gamma\Delta) = \frac{1}{3}\varepsilon(AB\Gamma) \cdot v$

Theorem 5.12. *The volume of the triangular pyramid $AB\Gamma\Delta$ is given by the formula*

$$o(AB\Gamma\Delta) = \frac{1}{3}\varepsilon(AB\Gamma)v,$$

where $\varepsilon(AB\Gamma)$ is the area of its base and v its altitude.

Proof. From the vertices B and Γ of the base draw parallels and equal to the edge $A\Delta$ and form the triangular prism $AB\Gamma Z\Delta E$ (See Figure 5.33). draw

5.7. VOLUME OF PYRAMIDS

next the diagonal ΓE. The created prism is divided into three pyramids: the initial $AB\Gamma\Delta$, $B\Gamma\Delta E$ and $\Gamma\Delta EZ$. These three pyramids have the same volume. $B\Gamma\Delta E$ and $\Gamma\Delta EZ$ with bases the congruent triangles $B\Gamma E$ and $E\Gamma Z$ and common vertex at Δ have equal bases and the same altitude hence the same volume. But also $\Gamma\Delta EZ$ has base equal to the initial $AB\Gamma\Delta$ and equal altitude. Since the three pyramids have the same volume, each one of them will have $1/3$ the volume of the prism.

Remark 5.5. It is interesting to observe in the last figure the position of the pyramid $B\Gamma\Delta E$ in the triangular prism. Its opposite edges $B\Gamma$ and ΔE are skew and the planes $AB\Gamma$ and ΔEZ are the unique (Proposition 3.13) parallel planes which contain these two skew lines. The common orthogonal of these two skew lines has length the distance between these two planes that is the altitude of the prism $AB\Gamma\Delta EZ$. This observation leads to the next proposition.

Proposition 5.7. *The volume of the pyramid with opposite edges $B\Gamma$ and ΔE is equal to*

$$\frac{1}{6}|B\Gamma||\Delta E|v\sin(\phi), \tag{5.1}$$

where v is the length of the common orthogonal of the (skew lines) $B\Gamma$ and ΔE and ϕ is the angle between $B\Gamma$ and ΔE.

Note that, in formula 5.1 the product $\frac{1}{2}|B\Gamma||\Delta E|v\sin(\phi)$ represents the volume of the parallelepiped which is circumscribed to the tetrahedron (Exercise 4.22) and shows that the pyramid has volume $1/3$ of that parallelepiped.

Corollary 5.8. *Given two skew lines α and β as well as line segments $B\Gamma$ and ΔE of fixed length and sliding respectively on these lines, the tetraheder $B\Gamma\Delta E$ has volume which is independent of their position on α and β.*

Remark 5.6. This section, relying on a corresponding by Legendre ([84, p.134]), is interesting, among others, in that it uses all the characteristics of the notion of limit without actually using its analytic determination. If we wanted to use limits, the proof of the formula for the volume of the pyramid would have finished almost immediately after the first proposition of the section. Indeed Proposition 5.5 reveals that the volume of the solid Π_2 (as well as of Π_1) is an approximation of the volume of the pyramid with error less than $\frac{\varepsilon v}{v}$. Therefore, increasing v and passing to the limit we have the requested formula. The calculation in Proposition 5.6 shows next that the volume of Π_2 is equal to

$$o(\Pi_2) = \frac{\mu\varepsilon v}{v} = \frac{(v+1)(2v+1)}{6v^2}\varepsilon v.$$

The conclusion follows by considering the limit of the fraction

$$\frac{(v+1)(2v+1)}{6v^2} = \frac{1}{6}\left(1+\frac{1}{v}\right)\left(2+\frac{1}{v}\right) \rightarrow \frac{1}{3}.$$

The method we followed is the one of *exhaustion of Eudoxus and Archimedes* applied to the case of these simple solids ([79, p.34]).

Theorem 5.13. *The volume of a pyramid is one third the product of the area of its base times its altitude.*

Proof. From one vertex, for example A of the base of pyramid $AB\Gamma\Delta$... draw all the diagonals towards the other vertices (See Figure 5.34). The pyramid is decomposed into triangular pyramids with the same altitude and the conclusion follows by summing the volumes of all the triangular pyramids relying on the preceding theorem

Corollary 5.9. *The volume of a truncated pyramid of altitude υ and with area of larger base ε_1 and smaller base ε_2 is equal to*

$$o = \frac{1}{3}\upsilon \cdot (\varepsilon_1 + \varepsilon_2 + \sqrt{\varepsilon_1 \varepsilon_2}).$$

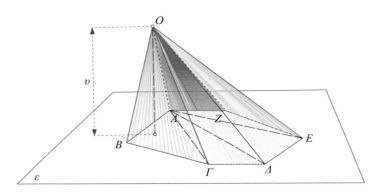

Fig. 5.34: Volume of pyramid

Proof. Follows directly from the preceding theorem, by considering the truncated pyramid as a difference (with respect to volume) of two pyramids with bases the two bases of the truncated pyramid (See Figure 5.35). This way, if η is the altitude of the pyramid with the smaller base, the wanted volume would be

$$o = \frac{1}{3}(\eta + \upsilon)\varepsilon_1 - \frac{1}{3}(\eta \varepsilon_2) = \frac{1}{3}(\upsilon \varepsilon_1 + \eta(\varepsilon_1 - \varepsilon_2)).$$

However the polygons on the two parallel planes are similar (Corollary 3.11), consequently the ratio of their areas will equal the ratio of the squares of their sides, which in turn is equal to the ratio of squares

$$\frac{\varepsilon_2}{\varepsilon_1} = \frac{\eta^2}{(\upsilon + \eta)^2} \quad \Rightarrow \quad \eta = \frac{\upsilon\sqrt{\frac{\varepsilon_2}{\varepsilon_1}}}{1 - \sqrt{\frac{\varepsilon_2}{\varepsilon_1}}}.$$

Substituting η in the preceding expression and performing calculations gives the wanted formula.

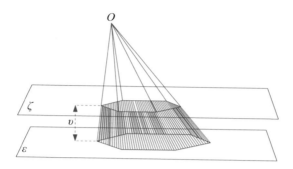

Fig. 5.35: Volume of truncated pyramid

Exercise 5.20. Show that the triangular pyramid, which has a triple orthogonal trihedral (whose three faces at the apex are right triangles) with edges of length δ has volume $\frac{\delta^3}{6}$.

Exercise 5.21. Show that the area of the lateral surface of a pyramid is greater than the area of its base.

Exercise 5.22. Show that for each convex polyhedron Π, circumscribed on sphere of radius r, the ratio $\lambda = \varepsilon(\Pi)/o(\Pi)$, of the area to its volume, depends only on the radius of the sphere.

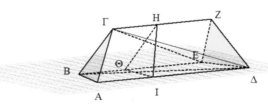

Fig. 5.36: Volume of truncated triangular prism

Exercise 5.23. Show that the volume of a *truncated* (§ 4.7) triangular prism $AB\Gamma\Delta EZ$ is equal to the sum of the volumes of the three pyramids which have as a base one triangle, for example $AB\Gamma$ and vertices, respectively, the three vertices of the other triangle ΔEZ.

Hint: Divide the prism into three pyramids $\pi_1 = AB\Gamma\Delta$, $\pi_2 = B\Delta E\Gamma$ and $\pi_3 = E\Delta\Gamma Z$ (See Figure 5.36). Moving the vertex Δ of π_2 parallel to its base

$B\Gamma E$, at A, we see that π_2 has volume equal to that of the pyramid $AB\Gamma E$. Moving the vertex E of π_3 parallel to its base $\Gamma\Delta Z$, at B, we see that it has volume equal to that of the pyramid $B\Gamma Z\Delta$. The last pyramid, similarly, has volume equal to that of the pyramid $B\Gamma AZ$ ([42, II, p.87]).

Exercise 5.24. Show that the volume of a *truncated* triangular prism $AB\Gamma\Delta EZ$ is equal to the product of the arithmetic mean of the lengths of its edges times the area of its orthogonal section (See Figure 5.36):

$$o(AB\Gamma\Delta EZ) = \frac{|A\Delta| + |BE| + |\Gamma Z|}{3} \cdot \varepsilon(H\Theta I).$$

5.8 Volume of cylinders

> You could not discover the limits of soul, even if you traveled every road to do so; such is the depth of its meaning.
>
> *Heraclitus*

The volume of a cylinder cannot be deduced directly from the basic properties of volume (§ 5.5), since these define completely the volume only for polyhedra. This way we resort again to limits and approximate the cylinder using inscribed in it prisms of the same height v as that of the cylinder. We limit ourselves to regular prisms inscribed in the cylinder and we define the **volume of the cylinder** as the limit of the volumes of the regular prisms inscribed in it, as their faces increase in number.

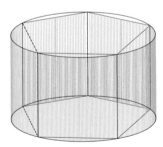

Fig. 5.37: Volume of cylinder

Theorem 5.14. *The volume of a cylinder is equal to the product of the area of its base times its altitude.*

Proof. For the calculation of the area we use the good old known sequence of regular polygons $\Pi_1, \Pi_2, \Pi_3,...$ (§ 1.2) which are inscribed in the base of the cylinder, beginning from the square (Π_1) and doubling in each step the number of vertices (See Figure 5.37). According to our calculations (Lemma

5.8. VOLUME OF CYLINDERS

1.5), the area of the inscribed in the circle of radius ρ polygon with μ sides is

$$\mu \left(\frac{\rho^2}{2} \sin\left(\frac{2\pi}{\mu}\right) \right).$$

It follows (Proposition 5.2) that the volume of the prism above this polygon with height v is

$$v\mu \left(\frac{\rho^2}{2} \sin\left(\frac{2\pi}{\mu}\right) \right).$$

Because the altitude remains fixed, it follows that the limit of the preceding expression, as $\mu \, (= 2^\nu)$ increases, will be the product of the height times the area of the circle.

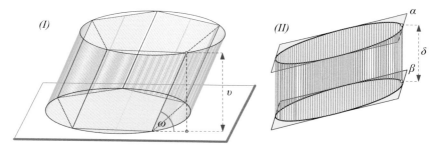

Fig. 5.38: Volume of oblique circular cylinder

Remark 5.7. The proof is valid also when the generators are not orthogonal to the plane of the circle of the base, but oblique. The only thing that changes in this case is the height of the cylinder, which is not coincident anymore with the length of the generator, but is given by the formula

$$v = \delta \sin(\omega),$$

where δ is the length of the generator of the oblique cylinder and ω is the angle formed between the generator and the plane of the base circle (See Figure 5.38-I). This way the preceding proposition, as it was written, holds as much for right circular cylinders as also for oblique ones.

Remark 5.8. The orthogonal section (using a plane orthogonal to the generators) of an oblique cylinder with base a circle *is not* a circle (it is an ellipse).

Exercise 5.25. We intersect a right circular cylindrical surface with two parallel planes $\{\alpha, \beta\}$ (See Figure 5.38-II). Compute the volume of the solid bounded by the cylindrical surface and the two planes in terms of the radius r of the cylindrical surface and the length δ of the generator between the planes.

Hint: Mimic the way of proposition 5.4.

5.9 Volume of cones

> ... the spirit of Geometry is ever present in almost the entire body of mathematics and it is a fundamental pedagogical mistake to try to eliminate it.
>
> *Rene Thom, Modern Mathematics. Do they exist?*

As with the cylinder, so also with the cone, the calculation of its volume is done by approximating the cone with inscribed regular pyramids. We have met the scenario several times. We inscribe regular polygons Π_1, Π_2, Π_3, ... in the circle of the base, beginning with the square and doubling the number of vertices successively for each next polygon. We define the **volume of the cone** as the limit of the volumes of the pyramids with base the polygons Π_ν and altitude the altitude of the cone.

Fig. 5.39: Volume of cone

Theorem 5.15. *The volume of the cone is equal to 1/3 the product of the area of its base times its altitude.*

Proof. The proof is the same as that of Theorem 5.14.

Remark 5.9. The proposition is true not only for right cones but also for oblique ones. The formulation of the preceding proposition is general enough to include these cases as well. The proof for the oblique cones is the same as that for the right ones.

The volume of a truncated cone is calculated by subtructing from the volume of the cone with the larger base the volume of the cone with the smaller base.

Proposition 5.8. *The volume of the truncated cone is equal to*

$$\frac{1}{3} \cdot \varepsilon \cdot Y \cdot (1 - \kappa^3),$$

where ε is the area of the larger base, Y is the altitude of the cone of the larger base and κ the ratio of the radius of the smaller to that of the smaller base.

5.9. VOLUME OF CONES

 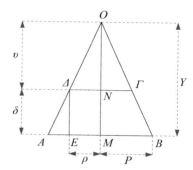

Fig. 5.40: Volume of truncated cone

Proof. If the ratio of the radii is κ, then the ratio of areas of the smaller circle to the larger is κ^2 (See Figure 5.40). Also the ratio of the altitude of the smaller cone to the altitude of the larger is (because of the similarity of triangles) also κ. The formula follows immediately writing the smaller altitude $\upsilon = \kappa Y$ and the smaller volume $\varepsilon' = \kappa^2 \varepsilon$ and substituting them into the formula of the difference of volumes between the two cones.

Next proposition gives the volume of the truncated cone as a function of its defining characteristics, which are the two radii of its bases P, ρ and its altitude δ.

Proposition 5.9. *The volume o of the truncated cone with larger base radius P, smaller base radius ρ and altitude δ is*

$$o = \frac{\pi \delta}{3}(P^2 + \rho^2 + P\rho).$$

Proof. The proof rests on the basic formula for the volume of a truncated cone, as a difference between the volumes of the two cones

$$o = \frac{\pi}{3}(P^2 Y - \rho^2 \upsilon),$$

where Y and υ are the altitudes of the cones with the larger and smaller base respectively. Figure-5.40 shows the characteristics of the triangle, which results by intersecting the truncated cone with a plane passing through its axis OM. The intersection is an isosceles trapezium $AB\Gamma\Delta$, which, by extending its sides leads to the isosceles ABO. From the similarity of the triangles $A\Delta E$ and ΔON we have

$$\frac{P-\rho}{\delta} = \frac{\rho}{\upsilon} \Rightarrow \upsilon = \frac{\rho \delta}{P-\rho}.$$

Also using the equation

$$Y = \upsilon + \delta,$$

we have

$$o = \frac{\pi}{3}(P^2 Y - \rho^2 v) = \frac{\pi}{3}(P^2(v+\delta) - \rho^2 v)$$
$$= \frac{\pi}{3}\left(P^2\left(\frac{\rho\delta}{P-\rho} + \delta\right) - \rho^2 \frac{\rho\delta}{P-\rho}\right)$$
$$= \frac{\pi}{3}\left(P^2\left(\frac{P\delta}{P-\rho}\right) - \rho^2 \frac{\rho\delta}{P-\rho}\right)$$
$$= \frac{\pi\delta}{3}\left(\frac{P^3 - \rho^3}{P-\rho}\right)$$
$$= \frac{\pi\delta}{3}(P^2 + \rho^2 + P\rho).$$

5.10 Volume of spheres

> The careless and thoughtless youth imagines that the world exists in order to be enjoyed; that it is the abode of a positive happiness; and that men miss this because they are not clever enough to take possession of it. .... The undeceiving comes too late.
>
> A. Schopenhauer, *Counsels and maximes*

As with the cylinder and the cone, so also with the sphere, the calculation of its volume is done by approximating it with the volume of solids which result from regular polygons. An approximation of a sphere of radius ρ is

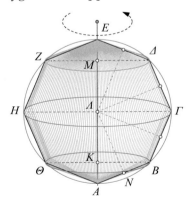

Fig. 5.41: Approximation of sphere

produced by rotating a regular polygon inscribed in the circle κ of radius ρ. Let us denote again by $\Pi_1, \Pi_2, \Pi_3, \ldots$ the regular polygons, which we used also in § 1.2. They begin with the square (Π_1) inscribed in the circle κ and

5.10. VOLUME OF SPHERES

each one of the others is also inscribed in κ and has double the number of sides of its predecessor. From the rotation of one such polygon about the diameter AE, which we suppose coinciding with one diagonal of the polygon, a solid is produced, the volume of which is used to approximate that of the sphere (See Figure 5.41). Next propositions lead to the calculation of this volume as well as the limit these volumes tend to, as the number of the corresponding polygon-sides increases.

Proposition 5.10. *The volume of the solid, which results through rotation of the triangle $AB\Gamma$ about its base AB, is equal to 1/3 the volume of the cylinder which results through rotation of the rectangle $AB\Delta\ E$ with the same base and the same altitude with the triangle.*

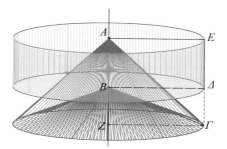

Fig. 5.42: Solid created through rotation of a triangle about a side

Proof. Let Z be the projection of the vertex Γ to the axis of rotation AB (See Figure 5.42). The solid produced from the triangle $AB\Gamma$ has volume the difference of volumes between the cones $A\Gamma Z$ and $B\Gamma Z$. The two cones have in common the base circle with radius $|Z\Gamma|$ and area $\varepsilon = \pi |Z\Gamma|^2$ and the difference of their volumes is

$$\frac{1}{3}\varepsilon |AZ| - \frac{1}{3}\varepsilon |BZ| = \frac{1}{3}\varepsilon |AB|.$$

Proposition 5.11. *The volume of the solid, which results through rotation of a triangle $AB\Gamma$ about an axis passing through its vertex A and not intersecting its base $B\Gamma$, is equal to 2/3 its area times the perimeter of the circle described by the middle M of its base.*

Proof. Let Δ be the intersection of $B\Gamma$ with the axis of rotation (See Figure 5.43). The volume o, produced by the triangle $AB\Gamma$, is equal to the difference of volumes produced by triangles $A\Delta\Gamma$ and $A\Delta B$ rotated about their common base $A\Delta$. According to the preceding proposition this volume is 1/3 the volume of the cylinder with the same base and same altitude of the respective triangles, therefore

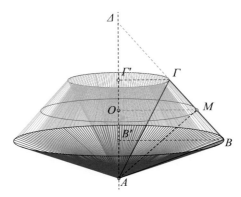

Fig. 5.43: Solid created through rotation of a triangle about a vertex

$$\begin{aligned}
o &= \frac{1}{3}(\pi |BB'|^2 |A\Delta|) - \frac{1}{3}(\pi |\Gamma\Gamma'|^2 |A\Delta|) \\
&= \frac{\pi |A\Delta|}{3}(|BB'|^2 - |\Gamma\Gamma'|^2) \\
&= \frac{\pi |A\Delta|}{3}(|BB'| - |\Gamma\Gamma'|)(|BB'| + |\Gamma\Gamma'|) \\
&= \frac{\pi}{3}(|A\Delta|(|BB'| - |\Gamma\Gamma'|))(|BB'| + |\Gamma\Gamma'|) \\
&= \frac{\pi}{3}(2\varepsilon(AB\Gamma))(2|OM|) \\
&= \frac{2}{3}\varepsilon(AB\Gamma)(2\pi|OM|).
\end{aligned}$$

Corollary 5.10. *The volume o of the solid, which results through rotation of an isosceles triangle $AB\Gamma$ about an axis passing through its vertex A and not intersecting its base $B\Gamma$ is equal to*

$$o = \frac{2}{3}\pi |AM|^2 |B'\Gamma'|,$$

where M is the middle of $B\Gamma$ and B', Γ' are the projections of B, Γ onto the axis of rotation (See Figure 5.44).

Proof. According to the preceding proposition, the volume of the aforementioned solid will be

$$o = \frac{2}{3}\varepsilon(AB\Gamma)(2\pi|OM|).$$

When $AB\Gamma$ is isosceles ($|AB| = |A\Gamma|$), then AM is orthogonal to $B\Gamma$ and the right triangles AMO and $N\Gamma'B'$, with N onto BB', such that $|BN| = |\Gamma\Gamma'|$, are similar. It follows that

$$\frac{|OM|}{|AM|} = \frac{|B'\Gamma'|}{|N\Gamma'|} = \frac{|B'\Gamma'|}{|B\Gamma|} \;\Rightarrow\; |OM| = \frac{|AM||B'\Gamma'|}{|B\Gamma|}.$$

5.10. VOLUME OF SPHERES

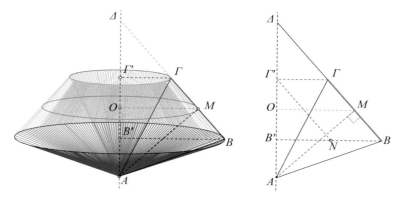

Fig. 5.44: Solid created through rotation of an isosceles triangle

Using this formula, the volume of the solid is written

$$o = \frac{2}{3}\varepsilon(AB\Gamma)(2\pi|OM|)$$
$$= \frac{2}{3}\left(\frac{1}{2}|B\Gamma||AM|\right)(2\pi|OM|)$$
$$= \frac{2}{3}\pi|AM|^2|B'\Gamma'|.$$

Remark 5.10. It is easy to see, that the formula of the last corollary holds also in the case where $B\Gamma$ does not intersect the axis of rotation, in other words when $B\Gamma$ is parallel to the axis $A\Delta$.

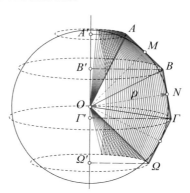

Fig. 5.45: Calculation of volume of solid by rotation

Proposition 5.12. *Consider the polygon consisting of the radii OA, $O\Omega$ and part of another polygon $AB\Gamma\Delta E...\Omega$ with equal sides and inscribed in the circle κ. Rotate this about an axis passing through the center O of κ and not intersecting the polygon. Then the solid produced has volume*

$$o = \frac{2}{3}\pi\rho^2 |A'\Omega'|,$$

where A', Ω' are the projections of A and Ω onto the axis of rotation and ρ is the radius of the inscribed circle of the polygon (See Figure 5.45).

Proof. The proof results by applying the preceding corollary to each one of the triangles OAB, $OB\Gamma$, $O\Gamma\Delta$, ... and summing up the formulas for the produced volumes.

Spherical sector is called the solid produced through rotation of a circular sector of circle κ about a line XY of the plane passing through the center O of the circle and not intersecting the circular sector.

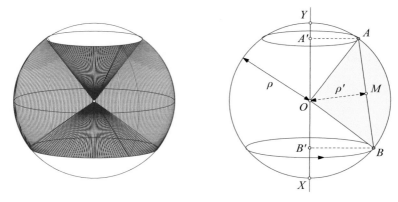

Fig. 5.46: Calculation of volume of spherical sector

Corollary 5.11. *The volume of the spherical sector of arc $\widehat{AB}$ of circle κ and radius ρ, which is rotated about a diameter of κ not intersecting the sector, is equal to*

$$o = \frac{2}{3}\pi\rho^2 |A'B'|,$$

where A' and B' are the projections of the endpoints of the arc $\widehat{AB}$ onto the axis of rotation (See Figure 5.46).

Proof. As volume of the aforementioned solid we consider the limit of the volumes produced in the way described in the preceding proposition from polygons $OAB\Gamma\Delta...\Omega$, which are inscribed in the circle κ, as the number of equal sides of the polygon $AB\Gamma\Delta...\Omega$ are increased. Then the radius ρ' of the inscribed to the polygon circle tends to the radius ρ of the circumscribed circle κ and the formula follows from the formula of the preceding proposition by substituting in ρ the radius of the circumscribed circle.

Theorem 5.16. *The volume of a sphere of radius ρ is*

5.10. VOLUME OF SPHERES

$$o = \frac{4}{3}\pi\rho^3.$$

Proof. We think of the sphere as generated by a circumference of radius ρ which is rotated about its diameter AE. In this circle we inscribe the polygons $\Pi_1, \Pi_2, \ldots$ such that AE becomes a diagonal. Let o_ν be the volume of the solid Σ_ν, which results by rotating Π_ν about the diagonal AE and suppose that ρ_ν is the radius of the circle inscribed to Π_ν. According to Proposition 5.12, we'll have (because $|A'E'| = |AE| = 2\rho$)

$$o_\nu = \frac{2}{3}\pi\rho_\nu^2|A'E'| = \frac{2}{3}\pi\rho_\nu^2|AE| = \frac{4}{3}\pi\rho\rho_\nu^2.$$

The volume of the sphere is the limit of the o_ν and this depends, according to the preceding formula, from the limit of the radii ρ_ν of the inscribed circles. As ν increases the radius of the inscribed circle ρ_ν tends to the radius of the circumscribed circle ρ, hence the result.

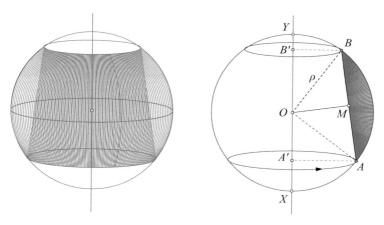

Fig. 5.47: Calculation of volume of spherical ring

Spherical ring is the solid which is enclosed between a truncated cone inscribed in the sphere and the sphere. It results from the rotation of a circular segment of circle κ about a diameter of circle XY not intersecting the circular segment.

Corollary 5.12. *The volume of the spherical ring produced from arc $\widehat{AB}$ of the circle of center O and radius ρ is*

$$o = \frac{2}{3}\pi|A'B'|(\rho^2 - |OM|^2) = \frac{1}{6}\pi|A'B'||AB|^2,$$

where A', B' are the projections of the endpoints of the arc $\widehat{AB}$ onto the axis of rotation and M is the middle of the arc (See Figure 5.47).

Proof. The proof results by calculating the difference $o_1 - o_2$ of two volumes. The first (o_1) is the volume of the solid which results from the rotation of the circular sector defined by the arc $\widehat{AB}$. The second (o_2) is the volume of the solid which results through rotation of the triangle OAB. The two volumes are calculated in corollaries 5.11 and 5.10 respectively.

Spherical zone is called the solid produced through rotation of the plane shape $AA'B'B$ defined by the arc $\widehat{AB}$ of the circle κ and the projections of its endpoints on a diameter XY not intersecting it. We rotate this shape about the diameter XY (See Figure 5.47). The solid produced can be also defined through the intersection of the sphere with two parallel planes at a distance $\upsilon = |A'B'|$. We consider the **volume of the spherical zone** as the sum of the volumes $o_1 + o_2$ of the truncated cone (o_1), produced from the rotation of the trapezium $AA'B'B$ and the volume of the spherical ring (o_2), produced from the rotation of the arc AB.

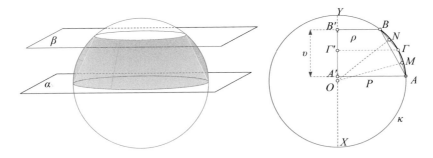

Fig. 5.48: Spherical zone

Proposition 5.13. *The volume of the spherical zone, produced from the arc $\widehat{AB}$ of the circle κ is equal to*

$$o = \frac{1}{2}\pi\upsilon(P^2 + \rho^2) + \frac{1}{6}\pi\upsilon^3,$$

where υ is the height of the spherical zone and P, ρ are the radii of the larger and smaller base respectively (See Figure 5.47).

Proof. The volume of the truncated cone is given in Proposition 5.9

$$o_1 = \frac{\pi\upsilon}{3}(P^2 + \rho^2 + P\rho),$$

where $\upsilon = |A'B'|$ is its altitude. Using also Corollary 5.12 we find that the volume o of the spherical zone is

5.10. VOLUME OF SPHERES

$$o = o_1 + o_2 = \frac{\pi v}{3}(P^2 + \rho^2 + P\rho) + \frac{1}{6}\pi v |AB|^2$$
$$= \frac{1}{6}\pi v (2(P^2 + \rho^2 + P\rho) + |AB|^2)$$
$$= \frac{1}{6}\pi v (2P^2 + 2\rho^2 + 2P\rho + (P-\rho)^2 + v^2)$$
$$= \frac{1}{6}\pi v (3P^2 + 3\rho^2 + v^2)$$
$$= \frac{1}{2}\pi v (P^2 + \rho^2) + \frac{1}{6}\pi v^3.$$

Remark 5.11. Let us note at this point, that all these calculations of volumes of solids, produced through rotation, are special cases of calculations of *integrals*, as they are defined in *Caclulus*. There it is developed a unified calculation method, which can be applied to many more solids than these we examined here. Spivak's Caclulus [86, p.326] gives all of our calculations and many more as exercises of application of the general methods of Calculus. These general methods however, are relying on the ideas we developed here and are due to Eudoxus and mainly to Archimedes.

Exercise 5.26. Show that for all polyhedra circumscribed in sphere of radius ρ (all their faces are tangent to the sphere) the ratio of their volume to their area is fixed and equal to $\frac{\rho}{3}$.

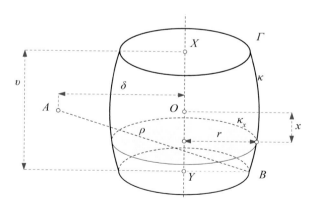

Fig. 5.49: Barrel

Remark 5.12. Often our endeavor, with simple questions, leads us to the limits of our knowledge and the need for transcendence. For example, the exact calculation of the volume (content) of a barrel, transcends the capabilities of this lesson. To face the problem we need knowledge of calculus. The problem is that the barrel (internal wall) is produced through rotation of an arc $\widehat{B\Gamma}$ of a circle κ of radius ρ. Unfortunately (or fortunately, for the chance to learn something new), however, the rotation of the arc $\widehat{B\Gamma}$ is not performed relative to an axis passing through the center A of the circle κ, but relative to another axis XY at a distance δ from the center of the arc (See Figure 5.49).

For those wishing to return here, after wandering a bit in calculus, the volume of the barrel is given by the integral ([88, p.269])

$$o = \pi \cdot \int_{-\frac{v}{2}}^{\frac{v}{2}} \left(\sqrt{\rho^2 - x^2} - \delta\right)^2 dx = \int_{-\frac{v}{2}}^{\frac{v}{2}} \varepsilon(\kappa_x) dx.$$

$r = \left(\sqrt{\rho^2 - x^2} - \delta\right)$ is the radius of the orthogonal to the axis circular section κ_x of the barrel, at distance $|x|$ from its center. $y = \pi \cdot r^2 = \varepsilon(\kappa_x)$ is the area of this section and the integral is the "sum" of all these areas, made precise with the use of limits.

From this method of volume calculations, results also the so called **Cavalieri principle** ([55, p. 383]). According to this, if two solids with the same height h, placed between two parallel planes $\{\alpha, \beta\}$ at distance h, have equiareal sections, for every plane γ parallel and lying between $\{\alpha, \beta\}$, then these solids have the same volume.

Note that Kepler (1571-1630) devoted an entire book on the subject ([83]), by calculating volumes of barrels made with different recipes, mainly by rotating arcs of conic sections around axes of their plane.

5.11 Comments and exercises for the chapter

> The major forms of beauty are order and symmetry and definiteness, which the mathematical sciences demonstrate in a special degree.
>
> *Aristotle, Metaphysics M 3,1.1078b*

Exercise 5.27. Show that the bisecting plane of a dihedral of a tetrahedron intersects the opposite edge at parts proportional to the areas of these two faces which define the faces of the dihedral.

Exercise 5.28. Given is the prismatic surface Σ, whose an orthogonal section is a square $AB\Gamma\Delta$. Is there an other plane section of the surface along a rhombus? How is determined such a section? Is it possible to find a plane such that the corresponding rhombus is similar to a given one? How is determined such a section?

Exercise 5.29. The tetrahedra $OAB\Gamma$, $OA'B'\Gamma'$ have in common the trihedral of the vertex O (See Figure 5.50-I). Show that the ratio of their volume is equal to the ratio $\frac{o(OA'B'\Gamma')}{o(OAB\Gamma)} = \frac{|OA'||OB'||OC'|}{|OA||OB||OC|}$.

5.11. COMMENTS AND EXERCISES FOR THE CHAPTER

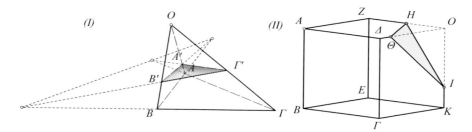

Fig. 5.50: Common trihedral vertex   Prism intersection

Exercise 5.30. In a rectangular parallelepiped with edges of length $a = |OZ|$, $b = |O\Delta|$ and $c = |OK|$, we define on them respectively points H, Θ, I, such that $\frac{|OH|}{|OZ|} = k_1$, $\frac{|O\Theta|}{|O\Delta|} = k_2$, $\frac{|OI|}{|OK|} = k_3$ (See Figure 5.50-II). Calculate the volume of the soild which remains if we delete the trihedral $OH\Theta I$.

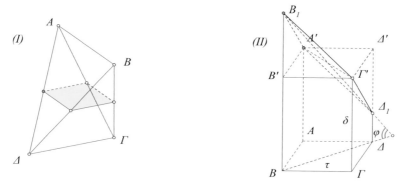

Fig. 5.51: Parallelogram intersection   Truncated square prism

Exercise 5.31. Show that a plane parallel to two opposite edges AB and $\Gamma\Delta$ of a tetrahedron intersects it along a parallelogram (See Figure 5.51-I). Also show that there exists one exactly such plane which intersects the tetrahedron along a rhombus. When does this rhombus become a square?

Exercise 5.32. Given is a prismatic surface Σ, whose one orthogonal section is a square $AB\Gamma\Delta$ of side τ. Let $A'B'\Gamma'\Delta'$ be a second orthogonal section parallel to the preceding one at distance δ. Draw a plane ε through $A'\Gamma'$, which intersects the plane of $AB\Gamma\Delta$ outside of this square by an angle ϕ. Calculate the area of the solid $AB\Gamma\Delta A'B_1\Gamma'\Delta_1$ as a function of τ, δ and ϕ (See Figure 5.51-II). Show that its volume is independent of ϕ.

Exercise 5.33. The tetrahedron $OAB\Gamma$ has the trihedral at O with all faces equal to ω and all edges running at O equal to x. Compute its volume in dependence of $\{\omega, x\}$ (See Figure 5.52-I).

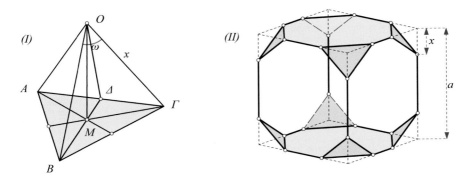

Fig. 5.52: Tetrahedron's volume   Truncated cube's volume

Exercise 5.34. From each vertex of a cube of edge-length a and on every edge, we define points at the same distance x and delete the resulting triply orthogonal tetrahedra. To find the area and the volume of the resulting solid (truncated cube) in dependence of $\{a,x\}$ (See Figure 5.52-II).

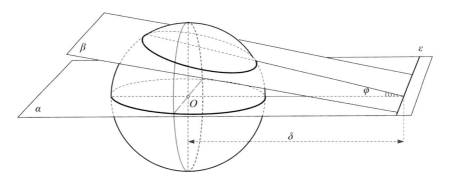

Fig. 5.53: Part of sphere between two planes

Exercise 5.35. The sphere $\Sigma(\rho)$ is intersected by two planes α, β passing through line ε at distance $\delta > \rho$ from its center. Plane α passes through the center of the sphere and forms with plane β a dihedral angle of measure ϕ. Find the volume and the area of the solid section of the sphere between the two planes (See Figure 5.53). Solve the same problem when plane α does not pass through the center of the sphere.

Exercise 5.36. Given are three spheres $\Sigma(O_1)$, $\Sigma(O_2)$, $\Sigma(O_3)$ of equal radii and pairwise externally tangent. Construct a hemisphere, which is tangent to all

5.11. COMMENTS AND EXERCISES FOR THE CHAPTER

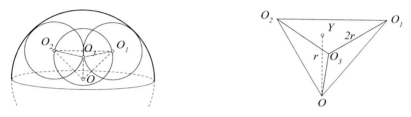

Fig. 5.54: Three spheres on a hemisphere

three, and such that the plane defining the hemisphere is also tangent to the spheres.

Hint: Calculate first the elements (edges) of the tetrahedron $OO_1O_2O_3$ which has as vertices the centers of the three spheres, as well as, the center of the requested hemisphere. The base of this tetrahedron is an equilateral with side $2r$, where r is the radius of the equal spheres, and its altitude is r (See Figure 5.54).

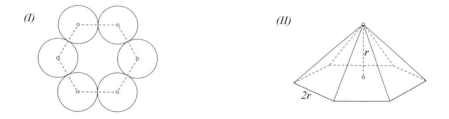

Fig. 5.55: N spheres in a hemisphere

Exercise 5.37. (Generalization of preceding) Given are N spheres of radius r, pairwise externally tangent, whose centers form a regular N-gon (See Figure 5.55). Construct a hemisphere which is tangent to all N spheres, and such that the plane defining the hemisphere is also tangent to the spheres.

Hint: Calculate first the elements of the pyramid with base a regular N-gon of side $2r$ and altitude r (See Figure 5.55-II).

Exercise 5.38. Calculate the area of the surface and the volume of the solid, which results from an isosceles trapezium with parallel sides of lengths $a < c$ and lateral sides of length b, onto which we have attached arcs of circles tangent to the lateral sides. The solid results by rotating the plane figure around its axis of symmetry (See Figure 5.56).

Exercise 5.39. Given is a line ε not intersecting the sphere Σ. Draw a plane passing through ε, which intersects the sphere into two parts, whose ratio of volumes/areas is λ.

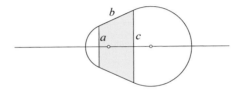

Fig. 5.56: Solid by rotation

Exercise 5.40. The tetrahedron $A\Gamma K\Lambda$ has two opposite edges on the diagonals of two opposite faces of a rectangular parallelepiped $AB\Gamma\Delta EZH\Theta$ (See

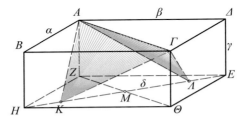

Fig. 5.57: Tetrahedron inside rectangular parallelepiped

Figure 5.57). One of its edges is coincident with the diagonal $A\Gamma$ and the other, of length δ, is contained in the diagonal HE and is symmetric relative to the center M of HE. Calculate the volume of the tetrahedron as a function of the lengths of the edges α, β, γ of the parallelepiped, as well as of the length δ.

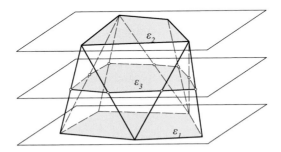

Fig. 5.58: Volume of a convex prism with two parallel faces

Exercise 5.41. Show that the volume of a convex prism whose vertices are contained in two parallel planes at a distance v (often such polyhedra are referred to as **prismatoids**) is equal to

5.11. COMMENTS AND EXERCISES FOR THE CHAPTER

$$o = \frac{v}{6}(\varepsilon_1 + \varepsilon_2 + 4\varepsilon_3),$$

where ε_1 and ε_2 are the areas of its parallel faces and ε_3 is the area of the its middle intersection, parallel to the planes (See Figure 5.58).

Hint: ([75, p.9]) Show first that the formula holds for tetrahedra. Subsequently divide the polyhedron into tetrahedra and add the volumes.

Exercise 5.42. Let ε be a line not intersecting the sphere Σ. Draw planes η_1, η_2,..., η_v passing through ε, which cut the sphere into $v+1$ parts which have the same volume.

Exercise 5.43. Show that the area ε of a spherical equilateral triangle of a sphere of radius ρ and its angle α are connected with the formula

$$\alpha = \frac{\varepsilon}{3\rho^2} + \frac{\pi}{3}.$$

Conclude that the angle of the spherical equilateral is always greater than 60 degrees (See also Exercise 4.76).

Platonic solid Σ	$\varepsilon(\Sigma)/a^2$	$o(\Sigma)/a^3$
Tetrahedron	$\sqrt{3}$	$\frac{1}{12}\sqrt{2}$
Cube	6	1
Octahedron	$2\sqrt{3}$	$\frac{1}{3}\sqrt{3}$
Dodecahedron	$3\sqrt{25+10\sqrt{5}}$	$\frac{1}{4}(15+7\sqrt{5})$
Icosahedron	$5\sqrt{3}$	$\frac{5}{12}(3+\sqrt{5})$

The preceding table records the ratio of the area of the surface $\varepsilon(\Sigma)/a^2$, as well as also the ratio of volume $o(\Sigma)/a^3$, respectively, to the square/cube of the edge of the Platonic solid.

Exercise 5.44. Show that the numbers of the first three lines of the preceding table are valid.

Exercise 5.45. Two spheres of different radii $\{\sigma_1(r_1), \sigma_2(r_2)\}$ are tangent externally and the cone κ is tangent to both. Compute the volume of the external domain of the spheres and internal to the cone included between the circles of contact of the spheres and the cone, in dependence of the radii $\{r_1, r_2\}$ (See Figure 5.59).

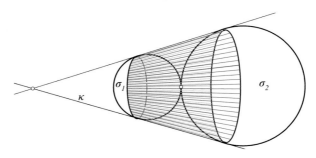

Fig. 5.59: Volume defined by two spheres

Exercise 5.46. Discuss the similar to the preceding problem for two spheres $\{\sigma_1(r_1), \sigma_2(r_2)\}$ with different radii, external to each other and the cone κ tangent to both. Compute the volume of the external domain of the spheres and internal to the cone included between the circles of contact of the spheres and the cone, in dependence of the radii $\{r_1, r_2\}$ and the distance d of the centers.

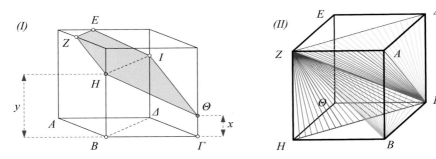

Fig. 5.60: Pentagonal section   Dihedral angle

Exercise 5.47. Let $AB\Gamma\Delta$ be the base of a cube. On the orthogonal edge from Γ we consider a line segment $\Gamma\Theta$ of length x (See Figure 5.60-I). From the point H of the orthogonal edge at B and at distance $|BH| = y$ we draw the parallel HI of the diagonal $B\Delta$. For some values of x, y the plane ΘHI intersects the cube along a pentagon. Given x, find for which other values of y this happens. Also calculate the area of the resulting pentagon and the volumes of the two parts of the cube.

Hint: One way is through Exercise 5.6.

Exercise 5.48. In a cube $AB\Gamma\Delta EZH\Theta$ draw the diagonals $\{A\Gamma, BZ\}$ (See Figure 5.60-II). Show that the dihedral angle between the planes $\{BZ\Gamma, ZH\Gamma\}$ are $60°$.

5.11. COMMENTS AND EXERCISES FOR THE CHAPTER

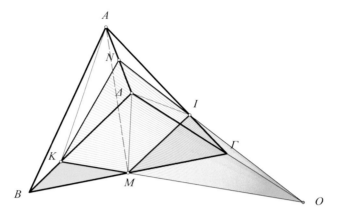

Fig. 5.61: Cutting with plane through the middles of opposite sides

Hint: The tetrahedra $\{B\Gamma ZH, AB\Gamma Z\}$ are symmetric relative to the plane $B\Gamma Z$ and the tetrahedra $\{AB\Gamma Z, A\Gamma\Delta Z\}$ are symmetric relative to the plane $A\Gamma Z$.

Exercise 5.49. The plane ε passes through the middles $\{M, N\}$ of the opposite edges of the tetrahedron $AB\Gamma\Delta$ (See Figure 5.61). Show that the intersection points $\{I, K\}$ defined on the two other edges, correspondingly, $\{A\Gamma, B\Delta\}$ divide them in proportional parts : $\frac{KB}{K\Delta} = \frac{I\Gamma}{IA}$.

Hint: Consider the intersection point O of the intersection plane with $\Delta\Gamma$ and apply the theorem of Menelaus (i) to the triangle $A\Delta\Gamma$ with secant line ON and (ii) to the triangle $AB\Gamma$ with secant line OM. It follows that

$$\frac{KB}{K\Delta} \cdot \frac{O\Delta}{O\Gamma} \cdot \frac{M\Gamma}{MB} = 1 = \frac{I\Gamma}{IA} \cdot \frac{O\Delta}{O\Gamma} \cdot \frac{NA}{N\Delta},$$

and the relation follows through simplification.

Exercise 5.50. Continuing the preceding exercise, show that a plane ε, passing through the middles $\{M, N\}$ of two opposite edges of the tetraheder $AB\Gamma\Delta$, divides it in two solids $\{\sigma_1, \sigma_2\}$ of equal volumes.

Hint: Each of the two solids $\{\sigma_1, \sigma_2\}$ is the union of a quadrangular pyramid τ_i and a tetrahedron δ_i. σ_1 consists of $\{\tau_1 = \Delta MINK, \delta_1 = \Delta M\Gamma I\}$ and σ_2 of $\{\tau_2 = AKMIN, \delta_2 = KBMA\}$. The pyramids $\{\tau_1, \tau_2\}$ have the same volume because they have the same base and the same height.

Tetrahedra $\{\delta_1, \delta_2\}$ have also equal volumes. This results by comparing first the volumes of the half of tetrahedra $o = o(\Delta ABM) = o(\Delta M\Gamma A)$.

$$\frac{o(\delta_1)}{o} = \frac{\varepsilon(\Delta I\Gamma)}{\varepsilon(\Delta\Gamma A)} = \frac{|I\Gamma|}{|A\Gamma|}, \qquad \frac{o(\delta_2)}{o} = \frac{\varepsilon(BMK)}{\varepsilon(BM\Delta)} = \frac{|BK|}{|B\Delta|}.$$

The conclusion results, using the preceding exercise, from the equality of the ratios of the segments ([32, p.241]).

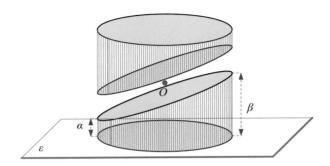

Fig. 5.62: Cylinder with oblique cut

Exercise 5.51. A right circular cylindrical surface of radius r is cut with a plane ε orthogonal to its axis and a plane ε' oblique to its axis, such that in the solid Σ, bounded by the cylinder and the two planes, the longest generator has length β and the shortest has length α (See Figure 5.62). Invent a method to calculate the lateral area of the solid as well as its volume in terms of $\{r, \alpha, \beta\}$. Show that these quantities do not change if $\{\alpha, \beta\}$ vary but their sum remains constant.

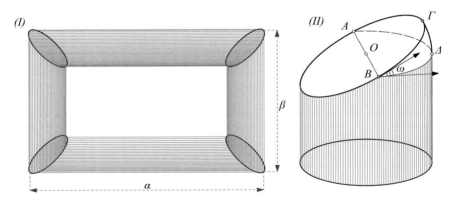

Fig. 5.63: Cylindrical frame

Exercise 5.52. Devise a method to construct a rectangular cylindrical frame of outer dimensions $\{\alpha, \beta\}$ (See Figure 5.63-I). The sides of the frame are parts of a cylindrical surface of radius r. Compute the total surface and volume of the frame in terms of $\{r, \alpha, \beta\}$.

Exercise 5.53. Suppose you would like to build a frame like the preceding one, but without these sharp corners. Devise a method to construct the same frame with rounded corners. You could use as a building element the one shown in figure 5.63-II. In this we use a solid Σ like the one of exercise 5.51, which we cut with a plane orthogonal to the axis at its intersection O with the oblique plane. Then we replace the upper cylindrical part with a spherical lune $AOB\Gamma\Delta$ (See Figure 5.63-II). Calculate the lateral surface and volume of such a solid building element, especially when the angle of the lune $\omega = 45°$. Then calculate the surface and volume of the corresponding frame with rounded corners. Note that, for the time being, we cannot compute the area of the oblique base of this building element, which consists of a half-circle and a half-ellipse glued together along the diameter AB of the half-circle.

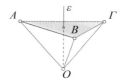

Fig. 5.64: Relation of inclinations relative to a triply orthogonal system

Exercise 5.54. Lines $\{OA, OB, O\Gamma\}$ form a triply orthogonal system at O, i.e. they are pairwise orthogonal. And line ε through O forms with these lines correspondingly the angles $\{\phi_A, \phi_B, \phi_\Gamma\}$. Show that

$$\cos^2(\phi_A) + \cos^2(\phi_B) + \cos^2(\phi_\Gamma) = 1 .$$

Hint: Intersect the lines with a plane $AB\Gamma$ orthogonal to ε (See Figure 5.64). By Pythagoras 3D theorem (4.9) $(AB\Gamma)^2 = (OAB)^2 + (OB\Gamma)^2 + (O\Gamma A)^2$. The triangles on the right side are the projections of $AB\Gamma$ on the planes spanned by two of the lines $\{OA, OB, O\Gamma\}$. According to exercise 5.6 the area of the projection of $AB\Gamma$ on the plane OAB is $(OAB) = \cos(\phi)(AB\Gamma)$, where ϕ the angle of the planes $AB\Gamma$ and ABO. But this is $\phi = \phi_\Gamma$, etc.

Exercise 5.55. Project a cube with edge of length d on a plane ε. Show that the sum of squares of the projections of its edges is $8d^2$.

Hint: Consider the three edges at a vertex A of the cube (See Figure 5.65). If the line ζ orthogonal to the projection plane ε makes angles $\{\phi, \chi, \psi\}$ with the edges of the cube then the projections of the three edges on ε have lengths $\{d\sin(\phi), d\sin(\chi), d\sin(\psi)\}$. Because of the relation of $\{\phi, \chi, \psi\}$ of the preceding exercise we have $\sin^2(\phi) + \sin^2(\chi) + \sin^2(\psi) = 2$. Hence for the edges of the cube at A, their projections squared sum up to $2d^2$. Projecting the edges at the 8 vertices of the cube we obtain analogously for the sum

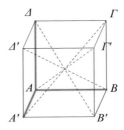

Fig. 5.65: Cube projection on a plane

of their squares $8 \cdot (2d^2) = 16d^2$. Since by this procedure each edge has been counted twice, the requested sum is actually $8d^2$.

Exercise 5.56. Is it possible that figure 5.65 represents actually the orthogonal projection of a cube on the plane of the paper?

References

79. M. Baron (1969) The Origins of the Infinitesimal Calculus. Dover, New Yorkk
80. C. Boyer (1991) A History of Mathematics, 2nd Edition. John Wiley, New York
81. R. Courant (1996) What is Mathematics. Oxford University Press, Oxford
82. T. Dantzig (1955) The bequest of the Greeks. George Allen and Unwin Ltd., London
83. J. Kepler (1596) Nova Stereometria doliorum vinariorum. Frankfurt
84. A. Legendre (1837) Elements de Geometrie suivis d' un traite de Trigonometrie. Langlet et compagnie, Bruxelles
85. D. Richeson (2008) Euler's gem. Princeton University Press, Princeton
86. M. Spivak (1994) Calculus, Third Edition. Publish or Perish, Houston
87. M. Spivak (1965) Calculus on Manifolds. Addison-Wesley, New York
88. V. Smirnov (1964) A course of higher Mathematics I,II,III. Pergamon Press, New York

Chapter 6
Conic sections

6.1 Conic sections

> If the Greeks had not cultivated Conic Sections, Kepler could not have superseded Ptolemy; if the Greeks had cultivated Dynamics, Kepler might have anticipated Newton.
>
> W. Whewell, *History of Inductive Science v.I, p.311*

Intersections of a right circular conical surface Σ with a plane ε, orthogonal to its axis and not passing through its apex O, are circles (Proposition 4.8). When the plane is not orthogonal to the axis of Σ, the resulting intersections are *ellipses, parabolas* and *hyperbolas*. The ellipses result, as we'll see, through intersections with planes cutting all the generators of Σ. An **ellipse** is also

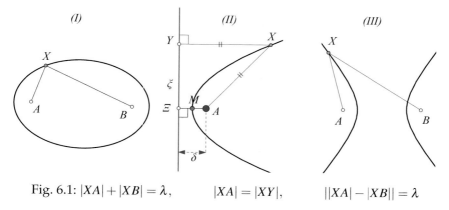

Fig. 6.1: $|XA|+|XB|=\lambda$, $\quad |XA|=|XY|, \quad ||XA|-|XB||=\lambda$

characterized as the geometric locus of points X, which have fixed sum of distances from two fixed points A and B (See Figure 6.1-I).

$$|XA|+|XB|=\lambda.$$

Points A and B are called **focal points** or **focals** of the ellipse. Ellipses also result when a right circular cylinder is intersected with a plane non-orthogonal (and not parallel) to its generators. Circles can be considered as a special case of ellipses, whose focal points A and B coincide.

Parabolas result through intersections with planes which are parallel to some plane tangent to the conical surface (§ 4.9). The **parabola** is characterized also as the geometric locus of the points X, which have the same distance from a fixed point A and fixed line ξ (See Figure 6.1-II). Point A is called **focus** or **focal point** of the parabola and the line ξ is called **directrix** of the parabola.

Finally hyperbolas result through intersections with planes, which are parallel to exactly two generators of the conical surface. The **hyperbola** is characterized also as the geometric locus of the points X, which have fixed difference of distances from two fixed points A and B (See Figure 6.1-III).

$$||XA| - |XB|| = \lambda.$$

Again the points A and B are called **focals** or *focal points* of the hyperbola.

These three kinds of curves (ellipses, parabolas, hyperbolas) are collectively called **proper conic sections** and result as intersections of a conical surface with a plane not passing through the vertex O. Ellipses and hyperbolas admit, as we'll see, two axes of symmetry as well as a center of symmetry and are called **conics with center** or **central conics**. The parabola however admits only one axis of symmetry but not a center of symmetry.

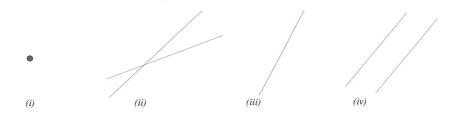

Fig. 6.2: The four kinds of singular (or degenerate) conic sections

When the plane ε, with which we intersect the conical surface, passes through the vertex O of the conical surface, then, depending on the inclination of ε, result the following shapes: (i) a point (plane ε contains then point O and no other point of Σ), (ii) two intersecting lines (ε intersects Σ along two generators) and (iii) a line (ε is tangent to Σ).

Collectively, these particular cases of sections are called **singular** or **degenerate** conic sections (See Figure 6.2). To these are added also pairs of parallel lines, considering them as intersections of a cylinder with a plane parallel to its generators.

6.1. CONIC SECTIONS

Remark 6.1. Usually the student meets conic sections, long before he meets them from the geometric viewpoint, in the form of graphs of simple functions. For example, the parabola as the graph of the function $y = ax^2 + bx + c$,

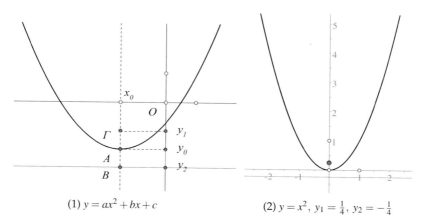

Fig. 6.3: Graphic representation of trinomial

whose focus is at point $x_0 = -\frac{b}{2a}$, $y_1 = \frac{4ac-b^2+1}{4a}$ and the directrix is the parallel to the x-axis from the point of the y-axis $y_2 = \frac{4ac-b^2-1}{4a}$ (See Figure 6.3).

The student meets also the hyperbola as the graph of the function $y = \frac{1}{x}$, which is a hyperbola passing through points $\pm(1,1)$ and with focals at points $\pm(\sqrt{2},\sqrt{2})$ (See Figure 6.4). More generally, the function $y = \frac{ax+b}{cx+d}$, where a, b, c and d are constants and $ad - bc \neq 0$, also represents a hyperbola

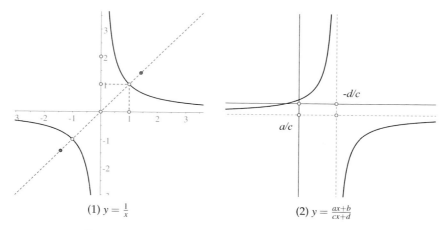

Fig. 6.4: Graphic representation of hyperbola

Exercise 6.1. Construct the points of the ellipse $\gamma(A,B,\lambda)$, which satisfy the equation $|XA|+|XB|=\lambda$ and which lie on the orthogonals of the line which joins the focals $\varepsilon = AB$ at points A and B.

Hint: If $x = |XA|$ is the distance of one such point of the orthogonal of AB at A, then $d^2 + x^2 = (\lambda - x)^2$, where $d = |AB|$.

Exercise 6.2. Construct the points of the parabola $\gamma(\Pi,\xi)$ with focus Π and directric ξ, by finding intersection points B of γ with lines orthogonal to ξ.

Exercise 6.3. Construct points of the parabola $\gamma(\Pi,\xi)$ with focus Π and directrix ξ, by finding intersection points B of γ with lines parallel to ξ and on the same side of ξ as point Π.

Hint: If ε is one such parallel of ξ and d the distance of the parallels $\{\xi,\varepsilon\}$, consider the intersections of ε with the circle $\kappa(\Pi,d)$.

Exercise 6.4. Construct the points of the hyperbola $\gamma(A,B,\lambda)$, which satisfy $||XA|-|XB||=\lambda$ and which are found on the orthogonals of the line through the focals $\varepsilon = AB$ at points A and B.

Hint: If $x = |XA|$ the distance of one such point of the orthogonal to AB at A, then $\lambda(2x+\lambda) = d^2$, where $d = |AB|$.

Remark 6.2. The hyperbolas of figure 6.4, called **rectangular**, are the only kind of conic sections which represent (relative to its symmetry axes) the graph of an *invertible* function. This function, which has the form $y = \frac{ax+b}{cx+d}$, is used for the definition of the *homographic relation* of $\{x,y\}$ (I-§ 5.21).

6.2 Dandelin's spheres

> There's music in the singing of a reed;
> There's music in the gushing of a rill;
> There's music in all things, if men had ears;
> Their earth is but an echo of the spheres.
>
> *Byron, Don Juan, Canto 15*

Dandelin's spheres (1794-1847) readily explain the relationship between conic sections and the geometric loci, which were mentioned in the preceding section. They are spheres inscribed in conical or cylindrical surfaces and simultaneously tangent to a plane ε. Next propositions show that every plane ε, which intersects such a surface defines two (or one in some cases, as we'll see in the next section) spheres called *Dandelin's spheres* relative to the plane ε. Of decisive importance in this section is Theorem 4.15, according to which every plane ε, not passing through the vertex O of a conical surface, intersects (i) either all generators, (ii) or all except two, (iii) or all except one.

6.2. DANDELIN'S SPHERES

Here we'll examine the first two cases. In the next paragraph we'll discuss the third as well.

Proposition 6.1. *Let ε be a plane, which intersects all the generators of the conical surface K. Then, there exist two spheres tangent to the conical surface and simultaneously tangent to the plane ε.*

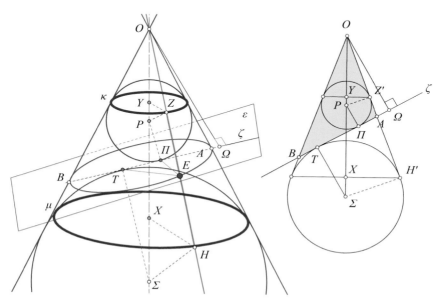

Fig. 6.5: Dandelin's spheres of conical surface intersected by plane

Proof. Consider the plane θ, which is defined from the axis OX of the conical surface K and the perpendicular $O\Omega$ from the apex O of the cone to the plane ε. Let also line ζ be the intersection of the planes ε and θ. By construction, line ζ is orthogonal to $O\Omega$ and intersects the two generators of the conical surface K which are contained in θ at points A and B (See Figure 6.5). This defines then on plane θ the triangle OAB and consequently its inscribed and escribed at the angle $\widehat{AOB}$ circles with centers, respectively, P and Σ. The spheres with centers P and Σ and radii, respectively, those of the inscribed and escribed circles, $P\Pi$ and ΣT, are the requested ones.

I show that one of them, $P(P\Pi)$ is tangent to the cone. Indeed, if OH is another generator and Z is the projection of P onto it, then the right triangles PZO and $PZ'O$, where Z' is the projection of P on OA, are congruent. This because P lies on the axis of the conical surface, therefore has the same distance from the generators, hence $|PZ| = |PZ'|$. Also the two right triangles have the hypotenuse OP in common, they are therefore congruent. This implies that OH is tangent to the sphere at Z and proves the claim. Similarly it is proved that the sphere $\Sigma(\Sigma T)$ is also tangent to the conical surface. That these two spheres are also tangent to the plane ε follows directly, e.g. for $P(P\Pi)$, follows from the fact that $P\Pi$ is also orthogonal to the plane ε, as parallel to $O\Omega$, which has this property. Consequently the plane ε is tangent to the sphere $P(P\Pi)$, being orthogonal at the endpoint of one of its radii.

Theorem 6.1. *The intersection of a conical surface K with a plane ε, which intersects all generators and does not pass through the apex, defines an ellipse with focals at the contact points of the plane with the corresponding two Dandelin's spheres.*

Proof. The proof results from the equality of the tangents from a point to a sphere (Theorem 4.17). If E is the point of intersection of ε and K (See Figure 6.5), then the generator OE is simultaneously tangent to both spheres. Also, if Π and T are the contact points of the two spheres $P(P\Pi)$ and $\Sigma(\Sigma T)$ with the plane ε, then $E\Pi$ and ET are respectively also tangent to these spheres. It follows that $|E\Pi| = |EZ|$ and $|ET| = |EH|$, where E and Z are the contact points of the generator OE with the spheres. Then however the sum

$$|ET| + |E\Pi| = |EZ| + |EH| = |ZH|,$$

is fixed and equal to the length of the generator of the truncated cone defined by the circles κ and μ, along which the two spheres are tangent to the conical surface.

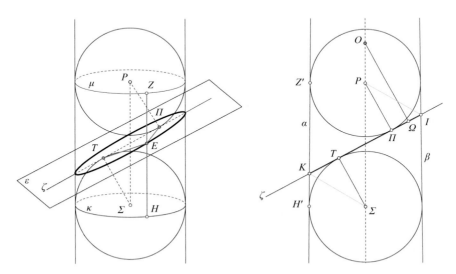

Fig. 6.6: Dandelin's spheres of cylindrical intersected by plane

6.2. DANDELIN'S SPHERES

Proposition 6.2. *Let ε be a plane intersecting all the generators of a cylindrical surface K. There exist then two spheres tangent to K and simultaneously tangent to the plane ε.*

Proof. The proof is similar to that of Proposition 6.1. The plane of the paper in figure-6.6, is the plane θ containing the axis η of the cylindrical surface and the perpendicular $O\Omega$ on the plane ε from an arbitrary point O of the axis. Plane θ intersects the cylindrical surface K along two (parallel) generators α and β. It also intersects plane ε along a line ζ. Drawing the bisectors $K\Sigma$, IP of the angles formed by ζ with α, β and their intersections P, Σ with the axis η, we define the circles $P(P\Pi)$ and $\Sigma(\Sigma T)$, which are simultaneously tangent to α, β and ζ. The spheres $P(P\Pi)$ and $\Sigma(\Sigma T)$ are the requested ones. The proof is similar to that of Proposition 6.1.

Theorem 6.2. *The intersection of a cylindrical surface K, with a plane which intersects all the generators of K, is an ellipse with focals at the contact points of the Dandelin's spheres with the plane ε.*

Proof. The proof results from the equality of the tangents from a point to a sphere (Theorem 4.17). If E is an intersection point of ε with K, then the generator EZ will be simultaneously tangent to both spheres (See Figure 6.6). Also, if Π and T are the points of contact of the spheres $P(P\Pi)$ and $\Sigma(\Sigma T)$ with the plane ε, then $E\Pi$ and ET are also respectively tangent to these spheres. It follows that $|E\Pi| = |EZ|$ and $|ET| = |EH|$, where E and Z are the points of contact of the generator OE with the spheres. Then however the sum

$$|ET| + |E\Pi| = |EZ| + |EH| = |ZH|,$$

is fixed and equal to the length of the generator of the cylinder which is formed from the circles κ and μ, along which the two spheres are tangent to the cylindrical surface.

Proposition 6.3. *Let the plane ε intersect all the generators of a conical surface K except two. There exist then two spheres tangent to K and simultaneously tangent to the plane ε.*

Proof. The proof is similar to that of Proposition 6.1. The plane of the paper in figure 6.7, is the plane θ containing the axis of the conical surface K and the perpendicular $O\Omega$ from the apex O of the conical surface to the plane ε. Let ζ be the line intersection of the planes ε and θ. By construction, ζ is orthogonal to $O\Omega$ and intersects the two generators of the conical surface contained in θ at points A and B. This defines then on plane θ the triangle OAB and consequently its escribed circles with centers P and Σ. The spheres with centers points P and Σ and radii those of the escribed circles respectively, $P\Pi$ and ΣT, are the requested ones.

I show that one of them, $P(P\Pi)$ is tangent to the conical surface. Indeed, if OH is another generator and Z is the projection of P to it, then the right triangles PZO and $PZ'O$, where Z' is the projection of P on OA, are congruent. This, because point P is on the axis of the cone, therefore has the same distance from the generators, consequently $|PZ| = |PZ'|$. Also the two triangles have the hypotenuse OP in common, they are therefore congruent. This implies that OH is tangent to the sphere at point Z and proves the claim. Analogously it is proved also that the sphere $\Sigma(\Sigma T)$ is tangent to the conical surface. The fact that these spheres are also tangent to the plane ε follows directly, e.g. for $P(P\Pi)$, from the fact that $P\Pi$ is also orthogonal to the plane ε, being parallel to $O\Omega$, which has this property. Consequently plane ε is tangent to the sphere $P(P\Pi)$, being orthogonal to the endpoint of one of its radii.

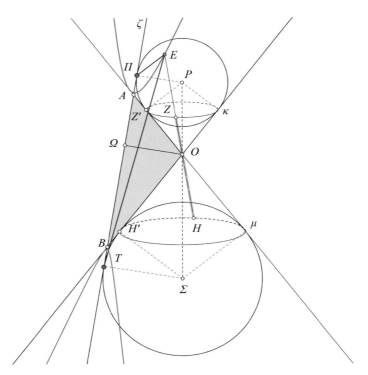

Fig. 6.7: Spheres of Dandelin II

Theorem 6.3. *The intersection points of a plane ε, meeting all the generators of a conical surface K except two, define a hyperbola on ε with focals at the contacts of the Dandelin's spheres with the plane ε.*

Proof. The proof results from the equality of the tangents from a point to a sphere (Theorem 4.17). If E is an intersection point of the plane ε and K, then the generator OE is simultaneously tangent to both spheres (See Figure 6.7). Also, if Π and T are the contact points of the spheres $P(P\Pi)$ and $\Sigma(\Sigma T)$ with the plane ε, then also $E\Pi$ and ET are respectively tangents to these spheres. It follows that $|E\Pi| = |EZ|$ and $|ET| = |EH|$, where E and Z are the contact points of the generator OE with the spheres. Then, however, the difference

$$|ET| - |E\Pi| = |EH| - |EZ| = |ZH|,$$

is constant and equal to the length of the generator of the truncated cone (union of the two cones with common vertex O) defined by the circles κ and μ, along which the spheres are tangent to the conical surface.

6.3 Directrices

> Man is a tool-using animal. Without tools he is nothing, with tools he is all.
>
> Thomas Carlyle, Sartor Resartus

Besides the parabola, whose directrix plays an important role for its usual definition as a geometric locus, directrices are defined also for the other kinds of conic sections with properties analogous to those of the directrix of the parabola. In this section we first relate the parabola directrix to the corresponding sphere of Dandelin and next we examine the generalization of the directrix for the other conics.

Proposition 6.4. *Let ε be a plane parallel to a tangent plane of a conical surface Σ. There exists then a sphere tangent to Σ and simultaneously tangent to ε.*

Proof. As in the propositions of the preceding section, we consider the plane θ containing the axis of the conical surface Σ and the perpendicular $O\Omega$ from the apex O of Σ to the given plane ε (in figure 6.8, θ coincides with the plane of the paper). Planes θ and ε intersect along a line ζ. This line is parallel to the generator β, which is contained in the tangent plane parallel to ε. Plane θ also contains the symmetric generator α of β relative to the axis. Also, because ζ and β are parallel, the triangle with sides α, ζ and the axis is isosceles. Thus, passing to the plane figure contained in θ, we find a circle $P(P\Pi)$ with center on the axis and simultaneously tangent to the two parallels ζ and β, as well as to the other generator α of Σ contained in θ. For this it suffices to consider the perpendicular to the axis OP from the intersection point of α and ζ. Latter bisects the angle at A of lines α and ζ and defines the circle $P(P\Pi)$, simultaneously tangent to α, β and ζ. The sphere $P(P\Pi)$ is the requested one. The proof of this claim is the same with that of Proposition 6.1.

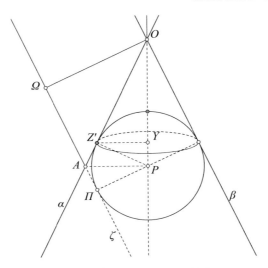

Fig. 6.8: Dandelin's sphere for a plane parallel to a tangent plane of Σ

Theorem 6.4. *The intersection of a conical surface Σ and a plane ε, parallel to a tangent plane of Σ, defines on ε a parabola with focus at the contact Π of Dandelin's sphere with ε and directrix the line ξ, along which ε intersects the plane η of the circle of the contact points of the sphere with the conical surface.*

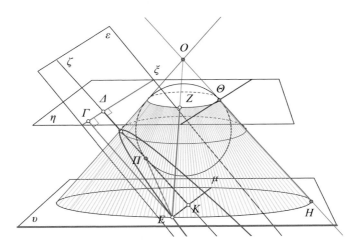

Fig. 6.9: Parabola and Dandelin's sphere

Proof. The proof results from the equality of the tangents from a point to a sphere (Theorem 4.17). To begin with, the circle, along which are tangent

6.3. DIRECTRICES

the sphere and the conical surface Σ, is contained in a plane η orthogonal to the axis of the conical surface, which intersects plane ε along a line ξ (See Figure 6.9). Line ξ is orthogonal to the plane θ which contains the axis and the generator $\beta = O\Theta$, which defines the tangent plane of Σ, which, by assumption, is parallel to ε. From an arbitrary point E of the intersection of the plane ε with Σ, passes a plane υ parallel to η and forming with that a truncated cone. Suppose EZ is the generator of this cone which passes through point E and $H\Theta$ the equal to EZ generator contained in plane θ. From the equality of the tangents from E to the sphere, follows that EZ and $E\Pi$ have the same length, therefore $E\Pi$ and $H\Theta$ also have the same length. It suffices then to show that $H\Theta$ and $E\Gamma$ have the same length. Here $E\Gamma$ is the perpendicular from E to ξ. Let line μ be the intersection of the planes ε and υ. Let also K be the intersection of μ with the plane θ. Because planes η and υ are parallel, the lines ξ and μ, which are intersected from them by the plane ε, are parallel. Because planes ε and η are orthogonal to θ, their intersection ξ will also be orthogonal to θ. Consequently, if Δ is the intersection point of ξ with θ, then $E\Gamma\Delta K$ is a rectangle and $K\Delta$ and $E\Gamma$ have the same length. Because however ζ and the generator $O\Theta$ are parallel, $K\Delta$ and $H\Theta$ will also have the same length, being segments of parallel lines between parallel planes.

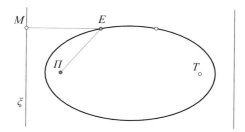

Fig. 6.10: Property of ellipse $\frac{|E\Pi|}{|EM|} = k < 1$

In what follows we show one more characteristic property of ellipses and hyperbolas in which appear two lines, called **directrices**, with properties similar to the directrix of the parabola.

Theorem 6.5. *Given an ellipse, to each focus Π corresponds a line ξ such that, for every point E of the ellipse, the ratio of distances*

$$\frac{|E\Pi|}{|EM|} = k < 1,$$

is fixed. M here dentotes the projection of E onto ξ.

328   CHAPTER 6. CONIC SECTIONS

Proof. For the proof we use the definition of the ellipse as the intersection of a conical surface Σ and a plane ε, as well as the Dandelin's sphere tangent to ε at Π (See Figure 6.11). The focus Π is the contact point of ε with the sphere and the arbitrary point E of the ellipse defines a generator EO of Σ as well as a plane υ passing through E and orthogonal to the axis of Σ and intersecting Σ along a circle μ. Also the points of contact between the sphere and the conical surface define a circle κ, contained in a plane η, parallel to υ. The two planes υ and η cut out from the conical surface a truncated cone. $E\Pi$ is equal to the generator EZ of the truncated cone. Let now ξ be the line along which planes ε and η intersect and M be the projection of E on ξ. We consider the plane θ containing the axis of Σ and the orthogonal from O to plane ε (this is the plane of the paper in the figure). Plane θ intersects

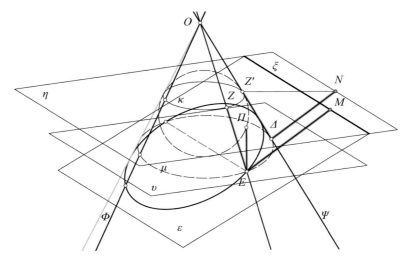

Fig. 6.11: The directrix ξ of the ellipse

the conical surface along two generators $O\Phi$ and $O\Psi$ lying symmetrically relative to the axis. Plane θ also defines the generator $\Delta Z'$, with Δ on the plane υ, of the truncated cone, whose length is

$$|\Delta Z'| = |EZ| = |E\Pi|.$$

Let N be the intersection of plane η with the parallel to EM from Δ. It is $|EM| = |\Delta N|$, since these are segments of parallels between parallel planes (Proposition 3.11). Triangle $Z'\Delta N$ depends on the position of E, however it remains similar to itself as E varies on the ellipse, hence the ratio of its sides

$$k = \frac{|\Delta Z'|}{|\Delta N|} = \frac{|E\Pi|}{|EM|}$$

6.3. DIRECTRICES

remains fixed. The fact that $k < 1$ follows easily from the inequality for angles $\widehat{\Delta NZ'} < \widehat{\Delta Z'N}$.

Theorem 6.6. *Given a hyperbola, to each focus Π corresponds a line ξ, such that for every point E of the hyperbola the ratio of distances*

$$\frac{|E\Pi|}{|EM|} = k > 1,$$

is fixed (See Figure 6.12). M denotes here the projection of E onto ξ.

Proof. For the proof we use the definition of the hyperbola as the intersection of a conical surface Σ and a plane ε as well as Dandelin's sphere tangent to ε

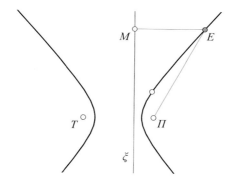

Fig. 6.12: Property of hyperbola $\frac{|E\Pi|}{|EM|} = \kappa > 1$

at Π (See Figure 6.13). The focus Π is the point of contact of ε and the sphere and the arbitrary point E of the hyperbola defines a generator EO of the cone Σ, as well as, a plane υ passing through E, orthogonal to the axis of Σ and intersecting Σ along a circle μ. Also the points of contact between the sphere and the conical surface define a circle κ contained in a plane η parallel to υ. The two planes υ and η cut out from Σ a truncated cone. $E\Pi$ is equal to the generator EZ of the truncated cone. Let now ξ be the intersection line of planes ε and η and M be the projection of E onto ξ. We consider the plane θ containing the axis of Σ and the perpendicular from O to the plane ε. Plane θ intersects the conical surface along two generators $O\Phi$ and $O\Psi$ lying symmetrically relative to the axis. Plane θ also defines the generator $\Delta Z'$ of the truncated cone, whose length is

$$|\Delta Z'| = |EZ| = |E\Pi|.$$

We draw from Δ the parallel to EM intersecting plane η at point N. It is $|EM| = |\Delta N|$, since these are segments on parallels between parallel planes

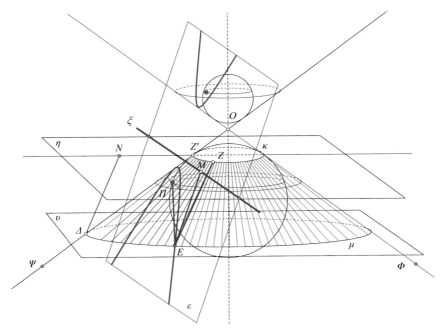

Fig. 6.13: The directrix ξ of the hyperbola

(Proposition 3.11). Triangle $Z'\Delta N$ depends on the position E, however it remains similar to itself as point E moves on the hyperbola, consequently the ratio
$$k = \frac{|\Delta Z'|}{|\Delta N|} = \frac{|E\Pi|}{|EM|}$$
remains fixed. The fact that $k > 1$ follows easily from the inequality for angles $\widehat{\Delta NZ'} > \widehat{\Delta Z'N}$.

The lines ξ, which are mentioned in the last two theorems, are called **directrices of the ellipse** and respectively **directrices of the hyperbola** while k is called **eccentricity** of the ellipse, respectively of the hyperbola. While in the parabola there exists a directrix (we often consider that the *line at infinity* is a second directrix for the parabola), in ellipses and hyperbolas there exist two and attention must be paid on how the fixed ratio of eccentricity is constructed
$$k = \frac{|E\Pi|}{|EM|}.$$

A directrix and the corresponding focus go together and we must be aware of this correspondence in order to define the ratio correctly. Otherwise this ratio will be not constant, while E varies on the conic (See Figure 6.14).

6.3. DIRECTRICES

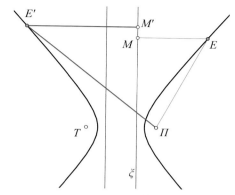

Fig. 6.14: Focus and corresponding directrix such that $\frac{|E\Pi|}{|EM|} = \frac{|E'\Pi|}{|EM'|}$

Corollary 6.1. *In every ellipse or hyperbola the directrices corresponding to the two focals are parallel.*

Proof. This is a direct consequence of the way they were defined through the intersection of their plane ε with the conical surface Σ. The directrices were defined, in the preceding theorems, as intersections of the plane ε with the planes η which contain the contact circles of Σ with corresponding spheres of Dandelin. These planes are parallel, being orthogonal to the axis of Σ, consequently their intersections with the plane ε of the conic will also be parallel lines (Proposition 3.9).

Remark 6.3. With all we said up to this point, we proved that every conic section is also described as a geometric locus as in § 6.1. We did not prove however the converse, for example that the geometric locus of points X for which the sum of distances from two fixed points $|XA| + |XB| = \lambda$ is fixed, is a conic section, in other words the specific locus results also as a an intersection of a conical surface with a plane. To show the converse, we must, given the locus, that is A, B and λ, construct a cone Σ and a plane ε, such that the resulting intersection coincides with given ellipse. This means that we must find the relations holding between the pair $(|AB|, \lambda)$, which defines the ellipse as a geometric locus, and the elements which define the conical surface Σ and the intersecting it plane ε. A similar remark holds also for the definition of the ellipse and the hyperbola through the directrix. Next exercise examines this converse in the simplest of cases, which is the parabola. For the cases of the ellipse and the hyperbola an alternative way is given in exercise 6.7.

Exercise 6.5. Given a parabola there exists a cone and a plane which intersect along the given parabola.

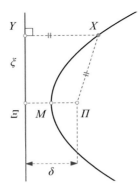

Fig. 6.15: A parabola is defined from a number δ

Hint: The parabola, defined as a geometric locus (§ 6.1), is completely determined from a number. This number is the distance δ of its focus from the directrix (See Figure 6.15). The middle M of the segment $\Pi\Xi$, where Ξ is the projection of the focus onto the directrix ξ, belongs to the parabola. To construct the requested conical surface we connect δ with its "opening" and with the sphere, which determines the position of its focus. Referring to the figure of Proposition 6.4, we consider its intersection with plane θ, relative to which the entire figure of this proposition is symmetric. We determine δ

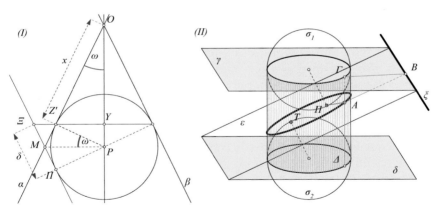

Fig. 6.16: Parabola as a conic section   Directrix ξ of ellipse

as a function of the opening ω of the conical surface and the length x of the tangent from the vertex O of the conical surface to the sphere of Dandelin (See Figure 6.16-I). Obviously it holds

$$\delta = 2|\Xi M| = 2|M\Pi| = 2|MZ'| = 2|PZ'|\tan(\omega) = 2x\tan^2(\omega).$$

Consequently, for every conical surface (opening ω), given δ we find, by solving the preceding equation, the position x on the generator, to which the sphere which defines the plane ε will be tangent and the corresponding conic section which will coincide with the parabola.

Exercise 6.6. Show that the directrix relative to the focus Π, of the ellipse which is defined as a section of a plane ε with a cylinder, coincides with the intersection line ξ of ε with the plane γ, which is orthogonal to the cylinder axis and passes through the center of the Dandelin's sphere σ_1 which defines the focus Π (See Figure 6.16-II).

Exercise 6.7. In figure 6.17, we have two circles in the plane ε external to each other. If we rotate the shape in space, about the axis ζ, then this produces: (i) a cone Σ_1 with vertex at O, (ii) a cone Σ_2 with vertex at Ω and (iii) two spheres κ_A, κ_B with centers respectively at A and B. The common tangent θ of the two circles will then be a generator of cone Σ_1 and the common tangent η will be a generator of cone Σ_2. Explain, how the tangent plane of Σ_1 along the generator θ will cut from the cone Σ_2 a hyperbola with focals

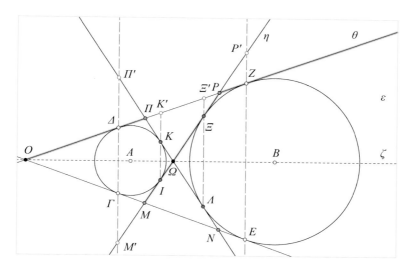

Fig. 6.17: Conics with given spheres of Dandelin

at Δ and Z, passing through points Π and P. Similarly, the tangent plane of cone Σ_2 along its generator η will cut from the cone Σ_1 an ellipse with focals at I and Ξ, passing through points M and P. Spheres κ_A and κ_B will then be spheres of Dandelin of these conic sections.

Exercise 6.8. Show that the ellipse of the preceding exercise has focal distance $2c = |I\Xi|$ and eccentricity $k = \frac{|IM|}{|MM'|}$. Points X of this ellipse satisfy the equation $|XI| + |X\Xi| = 2a$, where $2a = |MP|$.

Exercise 6.9. Show that the hyperbola of exercise 6.7 has focal distance $2c' = |\Delta Z|$ and eccentricity $k' = \frac{|\Pi \Delta|}{|\Pi K'|}$. Points X of this hyperbola satisfy the equation $||X\Delta| - |XZ|| = 2a'$, where $2a' = |\Pi P|$.

Exercise 6.10. Relate the characteristics of the conics of the two preceding exercises and show that $c = a'$, $a = c'$ and $k' = 1/k$. The lines orthogonal to the plane ε at $\{K', \Xi'\}$ are the directrices of the hyperbola, and the lines orthogonal to ε at $\{P', M'\}$ are the directrices of the ellipse.

6.4 General characteristics of conics

> Of our Thinking, we might say, it is but the mere upper surface that we shape into articulate Thoughts; - underneath the region of argument and conscious discourse, lies the region of meditation; here, in its quiet mysterious depths, dwells what vital force is in us; here, if ought is to be created, and not merely manufactured and communicated, must the work go on.
>
> T. Carlyle, Characteristics, Collected Works p. 211

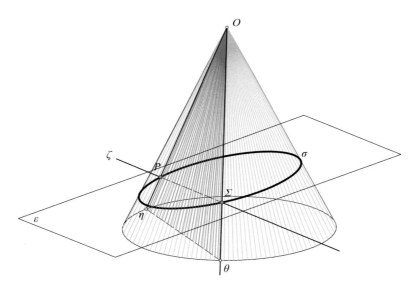

Fig. 6.18: Intersection of conic and line

There are some characteristics of the conic sections which are common for ellipses, parabolas and hyperbolas. Some other instead depend on the special nature of the conic section, determined from its eccentricity k. In this

6.4. GENERAL CHARACTERISTICS OF CONICS

section we'll examine some common properties resulting directly from the definition of these curves, as sections of a conical surface with a plane. All three kinds of conics have at least one focus, a corresponding directrix and eccentricities: $k < 1$, $k = 1$ and $k > 1$, which characterize respectively the ellipse, the parabola and the hyperbola.

For some properties eccentricity plays a general role independent of its precise magnitude. It is then simply used the fact that, the ratio of the distance of a point from the focus, to its distance from the corresponding directrix, is constant. In what follows we consider the conic σ defined by the intersection of a conical surface Σ with a plane ε, not passing through the vertex O of the conical surface. A fundamental property of conic sections is expressed with the next theorem.

Theorem 6.7. *Every line ζ on the plane ε of the conic intersects it in, at most, two points.*

Proof. Every line ζ on the plane ε of the conic defines, along with the vertex O of the cone, a plane which intersects from the conical surface two at most generators η and θ (See Figure 6.18). The common points of the conic and the line ζ will be on these generators, consequently they cannot be more than two (see also Exercise I-4.24).

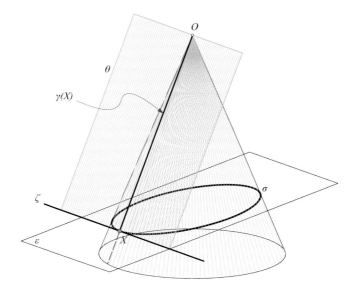

Fig. 6.19: Tangent ζ to conic at X

Tangent to the conic σ, at one of its points X, is called a line ζ defined as the intersection of the plane ε of the conic with the tangent plane of the conical surface along the generator $\gamma(X)$ defined by X.

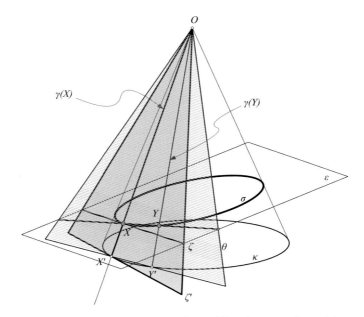

Fig. 6.20: Tangent as limiting position of line intersecting at two points

Theorem 6.8. *Through every point X of the conic, there is exactly one tangent to it (See Figure 6.19).*

Proof. The proof reduces to the corresponding property of the conical surface. Every point X on the conic, which is an intersection of the plane ε and the conical surface Σ, is contained in exactly one generator $\gamma(X)$ of the cone, which is common with exactly one tangent plane θ of the cone (§ 4.9). The intersection of planes θ and ε defines the unique line ζ through X which has the tangent property.

Proposition 6.5. *The tangent ζ of the conic σ at the point X of it coincides with the limiting position of line XY, passing through point X and a second point Y of the conic, as Y tends to coincide with X.*

Proof. The proof reduces to the corresponding property of the circle (See Figure 6.20). Every point X of the conic, which is the intersection of plane ε and the conical surface Σ, is contained in exactly one generator $\gamma(X)$. Two points X, Y of the conic and the corresponding generators $\gamma(X), \gamma(Y)$ define a plane θ which intersects ε along line XY. The plane χ trhough X and orthogonal to the axis of the conical surface intersects it along a circle κ and θ intersects this circle at points X and Y'. As Y tends to X, on the circle κ point Y' tends to X and $X'Y'$ tends to the tangent ζ of the circle κ at point X. Then, the plane θ tends to coincide with the tangent plane of the conical surface, which passes

6.4. GENERAL CHARACTERISTICS OF CONICS

through X and consequently its intersection with ε tends to coincide with the intersection of the tangent plane along $\gamma(X)$ with ε, which is the tangent ζ at X.

The common points of the interior of the conical surface and the plane ε, which intersects the surface along the conic κ are called **interior points** of the conic, and the other points not lying on the conic are called **exterior points** of the conic. The line segment AB defined from two points of the conic κ with line ε is called **chord** of the conic.

Proposition 6.6. *From every external point of the conic precisely two tangents to it can be drawn.*

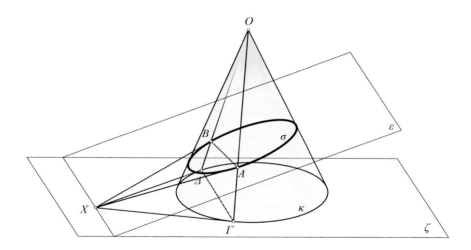

Fig. 6.21: Tangents to conic from an external point

Proof. If X is an external point on the plane ε of the conic σ, then we consider the orthogonal to the axis plane ζ passing through X (See Figure 6.21). Plane ζ intersects the conical surface along a circle κ and X is outside this circle, consequently two tangents $X\Gamma$ and $X\Delta$ to the circle can be drawn. These two tangents and the apex O define two planes $\{\alpha, \beta\}$ which intersect ε respectively along two lines $\{XA, XB\}$, which are tangent to the conic. Since every tangent to the conic is defined in a similar way (through a tangent plane of the cone), these are the unique tangents to σ from point X.

Proposition 6.7. *The middles of chords of a conic section, which are parallel to a line θ, are contained in a line ξ. If the conic has tangents parallel to θ, then ξ passes also through the contact points of these tangents.*

Proof. Suppose that the conic results as the intersection of a conical surface Σ with a plane ζ. Parallel chords A_1B_1 of the conic, define, together with the vertex O, planes η_1. These planes will intersect along a line θ, parallel to all these chords (Proposition 3.5) and passing through O (See Figure 6.22). Let Ω be the intersection point of θ and a plane ε, orthogonal to the axis of the conical surface, which intersects it along a circle κ. Planes η_1, which at ζ define the parallel chords, intersect plane ε along lines, which pass through the point Ω. Each one of these line defines also a corresponding chord of the circle κ. This way to the initial chords A_1B_1 of the conic section we correspond chords A_2B_2 of the circle, the extensions of which are lines passing through Ω. Let now $\Phi_2\Psi_2$ be the polar of Ω relative to circle κ. The lines $\Omega\Phi_2$, $\Omega\Psi_2$ are the tangents to the circle κ from Ω and the intersection of their planes (through O) defines the tangents of the conic which are parallel to the chords A_1B_1. Also, for every line through Ω, intersecting the circle at points A_2, B_2, and the polar at M_2, the cross ratio $(A_2B_2;M_2\Omega) = -1$, consequently the pencil of lines OA_2, OB_2, OM_2, $O\Omega$ which pass through these

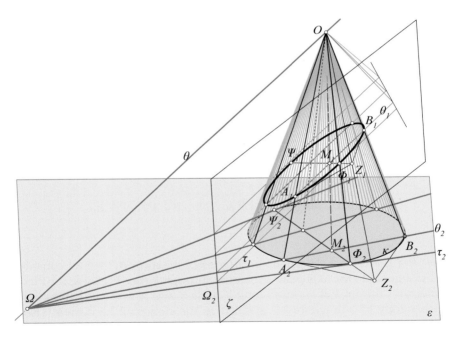

Fig. 6.22: The middles of parallel chords of conics

points will be harmonic, therefore it will intersect also every other line of the plane of these lines at pairs of harmonic conjugate points. However line θ_1 is parallel to $\theta = O\Omega$, therefore the other three lines will define on θ_1 two equal segments and point M_1 will be the middle of A_1B_1. However, points

6.4. GENERAL CHARACTERISTICS OF CONICS

M_2 and M_1 are contained in the plane of the three points O, Φ_2, Ψ_2, which intersects the plane ζ along the line $\Phi_1\Psi_1$. Consequently the middles of all these chords A_1B_1 are contained in the line $\Xi = \Phi_1\Psi_1$. The tangents of the conic at Φ_1, Ψ_1 are also parallel to the chords A_1B_1.

Remark 6.4. Using equations, the proof of the last proposition can be done in two lines ([92, p.151]). However figure 6.22, is interesting in many ways. Line θ is parallel to the plane ζ of the conic section. Similarly the lines like θ_1, which define chords A_1B_1 of the conic parallel to various directions, are contained all in a plane ζ' parallel to ζ and passing through the vertex O of the conical surface. This plane intersects plane ε of the figure, which contains the circle κ, along a line v. The pole N of v relative to circle κ, depending on the inclination of the plane ζ of the conic, is either an internal point of the circle or is on the circle or is external to the circle. The first case corresponds to ellipses, the second to parabolas and the third to the hyperbolas. It is proved easily that line ON intersects the plane of the conic at its center M or at infinity in the case of the parabola. The so called *diameters* of the conic, that is the chords through its center M, correspond to parallels to the lines θ which pass through M.

This way, in the case of the hyperbola, in which there are two generators σ and τ of the conical surface parallel to the plane ζ, these generators will define the plane ζ'. Then the line v of the intersection of ζ' and ε will intersect the circle κ at two points Γ and Δ and the pole N of v relative to κ will be outside the circle. From N then can be drawn two tangents to κ, which, together with point O will define two planes through O. These planes will intersect plane ζ by two lines which do not meet the conic (they are tangents to the conic at the points which correspond to the contact points Γ and Δ of the circle with the tangents from N. To Γ and Δ however correspond the points of the conic at infinity). These are the so called *asymptotes* of the hyperbola, about which we'll talk below.

Similar relations between plane characteristics of the conics and characteristics which are connected to space and the conical surface, through which the conic is defined, can be studied in many cases. Usually the description of these characteristics through equations is much simpler from their geometric description with shapes in space, for which we have a certain difficulty to draw and must rather imagine them than draw.

Remark 6.5. Figure 6.23, gives an image on the plane of the chords of the conic which are parallel to the line θ. Their middles are contained in a chord ξ. In the ellipse, for every line θ there exist tangents to it parallel to θ and the chord ξ passes through the points of contact of these tangents.

In the hyperbola, as can be seen in the figure, there do not always exist tangents to a given direction (θ in the figure). In the parabola, except the direction of its axis of symmetry (Proposition 6.11), for all other directions there exists exactly one tangent parallel to it. The figure also shows, that for conics with *center of symmetry*, which is called **center of the conic** (such a

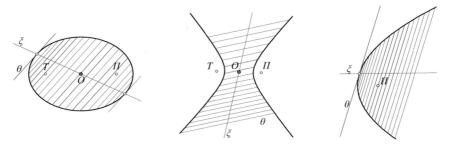

Fig. 6.23: Conjugate directions θ and ξ

point exists only for ellipses (Proposition 6.10) and hyperbolas (Proposition 6.14)), the chords ξ which pass through the middles of parallel chords pass also through the center of the conic. It is proved that the directions of θ and ξ have a symmetric relationship:

If the chords parallel to the direction θ have their middles on a chord parallel to direction ξ, then the chords parallel to ξ have their middles on a chord parallel to the direction θ.

Such pairs of directions, as θ and ξ, we say that they define pairs of **conjugate directions** of the conic. The parabola presents the peculiarity of having the special direction of its axis of symmetry, conjugate to every other direction. In the circle, which is a special case of ellipse, for every direction θ, the corresponding conjugate coincides with the orthogonal to θ direction ξ, since the middles of parallel chords of the circle are contained in a diameter orthogonal to the chords (Corollary I-2.3). Two chords of a conic which have conjugate directions are called **conjugate chords** of the conic. Two conjugate chords which pass through the center of a conic (if it exists) are called **conjugate diameters** of the conic.

Referring to figure 6.22 of proposition 6.7, let us remark that the conjugate diameters of the conic σ are perspectively projected from O to orthogonal diameters of the circle κ.

Theorem 6.9. *If the chord AB of the conic intersects the directrix α at point Z, then ΠZ, where Π is the corresponding focus of the directrix, is an external bisector of the **focal angle** $\widehat{A\Pi B}$.*

Proof. Because of the similarity of the right triangles it holds (See Figure 6.24-I)
$$\frac{|ZA|}{|ZB|} = \frac{|AX|}{|BY|} = \frac{|A\Pi|/k}{|B\Pi|/k} = \frac{|A\Pi|}{|B\Pi|},$$
where k is the eccentricity of the conic. The conclusion follows by applying Theorem I-3.3. The fact that the bisector is external stems from the fact that the diretrix does not intersects the conic.

6.4. GENERAL CHARACTERISTICS OF CONICS

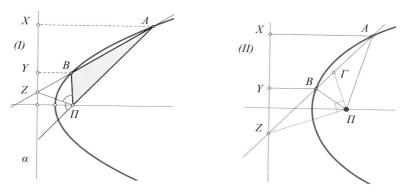

Fig. 6.24: Bisectors of *focal angle* of chord *AB*

Proposition 6.8. *The segment AZ of the conic tangent, from the contact point A to the intersection Z with the directrix, is seen from the corresponding focus under a right angle.*

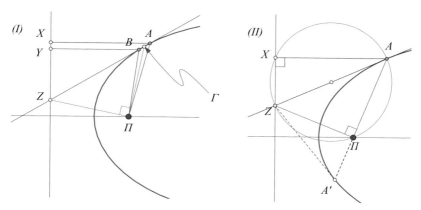

Fig. 6.25: Property of the conic's tangent

Proof. Let us consider two neighboring points A, B of a conic, their projections X, Y on the directrix of focus Π and the intersection point Z of AB with the directrix (See Figure 6.25-I). According to the preceding theorem, ΠZ is an external bisector of angle $\widehat{A\Pi B}$. The internal bisector of angle $\widehat{A\Pi B}$ is orthogonal to the external and is contained between ΠA and ΠB. Consequently, when B tends to coincide with A, the same will happen also with the trace Γ of the internal bisector and in the limiting position, in which AZ becomes the tangent at A (Proposition 6.5), angle $\widehat{Z\Pi A}$ will be a right one.

Corollary 6.2. *The circle which passes through point A of a conic, its projection X on directrix and the corresponding focus Π, also passes through the intersection point Z of the tangent at A and the directrix.*

Proof. According to the preceding proposition the quadrilateral $AXZ\Pi$ will be inscriptible (See Figure 6.25-II), having the angles at opposite vertices Π and X right.

Corollary 6.3. *The tangents of a conic at the endpoints of a chord AA', which passes through the focus Π, intersect at a point Z on the corresponding directrix, and ZΠ is an altitude of the triangle AA'Z (See Figure 6.25-II).*

Exercise 6.11. Show that, two chords AA' and BB' of a conic, which pass through its focus Π, define pairs of lines $(AB, A'B')$, $(AB', A'B)$ which intersect correspondingly at points Γ, Δ on the corresponding directrix ξ of Π (See Figure 6.26-I).

Hint: Point Δ is the intersection of ξ with the external bisector of angle $\widehat{B\Pi A'}$, which is also external bisector of the vertical angle $\widehat{A\Pi B'}$.

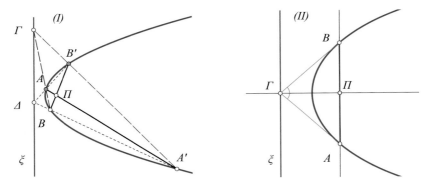

Fig. 6.26: Focal chords   Parameter $p = |\Pi\Gamma|$, latus rectum $\ell = |AB|$

Exercise 6.12. Show that the chord AB, which passes through the focus Π of a conic and is parallel to the corresponding directrix ξ, defines, through the tangents at its endpoints an isosceles triangle with vertex Γ on the directrix ξ. Show that the angle at Γ is right for the parabola, acute for the ellipse and obtuse for the hyperbola (See Figure 6.26-II).

Hint: The first claim is immediate consequence of the proposition 6.8. The second follows from the corresponding value of the eccentricity $k = \frac{|B\Pi|}{|\Gamma\Pi|}$.

The length $\ell = |AB|$ in the preceding exercise is called **latus rectum** and $|\Pi A| = \ell/2$ is called **semi latus rectum** of the conic. The distance $p = |\Pi\Gamma|$ is called **focal parameter** of the conic.

6.4. GENERAL CHARACTERISTICS OF CONICS

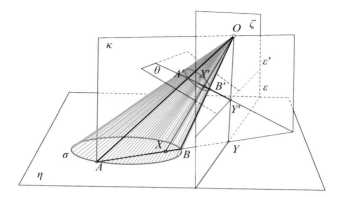

Fig. 6.27: Polar ε of point X relative to the conic σ

An important notion we met in circles, and can be generalized for conics, is that of the polar of a point relative to a conic. The definition is formally the same with the corresponding one for circles (I-§ 4.9). Indeed, if X is a point of plane η of a conic (See Figure 6.27) and a line through X intersects the conic at two points A and B, then this defines the harmonic conjugate $Y = X(A,B)$ of A, B relative to X. Yet again the locus of the Y's, as the line rotates about X is a line. This can be seen immediately, if we consider a right circular cone with vertex O, which contains the conic as the intersection with plane η. The intersection of the cone with the plane θ orthogonal to the axis is a circle and the lines OA, OB define a plane κ, which intersects the circle at points A', B', contained, respectively in generators OA, OB. Line OX intersects $A'B'$ at point X' and this defines the polar ε' of the circle relative to X'. Plane ζ, which contains O and ε', intersects plane η along a line ε, which contains all the harmonic conjugates $Y = X(A,B)$ of X relative to A, B, as AB rotates about X. This follows directly from the equality of the cross ratios $(AB;XY) = (A'B';X'Y') = -1$, on plane κ, where $Y' = X'(A',B')$ (Theorem I-5.29). We proved then the next property, through which the **polar** line $p(X)$, of point X relative to the conic is defined.

Theorem 6.10. *Given a conic σ and a point X of its plane, the geometric locus of the harmonic conjugates $Y = X(A,B)$ of X relative to the intersection points A, B of the conic with lines through X, is contained in a line $\pi(X)$.*

Remark 6.6. With the help of the same figure 6.27, we can see that, for a conic σ, as it happens with circles (Corollary I-4.25), if the polar $\pi(X)$ contains Y, then also the polar $\pi(Y)$ contains X. This implies, that the polars $\{\pi(X)\}$ of all points X of a line ε pass through a fixed point $\pi(\varepsilon)$, which is called the **pole** of ε relative to the conic σ.

A relatively easy exercise is the generalization and transfer of the properties valid for polars of circles (I-§ 4.9) to the polars of conics. Specifically,

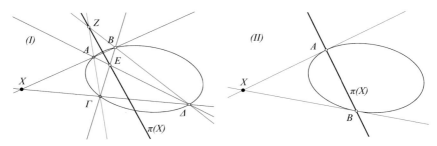

Fig. 6.28: Construction of polar $\varepsilon = p(X)$ of point X relative to conic

from the preceding observations and the properties of the complete quadrilateral (I-§ 5.17), follows the generalization of the construction of the polar for conics. In fact (Proposition I-5.17), to construct $\pi(X)$, we draw two secants AB and $\Gamma\Delta$ of the conic through X, and we connect the intersection points E, Z of the pairs of lines, respectively $(A\Delta, B\Gamma)$ and $(A\Gamma, B\Delta)$ (See Figure 6.28-I). For conics it also holds that the polar of a point X, from which one can draw tangents to the conic, is the chord of the points of contact (See Figure 6.28-II). The preceding properties, in combination with exercise 6.11 prove next property.

Corollary 6.4. *A directrix of a conic is the polar of its corresponding focus.*

Exercise 6.13. Given the focus Π and the corresponding directrix ξ and a point X (and no other points) of the conic κ, construct its tangent at X.

6.5 The parabola

> A parabola is approximately the path of a ball, a bullet, or a shell in the air. If the air offered no resistance the path would be exactly a parabola. Thus if warfare were conducted in a vacuum, as it should be, the calculations of ballistics would be much simpler than they actually are, and it would cost considerably less than the 25,000 or so of taxes which it is now necessary to shoot away in order to slaughter one patriot.
>
> T. Bell, *The Handmaiden of the Sciences*, p.24

The parabola can be considered to be the simplest case of a conic section. As it was mentioned also in a preceding section (Exercise 6.5), its shape is completely determined by a positive number δ, which gives the distance $\Pi\Xi$ of its focus from the directrix. The constant δ coincides with the focal parameter and is often mentioned as **parameter** of the parabola. The middle M of $\Pi\Xi$ belongs to the parabola and is called **vertex** of the parabola (See Figure 6.29).

6.5. THE PARABOLA

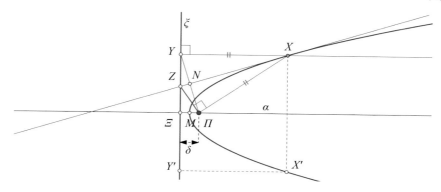

Fig. 6.29: First properties of parabola

Proposition 6.9. *Two parabolas are congruent, if and only if they have the same parameter.*

Proof. Here the congruence must be understood in the sense of placing the two shapes so that they are coincident. The proof is obvious and reduces to the fact that a system consisting of one line ξ and a point Π lying outside ξ, can be placed on top of another similar one, so that the two systems coincide, if and only if the distance of the points from the corresponding lines is the same.

In what follows Π denotes the focus and ξ the directrix of the parabola. We also consider the **axis of the parabola**, which is the line α which passes through the focus Π and is orthogonal to the directrix ξ. A line parallel to the axis is called **diameter** of the parabola.

Theorem 6.11. *Let Π be the focus and ξ the directrix of a parabola. The following properties are valid:*

1. *The parabola is symmetric relative to its axis $\Pi\Xi$ (See Figure 6.29).*
2. *For every point X on the parabola and its projection Y onto the directrix, point X lies on the medial line XN of the segment ΠY.*
3. *Line XN is tangent to the parabola at X.*
4. *The tangent and its orthogonal at X bisect the angles formed at X by the line $X\Pi$ and the parallel XY to the axis of the parabola.*
5. *The middle N of ΠY is contained in the tangent t_M of the parabola at its vertex M.*

Proof. (1) follows immediately from the definition of the parabola. If X is a point on the parabola and X' is its symmetric relative to its axis α, it will satisfy the definition of the same parabola.

(3): Because XN is the medial line of ΠY, triangles $Z\Pi X$ and ZYX are congruent, therefore angle $\widehat{Z\Pi X}$ is right and consequently ZX will coincide with the tangent from X (Proposition 6.8).

(2) and (4) follow directly from the fact that, by definition, triangle ΠXY is isosceles,

(5) follows from the fact that MN, joins the middles of $\Pi \Xi$ and ΠY.

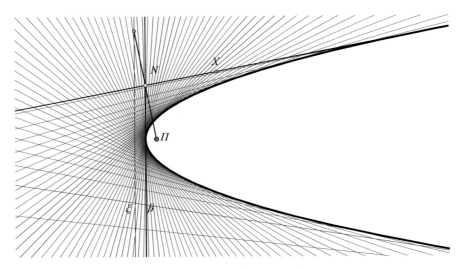

Fig. 6.30: Parabola as envelope

A consequence of the preceding proposition is that the parabola is, as we say, the **envelope** of the medial lines ε of the segments ΠY, where Π is fixed and Y moves on a line ξ. This means that, all these lines, which are produced for the various positions of Y on line ξ, are tangent to the parabola which has its focus at Π and directrix ξ.

Exercise 6.14. *The right angle $\widehat{\Pi NX}$ varies so that its vertex N moves on a line β and one of its sides passes through a fixed point Π. Show that its other side NX remains tangent to a parabola with focus at Π and the tangent at the vertex of the parabola is line β (See Figure 6.30).*

Archimedean triangle of the parabola is called a triangle, whose two sides are tangent to the parabola and its third side is the line (chord) which joins the contact points of these two tangents (See Figure 6.31). Archimedes, using an ingenious procedure to sum an infinite series of numbers, determined the area of the part of the interior of the parabola contained in such a triangle. In what follows we'll calculate this area using a different method ([66, p.239]), which gives the opportunity to study some fundamental properties of the parabola.

Theorem 6.12. *For every Archimedean triangle $AB\Gamma$ with contact points at A and B, the median ΓM is parallel to the axis of the parabola and intersects the parabola*

6.5. THE PARABOLA

at the middle Δ of ΓM, at which the tangent to the parabola is parallel to the base AB (See Figure 6.31).

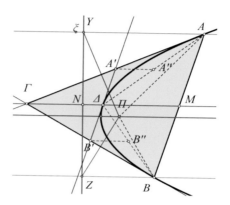

Fig. 6.31: Archimedean triangle

Proof. Draw from A and B the orthogonals AY and BZ to the directrix ξ. From the definition of the parabola, triangles ΠBZ and ΠAY are isosceli and $A\Gamma$ and $B\Gamma$ are tangent (Proposition 6.11) and simultaneously medial lines of ΠY and ΠZ respectively. In triangle ΠZY then, the medial lines $A\Gamma$ and $B\Gamma$ of ΠY and ΠZ meet at Γ. Therefore the medial line ΓN of the third side ZY will pass through Γ and will be parallel to the axis. In the trapezium $AYZB$, ΓN is parallel to its parallel sides and passes through the middle N of one of its lateral sides, therefore it will also pass through the middle M of its other lateral side. This proves the first claim of the proposition. For the second, consider the tangent at the intersection point Δ of the median ΓM and the parabola. Suppose that this tangent intersects $A\Gamma$ and $B\Gamma$ at points A' and B'. These define two other Archimedean triangles: $\Delta A'A$ and $\Delta B'B$, whose medians $A'A''$ and $B'B''$ will be parallel to the axis of the parabola. It follows that A' and B' are the middles of $A\Gamma$ and $B\Gamma$ respectively and $A'B'$ is parallel to AB.

Theorem 6.13 (Archimedes' quadrature of the parabola). *The region of the parabola contained in an Archimedean triangle $AB\Gamma$ has area equal to 2/3 the area $\varepsilon(AB\Gamma)$ of the triangle.*

Proof. The proof is a direct consequence of the preceding proposition. Indeed, triangle $AB\Delta$ is inside the region of the parabola and $A'B'\Gamma$ is outside and has area $\varepsilon(A'B'\Gamma) = \varepsilon(AB\Delta)/2$ (See Figure 6.31). I'll consider throwing triangle $A'B'\Gamma$ out and keeping $AB\Delta$. Next I repeat the same process using the new Archimedean triangles $AA'\Delta$ and $BB'\Delta$ throwing away again half the area of that I keep. The triangles I get rid of are these which do not contribute to the area of the region. The triangles I keep are those that contribute and I continue ad infinitum with smaller and smaller triangles throwing away part of them $\varepsilon(\pi)$ and keeping another, $\varepsilon(\kappa)$, always however in a ratio $\varepsilon(\pi) = \varepsilon(\kappa)/2$. In the limit of the infinite procedure I'll have exhausted the entire area of the initial triangle throwing part $\varepsilon(\Pi) = \varepsilon(K)/2$ of the part $\varepsilon(K)$ which I keep. The remaining area $\varepsilon(K)$ which will be left will be exactly the area of the region of the parabola which interests me. Because $\varepsilon(\Pi) + \varepsilon(K) = \varepsilon(AB\Gamma)$, it follows that $\frac{1}{2}\varepsilon(K) + \varepsilon(K) = \varepsilon(AB\Gamma)$ (generalization: [93]).

Theorem 6.14. *Let $AB\Gamma$ be an Archimedean triangle of the parabola κ and A', B' the middles, respectively, of $A\Gamma$, $B\Gamma$ and A_1, B_1 be the projections of the focus Π, respectively, to the sides $A\Gamma$, $B\Gamma$. The following properties are valid (See Figure 6.32):*

1. *Points Π, A_1, B_1, Γ are concyclic.*
2. *The triangles $\Pi\Gamma B$, $\Pi A\Gamma$, $\Pi A'B'$, $\Pi A_1 B_1$ are similar.*
3. *$|\Gamma\Pi|^2 = |\Pi A||\Pi B|$.*
4. *$\Gamma\Pi$ is a bisector of the angle $\widehat{B\Pi A}$ and a symmedian of the triangle $AB\Gamma$.*
5. *Angles $\widehat{B\Gamma\Pi}$ and $\widehat{M\Gamma A}$ are equal and angle $\widehat{A\Pi B}$ has double the measure of $\widehat{A\Gamma B}$.*
6. *The circumcircle $(\Gamma A'B')$ of triangle $\Gamma A'B'$ passes through the focus Π.*
7. *Line $A_1 B_1$ contains also the vertex of the parabola and is the Simson line of the tangential (circumscribed) triangle $A'\Gamma B'$ of the parabola relative to the point Π.*
8. *The Steiner line of the triangle $A'\Gamma B'$ relative to point Π coincides with the directrix of the parabola.*
9. *Every tangential (circumscribed) triangle $A'\Gamma B'$ of the parabola has its orthocenter at the directrix.*

Proof. (1) follows from the fact that, opposite angles of the quadrilateral $\Pi A_1 \Gamma B_1$ are right.

(2) follows from the fact that, because of (1), $\widehat{B\Gamma\Pi} = \widehat{B_1 A_1 \Pi}$. Also, because of the symmetry relative to $B\Gamma$, $\widehat{\Gamma B\Pi} = \widehat{\Gamma BZ} = \widehat{A_1 B_1 \Pi}$. The last equation holding because of the orthogonality of the corresponding sides of the angles. This shows that the triangles $\Gamma\Pi B$ and $A_1 B_1 \Pi$ are similar. Analogously is seen that also $\Gamma\Pi A$ and $A_1 B_1 \Pi$ are similar. The similarity of these triangles to $\Pi A'B'$ follows from the fact that $\Pi A'$, $\Pi B'$ are medians of similar triangles from respective vertices. Thus, we have $\widehat{A'\Pi B'} = \widehat{\Gamma\Pi B}$ and $\frac{|\Pi A'|}{|\Pi B'|} = \frac{|\Pi \Gamma|}{|\Pi B|}$. (3) is a direct consequence of (2).

(4) follows from the equality of angles $\widehat{\Gamma\Pi B} = \widehat{\Gamma\Pi A}$. Also, from the similarity of triangles of (2), follows that their altitudes are proportional to the corresponding sides, something which characterizes the points of the symmedian from A: $\frac{|\Pi B_1|}{|\Pi A_1|} = \frac{|\Gamma B|}{|\Gamma A|}$ (Theorem I-3.29).

6.5. THE PARABOLA

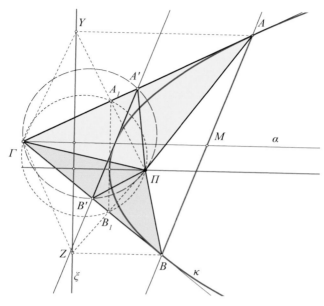

Fig. 6.32: Properties of the Archimedean triangle

(5) is a direct consequence of (4).

(6) follows also from (1) and (2), since angle $\widehat{A'\Pi B'} = \widehat{A_1 \Pi B_1}$ is supplementary to $\widehat{A\Gamma B}$.

(7) follows from theorem 6.11 and the definition of the Simson line, which contains the projections A_1, B_1, Γ_1 of Π on the sides of the triangle.

(8) is a consequence of (7) and theorem I-5.13, (9) is a direct consequence of (8).

Corollary 6.5. *A parabola is completely determined from one of its Archimedean triangles.*

Proof. From the preceding theorem and using figure 6.32, the Archimedean triangle $AB\Gamma$ determines the triangle $A'B'\Gamma$, whose circumscribed circle $\kappa = (A'B'\Gamma)$ contains the focus Π of the parabola. Π coincides with the intersection of κ with the symmedian of $AB\Gamma$ from Γ. Also the tangent to the parabola at its vertex coincides with the Simson line of $A'B'\Gamma$ relative to Π. Knowing the focus and the tangent at its vertex the parabola is considered to be known.

Theorem 6.15. *Let $AB\Gamma$ be an Archimedan triangle of the parabola κ with tangents ΓA, ΓB at A, B respectively. Let also $A'B'$ be another tangent to the parabola with A', B' on the sides, respectively, ΓA, ΓB and point of contact Δ. Then it holds*

$$\frac{|AA'|}{|A'\Gamma|} = \frac{|\Gamma B'|}{|B'B|}.$$

And conversely, if for two points $\{A', B'\}$ on the tangents holds the preceding relation, then $A'B'$ is also tangent to the parabola. The preceding ratio is equal to $\frac{|A'\Delta|}{|\Delta B'|}$.

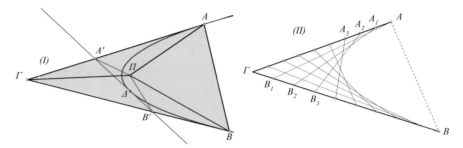

Fig. 6.33: Properties of ratios of Archimedean triangles

Proof. (See Figure 6.33-I) This is a consequence of (2) of the preceding theorem. In fact, applying the theorem to the Archimedean triangle $\Delta B'B$, we see that $\widehat{\Delta B'\Pi} = \widehat{B'B\Pi}$ and applying the theorem to the triangle $\Delta A'A$, we see that $\widehat{\Delta A'\Pi} = \widehat{A'A\Pi}$. Hence the triangles $A\Pi\Gamma$, $A'\Pi B'$ and $\Gamma\Pi B$ are similar. It follows then that $\Pi A'$ and $\Pi B'$ make the same angles $\widehat{A'\Pi\Gamma} = \widehat{B'\Pi B}$, which implies the property to prove.

Conversely, if the property holds and we draw from A' the tangent to κ, which intersects the other side at B'', then according to the first and proved part of the theorem, we'll have $\frac{|AA'|}{|A'\Gamma|} = \frac{|\Gamma B''|}{|B''B|}$. Latter, combined with the given relation gives $\frac{|\Gamma B''|}{|B''B|} = \frac{|\Gamma B'|}{|B'B|}$ and consequently point B'' will coincide with B'. Last claim is proved by applying theorem 6.14-(4) to the Archimedean triangle $AA'\Delta$, according to which $A'\Pi$ is a bisector of $\widehat{\Delta\Pi A}$ and consequently $\Pi\Delta$ divides $A'B'$ into the same ratio as A' divides $A\Gamma$.

Remark 6.7. The preceding theorem supports the practical construction of parabolas by dividing the sides of triangle $\widehat{A\Gamma B}$ into equal segments. According to it, we divide $A\Gamma$ into n equal segments starting from A, and ΓB also into n equal segments starting from Γ. By joining with line segments points with equal indices we construct tangents of a parabola, which has $A\Gamma B$ as an Archimedean triangle (See Figure 6.33-II). Details and related considerations are contained in exercise 6.45.

Theorem 6.16. *Three tangents $\{\alpha, \beta, \gamma\}$ of a parabola intersected by a variable fourth tangent ε at points, respectively $\{I, Z, E\}$, define a ratio $\kappa = \frac{IE}{IZ}$ independent of the position of the tangent ε (See Figure 6.34).*

6.5. THE PARABOLA

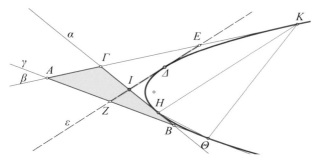

Fig. 6.34: Theorem of three tangents of parabola

Proof. We write the ratio

$$\kappa = \frac{IE}{IZ} = \frac{IE}{I\Delta} \cdot \frac{I\Delta}{IZ}, \quad \text{however} \quad \frac{IE}{I\Delta} = \frac{H\Gamma}{HI} \quad \text{and} \quad \frac{I\Delta}{IZ} = \frac{HI}{HB}.$$

The last two relations result by applying Theorem 6.15, respectively, to the Archimedean triangle ΓHK and to the Archimedean triangle $ZB\Delta$. Multiplying the relations by parts, we find $\kappa = \frac{H\Gamma}{HB}$, which is independent of the position of ε.

Exercise 6.15. Show that the middles M of parallel chords AB of a parabola lie on a line ε parallel to the axis and ε intersects the parabola at point Δ, at which the tangent is parallel to these chords (See Figure 6.35).

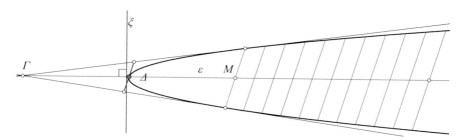

Fig. 6.35: Middles of parallel chords contained in diameter ε

Hint: Direct consequence of Theorem 6.12, according to which the middle M of the chord is on the median ΓM of the Archimedean triangle defined by the chord, and the median itself is parallel to the axis and passes through the point Δ at which the tangent is parallel to the chord. Consequently all the chords will have their middle on the line parallel to the axis passing through Δ. Point Δ is characterized by the fact, that there the tangent is parallel to the common direction of the chords. Now the fact that such a Δ exists, that is, that there exists only one point of the parabola at which the tangent has a given direction, follows from the fact that, by definition, the parabola has all its points on the same side of each tangent to it.

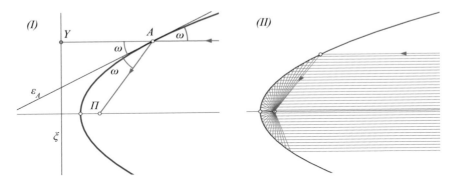

Fig. 6.36: Parabolic reflector

Exercise 6.16. Show that a ray incident to a parabola in direction parallel to its axis, reflected, passes through its focus.

Hint: Direct consequence of Theorem 6.11, according to which the tangent ε_A at A bisects angle $YA\Pi$ (See Figure 6.36).

Exercise 6.17. Show that the geometric locus of the centers of circles, tangent to a given line ξ and passing through a given point Π not lying on ξ, is the parabola with focus at Π and directrix ξ (See Figure 6.37).

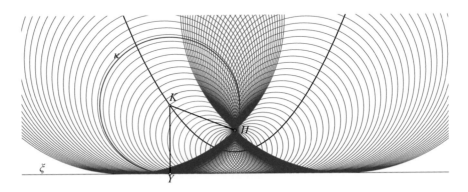

Fig. 6.37: Parabola as locus of centers of circles

Exercise 6.18. Find the intersection points of a line δ and the parabola κ, for which only the focus K and the directrix ξ is given.

6.5. THE PARABOLA

Hint: According to exercise 6.17, the two intersection points of δ with the parabola will also be centers of circles tangent to ξ but also passing through the focus K and its symmetric K' relative to δ. Apply then the results of Exercise I-4.1.

Exercise 6.19. Let us suppose that the shape of the parabola κ is given and that we have the capability to define points on it and its intersections with lines. How can we find, using these means, the positions of its focus and its directrix?

Hint: The middles of two parallel chords of it determine the direction of its axis and therefore also that of the directrix. Draw from the middle M of a chord AB parallel to the axis and find its intersection point N with the parabola. Extend MN towards N and find P: $|MP| = 2|MN|$. This defines the Archimedean triangle PAB and you can apply corollary 6.5.

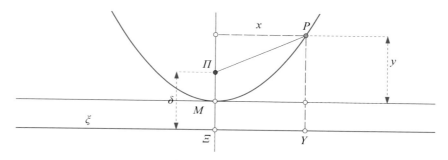

Fig. 6.38: Parabola through equation

Exercise 6.20. Show that the distances x, y of a point P of the parabola, from its axis and the tangent to its vertex respectively (See Figure 6.38), satisfy the equation
$$x^2 = 2\delta \cdot y,$$
where δ is the parameter of the parabola. And conversely, if (x, y), which are defined by projecting point P onto two orthogonal axes, satisfy this equation, then P is contained in a parabola.

Hint: $|P\Pi| = |PY| \Leftrightarrow \left(y + \frac{\delta}{2}\right)^2 = x^2 + \left(y - \frac{\delta}{2}\right)^2$.

Exercise 6.21. Prove the claims of remark-1 of § 6.1 which concerns parabolas. In other words, (i) that the point P of the plane is contained in a parabola, if and only if there exist two orthogonal axes OX, OY, such that the projections X', Y' of P on them, define oriented segments $x = OX'$, $y = OY'$, which satisfy the equation $y = ax^2 + bx + c$ and (ii) the aforementioned formulas for the focus and the directrix of the parabola hold.

Exercise 6.22. Show that there are no parallel tangents to a parabola. Conclude that each triple of different tangents of a parabola defines a triangle, as well as three corresponding Archimedean triangles.

6.6 The ellipse

> To read good books is like holding a conversation with the most eminent minds of past centuries and, more-over, a studied conversation in which these authors reveal to us only the best of their thoughts.
>
> R. Descartes

The shape of an ellipse is completely determined by two numbers: the **focal distance** which is defined as the distance between the two focals Π and T of the ellipse and the constant $\lambda > |\Pi T|$, which gives the sum of the distances

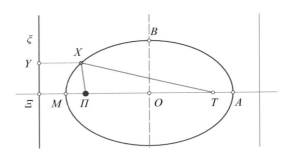

Fig. 6.39: Two ways to define an ellipse

of an arbitrary point X of the ellipse from the focals (See Figure 6.39)

$$|XT| + |X\Pi| = \lambda.$$

Two other numbers, which also determine completely the shape of an ellipse are: the distance $p = |\Pi \Xi|$ of one of its focals Π from the corresponding directrix ξ (the *focal parameter of the ellipse*) and the eccentricity $\kappa < 1$ for which holds

$$\frac{|X\Pi|}{|XY|} = \kappa,$$

where Y is the projection on the directrix of the arbitrary point X of the ellipse.

Proposition 6.10. *The ellipse is symmetric relative to the line ΠT of its focals and is also symmetric relative to the medial line of ΠT. The middle O of ΠT is a center of symmetry of the ellipse.*

6.6. THE ELLIPSE

Proof. A direct consequence of the definition (§ 6.1) of the ellipse. If the point X has distances from the focals which satisfy $|X\Pi| + |XT| = \lambda$, we easily see that the symmetric points of X relative to the two aforementioned lines, as well as point O will satisfy the same equality, therefore they will be points of the ellipse.

Line ΠT and the medial line of ΠT are called **axes** of the ellipse. The lengths of the two line segments OA and OB (but also often the segments themselves or/and the entire lines OA and OB), which are intersected by the ellipse on the axes, are respectively called **major axis** and **minor axis** of the ellipse. Their point of intersection O is called **center of the ellipse**. Chords of the ellipse which pass through its center are called **diameters of the ellipse**.

Proposition 6.11. *The directrices are orthogonal to the line of the focals ΠT and lie symmetrically relative to the center of the ellipse. The intersection Ξ of the directrix with the axis and the corresponding focus Π are harmonic conjugate points of A and M at which the ellipse intersects the axis ΠT (See Figure 6.39).*

Proof. The orthogonality results from the symmetry relative to the axis OA. If X belongs to the ellipse then its symmetric X' relative to this axis also belongs to the ellipse. Because $X\Pi$ and $X'\Pi$ have the same length, XY and $X'Y'$ will be orthogonal to the same line and they will have the same length, therefore $XYY'X'$ will be a rectangle. The second claim follows directly from the definition through the eccentricity κ, since points M and A will satisfy

$$\frac{|M\Pi|}{|M\Xi|} = \frac{|A\Pi|}{|A\Xi|} = \kappa,$$

therefore A and M will be harmonic conjugate relative to Ξ and Π. Consequently (Exercise 1.77) points Ξ, Π will be harmonic conjugate of A and M.

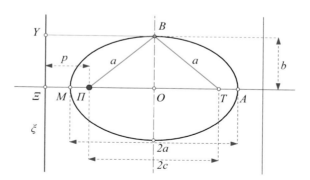

Fig. 6.40: Basic formulas for the ellipse

Next formularium contains some relationships of the various segment lengths which characterize the ellipse (See Figure 6.40). We use the notation:

$$a = |OA|, \ b = |OB|, \ c = |O\Pi| = \frac{|\Pi T|}{2}, \ p = |\Pi \Xi|.$$

Proposition 6.12.

$$a = |OA| = |B\Pi| = |BT| = \frac{\lambda}{2} \ \Leftrightarrow \ \lambda = 2a = |AM|, \quad (6.1)$$

$$c = \frac{|\Pi T|}{2}, \ b = |OB| \ \Rightarrow \ a^2 = b^2 + c^2, \quad (6.2)$$

$$|M\Pi| = a - c, \ |A\Pi| = a + c \ \Rightarrow \ |M\Xi| = \frac{a-c}{\kappa}, \ |A\Xi| = \frac{a+c}{\kappa}, \quad (6.3)$$

$$p = |\Pi\Xi| = |\Pi M| + |M\Xi| = (a-c) + \frac{a-c}{\kappa} = \frac{1+\kappa}{\kappa}(a-c), \quad (6.4)$$

$$\kappa = \frac{c}{a} \ \Rightarrow \ p = \frac{1-\kappa^2}{\kappa}a = \frac{a^2-c^2}{c} = \frac{b^2}{c}, \quad (6.5)$$

$$a = \frac{\kappa}{1-\kappa^2}p, \ c = \frac{\kappa^2}{1-\kappa^2}p. \quad (6.6)$$

Proof. Trivial consequences of the definitions. The first of 6.5 follows from the eccentricity $\kappa = \frac{|B\Pi|}{|BY|}$ (See Figure 6.40), using the preceding ones. The last two lines give formulas for the pairs of numbers (p, κ) and (λ, c), which define the ellipse in the two usual ways.

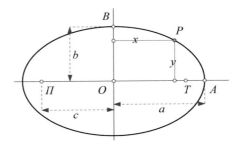

Fig. 6.41: Ellipse equation referred to its axes

Proposition 6.13. *If x, y are the distances of a point P of the ellipse from the minor and major axis, respectively OY and OX, then the following relation holds*

$$\frac{x^2}{a^2} + \frac{y^2}{b^2} = 1, \quad (6.7)$$

where a, b, are, respectively, the major and minor axes of the ellipse. And conversely, if the point P has distances x, y from two orthogonal axes OY and OX (See Figure

6.6. THE ELLIPSE

6.41), which satisfy this equation, then it is contained in the ellipse of center O, major axis a along OX and minor axis b along OY.

Proof. The equality $|P\Pi| + |PT| = 2a$ is expressed as a function of x and y with
$$\sqrt{(c+x)^2 + y^2} + \sqrt{(c-x)^2 + y^2} = 2a. \qquad (6.8)$$
Squaring both sides, isolating the root and squaring again so that it is eliminated, we are led to the aforementioned formula, taking into account also the formula $a^2 = b^2 + c^2$. The calculations are reversed and through them, from formula 6.7 follows 6.8, which shows the converse.

Exercise 6.23. Show that the half parameter of the ellipse is $\frac{b^2}{a} = \kappa \cdot p$, where κ is the eccentricity of the ellipse and p is its focal parameter.

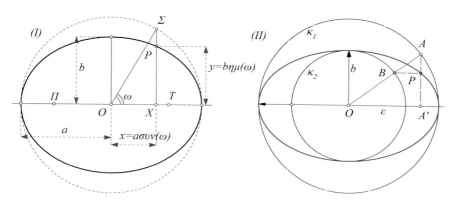

Fig. 6.42: Eccentric angle   Major and minor circle

Primary or also **major circle** of the ellipse is called the circle κ_1, with center the center O of the ellipse and radius the major axis a. This circle is tangent to the ellipse at its vertices contained in the major axis. **Minor circle** or **secondary** of the ellipse is called the circle κ_2 with center the point O and radius the minor axis b of the ellipse. This circle is concentric with the major one and is tangent to the ellipse at its vertices contained in the minor axis (See Figure 6.42-II). Last figure suggests a construction of the ellipse which relies on these two auxiliary circles and is examined in exercise 6.24.

Eccentric angle at the point P of the ellipse is called the angle $\omega = \widehat{XO\Sigma}$, which is formed from the major axis and the radius $O\Sigma$ of the major circle at point Σ, at which the orthogonal from P to the major axis intersects this circle (See Figure 6.42-I).

Theorem 6.17. *Let P be a point on the ellipse and X be the projection of P on the major axis. Let also Σ be the intersection point of XP with the major circle, lying on* the same side as P relative to the major axis and ω be the eccentric angle of P. Then the following properties hold (See Figure 6.42-I):

1. $x = |OX| = a\cos(\omega)$, $y = |PX| = b\sin(\omega)$.
2. $\frac{|PX|}{|\Sigma X|} = \frac{b}{a}$.
3. Conversely, if to every point Σ of a circle κ with diameter $2a$ we correspond point P, such that the preceding relation holds, where X is the projection of P to a fixed diameter of the circle, then P lies in the ellipse with major axis a on the fixed diameter, minor axis b and major circle κ.

Proof. (1) follows from the fact that $x = a\cos(\omega)$ and Proposition 6.13, according to which $\frac{x^2}{a^2} + \frac{y^2}{b^2} = 1$ must also hold. By substituting x, we have $\cos^2(\omega) + \frac{y^2}{b^2} = 1$, which shows the wanted.

(2) is a direct consequence of (1), since $\frac{|PX|}{|\Sigma X|} = \frac{b\sin(\omega)}{a\sin(\omega)}$.

(3) holds because, then, for angle $\omega = \widehat{XO\Sigma}$ hold $x = a\cos(\omega)$, $y = b\sin(\omega)$ and equation 6.7 is satisfied.

Exercise 6.24. Show that the points of the ellipse can be constructed using the following method (See Figure 6.42-II). We consider two concentric relative to the ellipse circles κ_1, κ_2, with radii, respectively, a and b, as well as line ε through their centers O. The arbitrary point A of κ_1 defines through the radius OA the point B on κ_2. The perpendicular AA' on ε from A and the perpendicular BP from B to AA' define the point P, which describes an ellipse having κ_1, κ_2, respectively, as major and minor circles.

Besides the major and minor circles there exist two more (congruent) circles with centers at the focals of the ellipse and radius $2a$, called **auxiliary circles** of the ellipse (See Figure 6.43). Next theorem shows, that all these circles are connected with important properties of the ellipse.

Theorem 6.18. Let Σ be an arbitrary point of the auxiliary circle σ_Π of the ellipse with center at the focus Π and second focus at T. The following properties are valid (See Figure 6.43):

1. The medial line ε of the segment ΣT intersects $\Sigma \Pi$ at a point X of the ellipse.
2. Line ε is tangent to the ellipse and X is the point of contact of ε.
3. The ellipse is contained whole in the side of its tangent which contains also the focals of the ellipse.
4. The ellipse is contained whole in the interior of its major circle σ.
5. The tangent ε at X bisects the supplementary of angle $\widehat{\Pi X T}$.
6. The projections Y and Y' of the focals to the tangent ε are contained in the major circle of the ellipse.
7. The product $|TY||\Pi Y'|$ is fixed and equal to b^2.
8. The tangent ε intersects the directrix β of the focus Π at point B, for which $\widehat{B\Pi X}$ is right.

6.6. THE ELLIPSE

9. Line ε intersects the directrix γ of the focus T at a point Γ, for which $\widehat{X T \Gamma}$ is right and $\Sigma \Gamma$ is tangent to σ_Π. Also $\Sigma \Gamma$ is parallel to $B\Pi$.

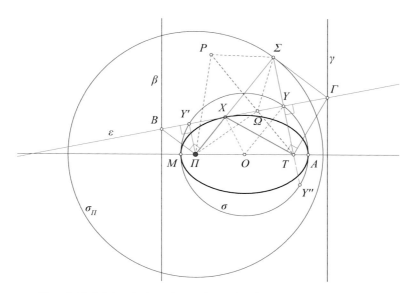

Fig. 6.43: Major (σ) and auxiliary (σ_Π) circles of the ellipse

Proof. (1) is a consequence of the definition, since $|\Pi X| + |XT| = |\Pi X| + |X\Sigma| = 2a$.

For (2) let us consider a point Ω of ε, different from X. Because of the medial line
$$|\Omega\Pi| + |\Omega T| = |\Pi\Omega| + |\Omega\Sigma| > |\Pi\Sigma| = 2a.$$
The last inequality is the triangle inequality for the triangle $\Pi\Sigma\Omega$. The inequality shows that point Ω does not belong to the ellipse, therefore X is the unique point of ε in common with the ellipse.

For (3) let us consider an arbitrary point P on the other side of ε from the one containing the focals. Let Ω be the intersection of PT with ε. It holds
$$|P\Pi| + |PT| = |P\Pi| + |P\Omega| + |\Omega T|$$
$$= |P\Pi| + |P\Omega| + |\Omega\Sigma| \ > \ |P\Pi| + |P\Sigma| \ > \ |\Pi\Sigma| \ = \ 2a.$$

Therefore P does not belong to the ellipse. The last two are the triangle inequalities for triangles $P\Sigma\Omega$ and $\Pi P\Sigma$.

(4) holds because OX is a median of the triangle ΠXT and consequently $|OX| < \frac{1}{2}(|X\Pi| + |XT|) = a$ (Exercise I-1.48).

(5) holds because the triangle $TX\Sigma$ is isosceles, therefore the medial line ε of ΣT bisects its apical angle.

(6) for point Y, holds because OY joins the middles of the sides of triangle $\Sigma\Pi T$, therefore $|OY| = \frac{|\Pi\Sigma|}{2} = a$. The property for the projection Y' of Π on ε is proved analogously.

(7) follows by extending TY until it intersects the circle σ at the second point Y''. Because of the symmetry of the ellipse relative to O, TY'' and $\Pi Y'$ are equal and the product $|TY||\Pi Y'| = |TY||TY''|$ is equal (modulo sign) with the power of the focus T relative to the major circle σ.

(8) and the second claim of (9) follow from the corresponding general property of conics (Proposition 6.8). (9) follows from (8) and the symmetry of triangles $XT\Gamma$ and $X\Sigma\Gamma$ relative to the line ε.

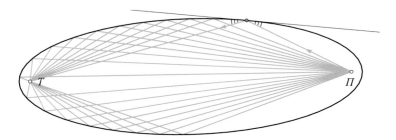

Fig. 6.44: Reflecting property of the ellipse focals

Remark 6.8. Property (5) implies that the reflected rays of light, which initially emanate from one focus Π and reflect at the ellipse so that the incident and reflected rays form equal angles with the tangent, pass through the other focus T (See Figure 6.44).

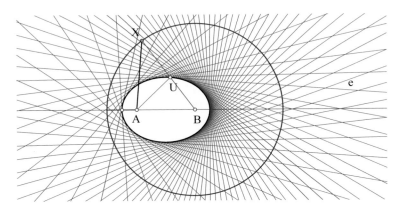

Fig. 6.45: Envelope of medial lines of segments AX

Exercise 6.25. Show that the medial line of a segment AX, with A fixed and X variable on the circle $\kappa(B,\rho)$, is a tangent to ellipse with focals at A and

6.6. THE ELLIPSE

at the center of the circle B and great axis the radius ρ of the circle (Figure 6.45).

Hint: Consider the intersection Y of the medial line with BX. The sum $|XA| + |XB|$ is fixed and equal to the radius ρ of the circle κ.

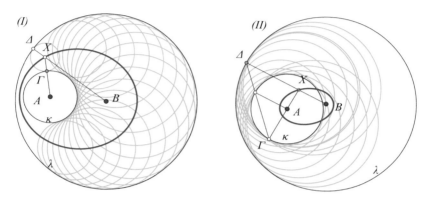

Fig. 6.46: Ellipse as a locus of centers of circles tangent to two others

Exercise 6.26. Given are two non concentric circles $\kappa(A,\rho)$ and $\lambda(B,\rho')$, of which the first is contained whole inside the second. Show that the geometric locus of the centers X of circles, which are tangent simultaneously to κ and λ and do not contain/contain κ, is an ellipse with focals at points A and B and major axis the half sum/difference of the radii of the two circles.

Hint: The point of contact Γ of a circle tangent to κ lies on the center-line AX (See Figure 6.46-I) (analogous is figure 6.46-II in the case the circles contain κ). The same thing happens also with the contact point Δ of the circle with the other circle λ. Then, we have

$$|XA| + |XB| = |A\Gamma| + |\Gamma X| + |XB| = |A\Gamma| + |XB| + |X\Delta| = \rho + \rho'.$$

Exercise 6.27. From points A, A' of the major axis of an ellipse and to the same side of this axis draw orthogonal half lines, which intersect, respectively, the ellipse and the auxiliary circle at points B, B' and Γ, Γ' (See Figure 6.47-I). Show that the lines BB' and $\Gamma\Gamma'$ intersect on the major axis. Conclude also that the tangents, respectively, of the ellipse at B and the auxiliary circle at Γ intersect at the same point of the major axis (See Figure 6.47-II).

Hint: The first property results from Theorem 6.17, by applying the theorem of Thales (See Figure 6.47-I). The second, for the tangents, follows from the first, by letting A' converge towards A, in which case points B', Γ' converge respectively towards B, Γ and the secants BB', $\Gamma\Gamma'$ converge towards the

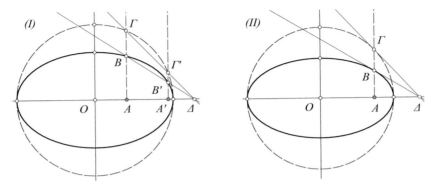

Fig. 6.47: Intersection of respective chords/tangents on the major axis

corresponding tangents at B and Γ (See Figure 6.47-II). The proof can also be done directly, without limits, using the formulas of Theorem 6.17.

Remark 6.9. This exercise shows that the polars of points on the major axis, relative to the conic, coincide with the polar lines of the same point relative to the major circle of the ellipse. A corresponding property holds also for the minor axis and the respective minor circle of the ellipse.

Theorem 6.19. *Let M be the intersection point of two tangents MA, MB of the ellipse, which are parallel to conjugate directions, so that $OAMB$ is a parallelogram, where O is the center of the ellipse (See Figure 6.48-I). From A and B we draw perpendiculars AA_0, BB_0 to the major axis, which intersect the major circle at points, respectively, A_1 and B_1. Then the angle $\widehat{A_1OB_1}$ is right. The converse also holds. If angle $\widehat{A_1OB_1}$ is right, then OA, OB are conjugate directions.*

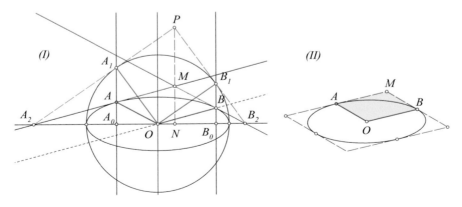

Fig. 6.48: Conjugate tangents

6.6. THE ELLIPSE

Proof. Follows directly from exercise 6.27, which guarantees the intersection of the tangents at A, A_1 at point A_2 of the axis (See Figure 6.48-I), and the equality of ratios:

$$\frac{A_1A_0}{A_2A_0} = \frac{A_1A_0}{AA_0} \cdot \frac{AA_0}{A_2A_0} = \frac{B_1B_0}{BB_0} \cdot \frac{BA_0}{OB_0} = \frac{B_1B_0}{OB_0}.$$

The second equality holds because of Theorem 6.17, according to which $\frac{A_1A_0}{AA_0} = \frac{B_1B_0}{BB_0} = \frac{a}{b}$ and the similarity of triangles A_2A_0A, OB_0B. This equality implies that the right triangles $A_2A_0A_1$ and OB_0B_1 are similar, therefore angle $\widehat{A_1OB_1}$ is right. The converse is proved by reversing the sequence of arguments.

Corollary 6.6. *All the parallelograms circumscribed to an ellipse, whose sides define conjugate directions, have fixed area equal to 4ab, where a, b are the axes of the ellipse.*

Proof. The parallelogram $AOBM$ of figure (See Figure 6.48-I) is one fourth (regarding the area) of a typical such circumscribed parallelogram (like that of figure 6.48-II). However, by decomposing the areas into differences of triangles with common bases like

$$\varepsilon(OAMB) = \varepsilon(MA_2B_2) - \varepsilon(AA_2O) - \varepsilon(OBB_2),$$
$$\varepsilon(PA_1OB_1) = \varepsilon(PA_2B_2) - \varepsilon(A_1A_2O) - \varepsilon(OB_1B_2),$$

we see that $AOBM$ and the square A_1OB_1P have ratio of areas b/a, while the square has area a^2. Therefore the area of $AOBM$ will be ab.

Corollary 6.7. *For two conjugate half diameters OA, OB, holds $|OA|^2 + |OB|^2 = a^2 + b^2$.*

Proof. It holds (See Figure 6.48-I):

$$|OA|^2 + |OB|^2 = (|OA_0|^2 + |A_0A|^2) + (|OB_0|^2 + |B_0B|^2)$$
$$= (|OA_0|^2 + |OB_0|^2) + \frac{b^2}{a^2}(|A_0A_1|^2 + |B_0B_1|^2) = a^2 + b^2.$$

In the last equation holds $|OA_0|^2 + |OB_0|^2 = a^2$, because of the Theorem 6.17 and the orthogonality between OA_1, OB_1, which implies that the eccentric angles satisfy $\widehat{B_0OA_1} = \widehat{B_0OB_1} + \frac{\pi}{2}$. Also holds $|A_0A_1|^2 + |B_0B_1|^2 = a^2$, since the right triangles OA_1A_0 and B_1OB_0 are congruent.

Exercise 6.28. Show that every parallelogram circumscribed to an ellipse defines through its diagonals a pair of conjugate directions (See Figure 6.49-I). Also every parallelogram inscribed in an ellipse defines through its sides a pair of conjugate directions (See Figure 6.49-II).

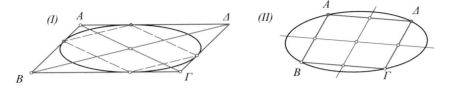

Fig. 6.49: Conjugate directions: circumscribed/inscribed parallelograms

Exercise 6.29. Show that from all the parallelograms circumscribed to an ellipse, the ones whose sides define conjugate directions have the least possible area.

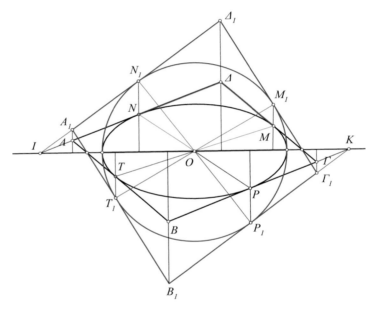

Fig. 6.50: Minimization of area of circumscribed parallelograms

Hint: Let $p = AB\Gamma\Delta$ be a circumscribed parallelogram and $p_1 = A_1 B_1 \Gamma_1 \Delta_1$ be the parallelogram circumscribed to the major circle and having contact points the projections of the respective points of p on the major circle (See Figure 6.50). By decomposing the areas into triangles with the same base, as in corollary 6.6, we see that the ratio of areas is fixed, as $\varepsilon(p_1)/\varepsilon(p) = a/b$. Therefore p is minimized, regarding the area, simultaneously with p_1. However all p_1 are rhombuses circumscribed to a fixed circle and the least one

6.6. THE ELLIPSE

such is the circumscribed square (Exercise I-3.142). This, implies, according to Theorem 6.19, the claim.

Exercise 6.30. Show that all parallelograms inscribed in an ellipse, whose diagonals define conjugate directions, have the same area $2ab$. Also show, that from all the inscribed in an ellipse parallelograms, the ones whose sides define conjugate directions have the maximum possible area.

Next exercise is, essentially, proposition 21 from the first book "Conics" of Apollonius ([97, p.19]). For its solutions one can use the affine projection of the major cirlce onto the ellipse (see remark 6.14).

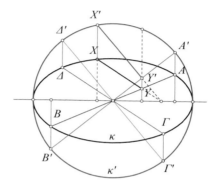

Fig. 6.51: A theorem of Apollonius

Exercise 6.31. Let $\{AB, \Gamma \Delta\}$ be conjugate diameters of the ellipse κ. An arbitrary point X of κ is projected parallel to $\Delta \Gamma$ onto the point Y of AB. Show that the ratio $\frac{XY^2}{YA \cdot YB}$ is constant.

Hint: Consider the points $\{A', B', \Gamma', \Delta', X'\}$ of the major circle κ', which are orthogonally projected onto the major axis to the corresponding points without primes (See Figure 6.51). Let Y' be the intersection of $A'B'$ with the parallel to XX' from Y. Show that, for constant conjugate diameters $\{AB, \Gamma \Delta\}$, the ratios $\{\frac{XY}{X'Y'}, \frac{YA}{Y'A'}, \frac{YB}{Y'B'}\}$ remain constant and independent of the position of X. Use the power of Y' relative to κ', which is equal to $|X'Y'|^2 = |Y'A'||Y'B'|$.

6.7 The hyperbola

> All things are strange. One can always sense the strangeness of a thing once it ceases to play any part; when we do not try to find something resembling it and we concentrate on its basic stuff, its intrinsicality.
>
> P. Valery, Tel Quel

Hyperbola's shape, like that of the ellipse, is completely determined by two numbers: the **focal distance**, defined by the distance between the two focals Π and T of the hyperbola $|\Pi T| = 2c$ and the constant $2a < |\Pi T| = 2c$ giving the difference of distances of the arbitrary point X of the hyperbola from the focals (See Figure 6.52):
$$||XT| - |X\Pi|| = 2a.$$

Two other numbers, which also completely determine hyperbola's shape are: the distance $p = |\Pi\Xi|$ of one of its focals Π from the corresponding directrix ξ (or **focal parameter** of the hyperbola) and the eccentricity $\kappa > 1$ for which holds
$$\frac{|X\Pi|}{|XY|} = \kappa,$$

where Y is the projection to the directrix of the arbitrary point X of the hyperbola.

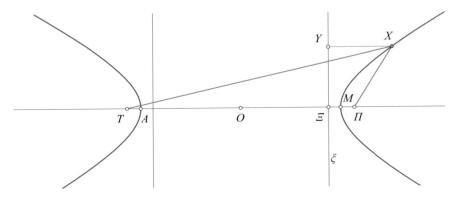

Fig. 6.52: Two ways to define a hyperbola

Proposition 6.14. *The hyperbola is symmetric relative to the line ΠT of the focals and also symmetric relative to the medial line of ΠT. The middle O of ΠT is a center of symmetry of the hyperbola.*

Proof. This is a direct consequence of the definition (§ 6.1) of the hyperbola. If point X has distances from the focals which satisfy $||X\Pi| - |XT|| = 2a$, we

6.7. THE HYPERBOLA

easily see that the symmetric points of X relative to the aforementioned two lines, as well as the symmetric of X relative to the point O will satisfy the same equation, therefore they will be points of the hyperbola.

TLine ΠT of the focals and the medial line of ΠT are called **axes** of the hyperbola. Their intersection point O is called **center** of the hyperbola. Chords of the hyperbola, which pass through its center, are called **diameters** of the hyperbola.

Proposition 6.15. *The directrices are orthogonal to the line of the focals ΠT and lie symmetrically relative to the center of the hyperbola. The intersection point Ξ of the directrix with the axis and the corresponding focus Π are harmonic conjugate points of $\{A, M\}$, at which the hyperbola intersects the axis ΠT.*

Proof. The orthogonality results from the symmetry relative to the axis OA. If X belongs to the hyperbola, then its symmetric X', relative to this axis, will also belong to the hyperbola. Because $X\Pi$ and $X'\Pi$ have the same length, XY and XY' will also be orthogonal to the same line and will have the same length, therefore $XYY'X'$ will be a rectangle. The second claim follows directly from the definition, since points M and A satisfy

$$\frac{|M\Pi|}{|M\Xi|} = \frac{|A\Pi|}{|A\Xi|} = \kappa,$$

therefore $\{A, M\}$ will be harmonic conjugate to $\{\Xi, \Pi\}$. Consequently (Exercise 1.77) latter will also be harmonic conjugate to $\{A, M\}$.

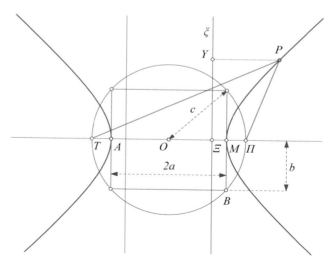

Fig. 6.53: Basic formulas for the hyperbola

Next formularium establishes relations between various segment lengths which characterize the hyperbola, using following notation (See Figure 6.53)

$$a = |OA|, \quad b = |MB|, \quad c = |O\Pi| = \frac{|\Pi T|}{2}.$$

Proposition 6.16.

$$a = |OA| \Leftrightarrow 2a = |AM|, \tag{6.9}$$

$$c = \frac{|\Pi T|}{2}, \quad b = |MB| \Rightarrow c^2 = a^2 + b^2, \tag{6.10}$$

$$|M\Pi| = c - a, \quad |A\Pi| = c + a \Rightarrow |M\Xi| = \frac{c-a}{\kappa}, \quad |A\Xi| = \frac{c+a}{\kappa}, \tag{6.11}$$

$$|\Pi\Xi| = |\Pi M| + |M\Xi| = (c-a) + \frac{c-a}{\kappa} = \frac{1+\kappa}{\kappa}(c-a), \tag{6.12}$$

$$\kappa = \frac{c}{a} \Rightarrow |\Pi\Xi| = \frac{\kappa^2 - 1}{\kappa}a = \frac{c^2 - a^2}{c} = \frac{b^2}{c}, \tag{6.13}$$

$$a = \frac{\kappa}{\kappa^2 - 1}p, \quad c = \frac{\kappa^2}{\kappa^2 - 1}p, \quad p = |\Pi\Xi|. \tag{6.14}$$

Proof. Trivial consequences of the definitions. The first of (6.13) follows from the ratio $\kappa = \frac{|A\Pi|}{|A\Xi|}$, using the preceding it relations. The last two lines give relations between the pairs of numbers $(|\Pi\Xi|, \kappa)$ and $(a, |\Pi T|)$, which define the hyperbola in the two usual ways.

A characteristic of the hyperbola we do not meet in other kinds of conic sections is the existence of **asymptotes** (See Figure 6.54). These are two lines, which separate all the lines through the center O of the hyperbola into two groups. The lines of the first group intersect the hyperbola, while these of the second group do not.

Proposition 6.17. *For every hyperbola there exists an angle ω with vertex at its center O, such that every line through O contained in that angle will intersect the hyperbola, while every line through O and not contained in that angle will not intersect it.*

Proof. For the proof we use Proposition I-3.7, according to which, points P of a line ε through O are characterized from a fixed ratio $\mu = \frac{|PX|}{|PY'|}$ (See Figure 6.54), where X and Y' are the projections of P onto two orthogonal lines (the axes of the hyperbola). For one point of such a line, then, the requirement for it to belong to the hyperbola leads to the equation

$$\kappa = \frac{|P\Pi|}{|PY|} = \frac{\sqrt{|PX|^2 + |X\Pi|^2}}{|PY'| - |YY'|} \Leftrightarrow |PX|^2 + |X\Pi|^2 = \kappa^2(|PY'| - |YY'|)^2.$$

6.7. THE HYPERBOLA

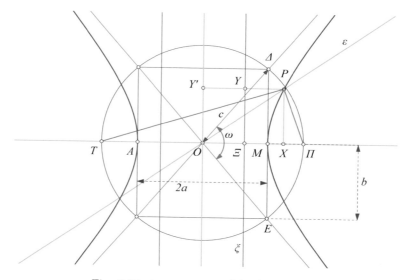

Fig. 6.54: Asymptotes of the hyperbola

If we set $|PY'| = x$, then we have the following relations

$$|PX| = \mu \cdot x, \quad |X\Pi|^2 = (c-x)^2, \quad |YY'| = \frac{a}{\kappa}$$

and the preceding equation is translated into a relation for x.

$$x^2(\mu^2 + 1 - \kappa^2) + b^2 = 0.$$

The last equation then has exactly a solution, when it holds

$$\mu^2 + 1 - \kappa^2 < 0 \quad \Leftrightarrow \quad \mu^2 < \frac{b^2}{a^2},$$

which is equivalent to the fact that the line ε intersects the orthogonal ΔE of AM at a distance from M less than b.

The constant a, often also the line segment OM or/and the line OM itself are called **transverse axis** of the hyperbola. Respectively the length b or/and the axis of symmetry orthogonal to OM is called **conjugate axis** of the hyperbola. The points (A, M) at which the transverse axis intersects the hyperbola are called **vertices** of the hyperbola.

Auxiliary circles (also **Circular directrix**) of the hyperbola are called the two circles with centers at its focals with radius $2a$, where a is the transverse axis of the hyperbola. The circle with center O and radius a is called **primary circle** or also **major circle** of the hyperbola.

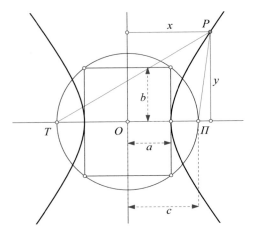

Fig. 6.55: Hyperbola through equation

Proposition 6.18. *If x, y are the distances of a point P of a hyperbola from the conjugate and transverse axis respectively, then the following equation holds*

$$\frac{x^2}{a^2} - \frac{y^2}{b^2} = 1,$$

where a, b are respectively the transverse and the conjugate axes of the hyperbola. And conversely, if for the projections of P to two orthogonal axes the preceding equation holds, then the point is contained in a hyperbola with transverse and conjugate axes respectively a and b.

Proof. (See Figure 6.55) The equation $|PT| - |P\Pi| = 2a$ is expressed as a function of x and y with

$$\sqrt{(x-c)^2 + y^2} - \sqrt{(x+c)^2 + y^2} = 2a.$$

Squaring both sides, isolating the root and re-squaring, so that it is eliminated, we are led to the aforementioned formula, taking also into account the equation $c^2 = a^2 + b^2$. The calculations in reverse order give also the proof of the converse.

Proposition 6.19. *Let Σ be an arbitrary point of the auxiliary circle σ_T of the hyperbola with center the focus T and second focus Π. The following properties hold (See Figure 6.56):*

1. *The medial line ε of the segment $\Sigma\Pi$ intersects ΣT at a point X of the hyperbola.*
2. *ε is tangent to the hyperbola and X is the contact point of ε.*
3. *The hyperbola is contained whole in the exterior of its major circle σ.*
4. *The tangent ε at X bisects the angle $\widehat{\Pi X T}$.*

6.7. THE HYPERBOLA

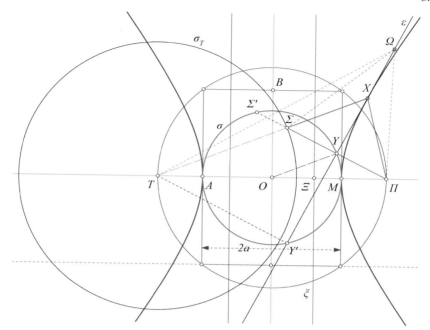

Fig. 6.56: Basic properties of hyperbola

5. The projections Y and Y' of the focals on the tangent ε are contained in the major circle of the hyperbola.
6. The product $|\Pi Y||TY'|$ is fixed and equal to b^2.

Proof. (1) is a consequence of the definition, since $|TX| - |X\Pi| = |TX| - |X\Sigma| = 2a$.

For (2) let us consider a point Ω of ε, different from X. Because of the medial line
$$|\Omega T| - |\Omega \Pi| = |\Omega T| - |\Omega \Sigma| < |T\Sigma| = 2a.$$
The last one is the triangle inequality for triangle $T\Sigma\Omega$. The inequality shows that Ω does not belong to the hyperbola, therefore X is the only point of ε in common with the hyperbola.

(3) is due to the fact that OX is a median of triangle ΠXT and consequently $|OX| > \frac{1}{2}(|XT| - |X\Pi|) = a$ (Exercise 1.48).

(4) is due to the fact that triangle $\Pi X\Sigma$ is isosceles, therefore the medial line ε of $\Sigma\Pi$ bisects the angle at its apical vertex.

(5) for the point Y, is due to the fact that OY joins the middles of the sides of triangle $\Sigma\Pi T$, therefore $|OY| = \frac{|T\Sigma|}{2} = a$. The property for the projection Y' of T on ε is proved analogously.

(6) follows by extending ΠY until it intersects the circle σ at point Σ'. Because of the symmetry of the hyperbola relative to O, $\Pi\Sigma'$ and TY' are

equal and the product $|\Pi Y||TY'| = |\Pi Y||\Pi \Sigma'|$ is equal to the power of the focus Π relative to the major circle σ.

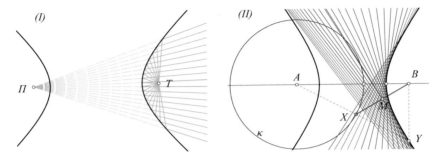

Fig. 6.57: Reflecting property   Hyperbola as envelope of medial lines

Figure 6.57-I shows the consequence of (4) of the preceding proposition, according to which, rays emanating from a focus T of the hyperbola and reflected on it converge to the other focus Π of the hyperbola.

Exercise 6.32. Given a circle $\kappa(A, \rho)$ and a point B outside it, show that the medial lines of the segments BX, where X is a point of the circle, are tangents to a hyperbola with focals at points A and B and transverse axis ρ.

Hint: Extend AX until it intersects the medial line at Y (See Figure 6.57-II) Then $|YA| - |YB| = |YA| - |YX| = \rho$.

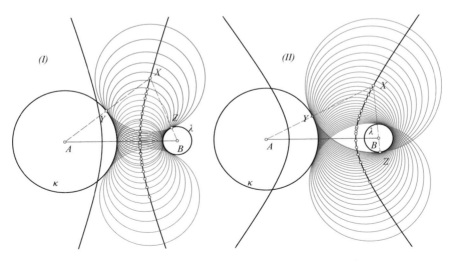

Fig. 6.58: Hyperbola as locus of centers of circles

6.7. THE HYPERBOLA

Exercise 6.33. Given the circles $\kappa(A,\rho)$ and $\lambda(B,\rho')$ show that the centers of variable circles tangent to these lie on a hyperbola with focals at the centers A, B of the circles and transverse axis equal to $|\rho \pm \rho'|$.

Hint: Let Y, Z be the contact points of a third circle, which is tangent to the two given ones externally (See Figure 6.58-I) (for the other contact possibilities hold analogous properties). X, Y and A are collinear and similarly, X, Z, B are collinear. Consequently, $||XA| - |XB|| = |(|XY| + |YA|) - (|XZ| + |ZB|)| = ||YA| - |ZB|| = |\rho - \rho'|$. Give a similar proof for circles tangent to κ externally and to λ internally (See Figure 6.58-II). Are the centers contained in only one branch of the hyperbola? What happens with the other branch?

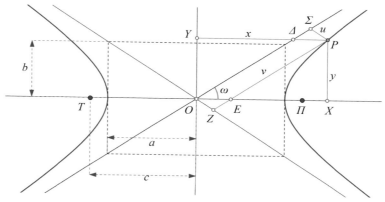

Fig. 6.59: Hyperbola relative to asymptotes

Proposition 6.20. *If $u = |P\Sigma|$, $v = |PZ|$ are the lengths of the sides of a parallelogram $P\Sigma OZ$, which is defined from the point P of hyperbola, by drawing parallels to the asymptotes, then it holds:*

$$u \cdot v = \frac{c^2}{4},$$

where $2c$ is the focal distance of the hyperbola.

Proof. A calculation of $\{u,v\}$ as functions of $\{x,y\}$ of the preceding exercise suffices. Assuming that the angle between the asymptotes is 2ω and using figure 6.59, we have:

$$\tan(\omega) = \frac{b}{a} = \frac{|OY|}{|Y\Delta|}, \quad |Y\Delta| = y\cot(\omega), \quad |O\Delta| = \frac{y}{\sin(\omega)} \tag{6.15}$$

$$|\Delta P| = x - |Y\Delta| = x - y\cot(\omega) \tag{6.16}$$

$$\frac{|\Delta P|}{\sin(2\omega)} = \frac{u}{\sin(\omega)} \tag{6.17}$$

$$u = \frac{x\sin(\omega) - y\cos(\omega)}{\sin(2\omega)} \tag{6.18}$$

$$v = \frac{x\sin(\omega) + y\cos(\omega)}{\sin(2\omega)} \tag{6.19}$$

The third equation follows from the sine formula for triangle $\Delta P\Sigma$ and the rest, as well as the requested equation, follow using simple calculations.

Exercise 6.34. Let α, β be two lines intersecting at O under the angle ω. Parallelograms $OZP\Sigma$ are constructed with a vertex at O, vertices Z and Σ respectively on α and β and such that the area of the parallelogram is fixed. Show that the fourth vertex P of the parallelogram is contained in a fixed hyperbola, which has its center at point O and its asymptotes are the lines α and β.

Hint: Applying the preceding proposition, find the characteristics of the hyperbola from the given data of the problem and identify the hyperbola with the requested geometric locus.

Asymptotic triangle of a hyperbola, is called a triangle like the triangle AOB of figure 6.60, i.e. a triangle, whose two sides are the asymptotes and the third is a tangent to the hyperbola.

Proposition 6.21. *Let AOB be an asymptotic triangle of a hyperbola with A, B on the corresponding asymptotes α, β and P be the point of contact of side AB (See Figure 6.60). Let also Z and Σ be respectively the projections of P on the asymptotes α and β. The following properties hold:*

1. *The tangent side AB is parallel to $Z\Sigma$ and has twice its length.*
2. *The point of contact P is the middle of side AB.*
3. *Triangle AOB has area equal to ab, where a and b are the axes of the hyperbola.*
4. *Conversely, every triangle AOB with A, B on the asymptotes and AB on the same angle of a hyperbola branch and area equal to ab is an asymptotic triangle of the hyperbola.*

Proof. (1) and (2) follow from Exercise 6.20 in combination with Exercise I-3.85. Indeed suppose that for every point X we define the positive numbers (u,v) through their parallel projections to the asymptotes and that P is characterized from $u = s$, $v = t$. Then by Exercise I-3.85 segment $Z\Sigma$ will be characterized by the fact that for its points will hold $\frac{u}{s} + \frac{v}{t} = 1$, while for the points of its parallel AB from P will hold $\frac{u}{s} + \frac{v}{t} = 2$. For the proof then it suffices to show that of the points of AB characterized by $\{(u,v) : \frac{u}{s} + \frac{v}{t} = 2\}$ the

6.7. THE HYPERBOLA

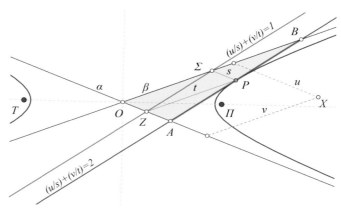

Fig. 6.60: Asymptotes and tangents of hyperbola

only point which satisfies also $u \cdot v = s \cdot t$ is P. The latter however is easy to show and implies that line AB has only one common point with the hyperbola, therefore it is a tangent to it.

(3) follows from Proposition 6.20, according to which st, which is half the area of AOB is fixed. The fact that the fixed area is ab is seen by taking point P at the vertex of the hyperbola.

(4) follows by applying Proposition 6.20.

Remark 6.10. The preceding proposition suggests an easy way to find the tangent at P: the parallelogram $OZP\Sigma$ is first constructed and subsequently from P one draws the parallel AB to its diagonal $Z\Sigma$. Next exercise gives a different view of the asymptotic triangle by defining it through a circle which passes through the two focals of the hyperbola.

Exercise 6.35. Show that every circle passing through the focals of a hyperbola defines through its intersections with the asymptotes two asymptotic triangles OAB and $OA'B'$ (See Figure 6.61). Also show that triangles $OA\Pi$, $O\Pi B$ and ATB are similar.

Hint: The area of triangle AOB is equal to $\frac{1}{2}|OA||OB|\sin(\omega)$, where ω is the angle of the asymptotes. The product of the segments expresses the power of point O relative to the circle and is equal to the product $|OB'||OA| = |OT|^2$. Then

$$\frac{1}{2}|OA||OB|\sin(\omega) = \frac{1}{2}|OT|^2 \left(2\sin\left(\frac{\omega}{2}\right)\cos\left(\frac{\omega}{2}\right)\right) = ab.$$

This shows that triangle AOB is asymptotic (Proposition 6.21). The similarities between the triangles follow from easily detected equalities of the corresponding angles.

Hyperbolas among the conics, except their asymptotes, have one more peculiarity. Every hyperbola γ with axes a, b defines another hyperbola γ'

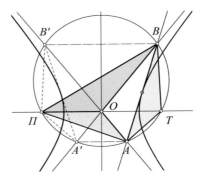

Fig. 6.61: Asymptotic via a circle

with axes $a' = b$ and $b' = a$ and the same asymptotes (See Figure 6.62). Hyperbola γ' is called **conjugate** of γ. From the definition follows that the conjugate of the conjugate is the original hyperbola. Also γ' has the same focal distance $2c$ with γ and the focals of the two hyperbolas lie on the same circle of radius c. Also, according to Proposition 6.18, γ will satisfy the equation $\frac{x^2}{a^2} - \frac{y^2}{b^2} = 1$ and, relative to the same axes γ' will satisfy equation $\frac{x^2}{b^2} - \frac{y^2}{a^2} = 1$.

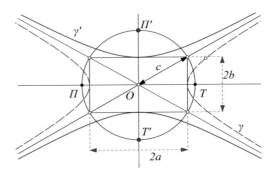

Fig. 6.62: Conjugate hyperbola γ' of γ

Exercise 6.36. Show that two lines α, β intersecting at point O under the angle ω and a circle $\kappa(O,c)$ determine two conjugate hyperbolas with focals on the bisectors of ω and focal distance $2c$.

A kind of converse of Proposition 6.20, with proof similar to that of Exercise 6.34 is formulated in the following exercise.

Exercise 6.37. Let α, β be two lines intersecting at O under an angle ω. Parallelograms $OZP\Sigma$ with one vertex at O are constructed, having their sides parallel to α, β and such that the area of the parallelogram is fixed and equal

6.7. THE HYPERBOLA

to c^2. Then vertex P is contained in one of the two conjugate hyperbolas with asymptotes α, β and focal distance $2c$.

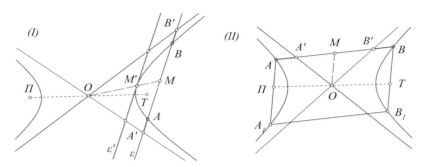

Fig. 6.63: Equal segments AA' and BB'

Proposition 6.22. *A hyperbola chord AB, which extended intersects the asymptotes at points A', B', defines segment $A'B'$ with middle coincident to the middle M of AB. Consequently segments AA' and BB' are equal.*

Proof. In the case of figure 6.63-I, the tangent parallel to the secant has its middle M' on the conjugate direction OM of the secant (Proposition 6.21). By Thales, the middle of $A'B'$ will also be on OM, therefore it will coincide with M. In the case of figure 6.63-II, the sides of the parallelogram ABB_1A_1 define conjugate directions and the result follows using a similar argument.

Rectangular hyperbola is called a hyperbola whose asymptotes are orthogonal. This is equivalent to $a = b$ and to the fact that the equation of the hyperbola relative to its axes has the form

$$\frac{x^2}{a^2} - \frac{y^2}{a^2} = 1.$$

Proposition 6.23. *The angle $\widehat{BAI}$ of two chords AB and AI of a rectangular hyperbola is equal to the angle $\widehat{\Gamma OZ}$ of the conjugate directions of the chords.*

Proof. If AB, extended, intersects the asymptotes at A', B', then triangle $A'OB'$ is right (See Figure 6.64-I) and triangle $O\Gamma A'$ is isosceles, where Γ is the middle of $A'B'$, which is simultaneously also the middle of AB. Similarly ΔOE, where O, E are the intersection points of AI with the asymptotes, is right and the triangle $OZ\Delta$ is isosceles. It follows that angle $\widehat{\Gamma OZ}$ is equal to angle $\widehat{\Gamma AZ}$.

Theorem 6.20. *For every triangle ABΓ inscribed in a rectangular hyperbola (i.e. its vertices are on the hyperbola), the following properties are valid (See Figure 6.64-II):*

1. *The Euler circle of ABΓ passes through the center O of the hyperbola.*
2. *The orthocenter H is also contained in the hyperbola.*
3. *The fourth intersection point Θ of the circumcircle of ABΓ and the hyperbola is the symmetric of H relative to O.*

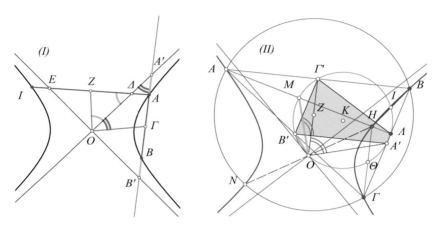

Fig. 6.64: Conjugate isogonals   Orthocenter of inscribed

Proof. The key to the proof is Proposition 6.23, according to which $\widehat{\Gamma'BA'} = \widehat{A'O\Gamma'} = \widehat{\Gamma'B'A'}$, where A', B', Γ' are the respective middles of the sides. This proves (1). If $A\Lambda$ is the altitude from A, define H as the intersection point of $A\Lambda$ with the hyperbola. We'll show that H coincides with the orthocenter of the triangle. For this, consider the middles M, I, Θ, respectively, of HA, HB, $H\Gamma$. According to Proposition 6.23, $\widehat{B'OM} = \widehat{B'\Lambda M} = \widehat{MAB'}$. The last two equalities are satisfied however also by the intersection point M' of the circle of Euler with the altitude $A\Lambda$. It follows that M coincides with M' and consequently H is the orthocenter of the triangle. (3) follows from the preceding properties, since the circumcircle is homothetic to the Euler circle relative to H with ratio $\lambda = 2$.

6.8 Comments and exercises for the chapter

> For us, there is only the trying. The rest is not our business.
>
> T.S. Eliot, Four Quartets

Exercise 6.38. Construct the tangents of a conic section from a point P, as well as their contact point. For the conic we know only the position of its focals and the directrices and not its entire set of points.

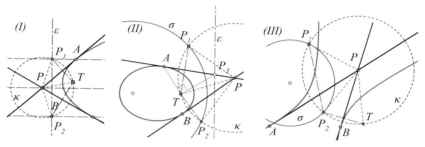

Fig. 6.65: Tangents from a point P to a parabola, ellipse, hyperbola

Hint: In all three cases the key ingredient is the circle $\kappa(P,|PT|)$, where T is one focus and ε is the associated directrix of it.

In the case of the parabola, we define the points of intersection P_1, P_2 of κ with the directrix ε (See Figure 6.65-I). From P_1, P_2 we draw respectively orthogonals to ε. The intersection points A, B of these orthogonals with the medial lines respectively of P_1T, P_2T are the wanted points of contact (Theorem 6.11).

For ellipses and hyperbolas, we consider the intersection points P_1, P_2 of κ with the major circle σ with center the other than T focus (See Figure 6.65-II). Yet again the tangents are the medial lines of P_1T, P_2T. For example, the contact point of PA is the intersection point A of the medial line of P_1T and the orthogonal to O_3T at T, where P_3 is the intersection of P_1A with the directrix ε (Theorem 6.18). The construction of the tangents of the hyperbola is done analogously (See Figure 6.65-III).

Exercise 6.39. Referring to figure 6.65 show that the angles $\widehat{PTA} = \widehat{PTB}$ or $\widehat{PTA} + \widehat{PTB} = \pi$.

Exercise 6.40. The rhombus $AB\Gamma\Delta$ has fixed side-length δ and fixed also the position of its vertex Γ (See Figure 6.66). The opposite to it vertex A is moving on a fixed circle $\kappa(E,r)$. Show that the intersection point Z of the diagonal $B\Delta$ of the rhombus with the line EA lies on a conic with focal points at $\{E,\Gamma\}$ and that the diagonal $B\Delta$ is tangent to this conic at Z.

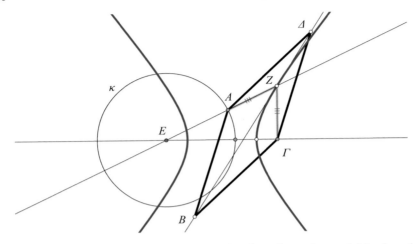

Fig. 6.66: Conic drawing linkage with a fixed circle and a variable rhombus

Hint: In the case point Γ lies outside the circle, $B\Delta$ is the medial line of the segment $A\Gamma$ hence $|ZA| = |Z\Gamma|$ and $|ZE| - |Z\Gamma| = r$ (Proposition 6.19). Analogous argument is valid also for the other positions of Γ with respect to the circle κ ([99]).

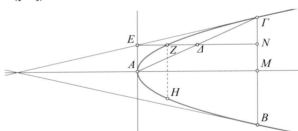

Fig. 6.67: Parabola from three special points

Exercise 6.41. Let point A be the vertex and B, Γ two points of a parabola lying symmetrically relatively to its axis. Find two additional points of the parabola.

Hint: Point N on $B\Gamma$ and Z on EN are defined from the ratios $|NB|/|N\Gamma| = |ZN|/|ZE| = 3$ (See Figure 6.67). Point Z and its symmetric H relative to the axis AM are points of the parabola (Theorem 6.12).

Exercise 6.42. Show that four lines in general position define exactly one parabola tangent to these lines (See Figure 6.68).

Hint: Four lines in general position define four triangles $\tau_1 = ABZ$, $\tau_2 = BE\Delta$, $\tau_3 = A\Delta\Gamma$ and $\tau_4 = \Gamma ZE$ and the corresponding Miquel point Π, which is

6.8. COMMENTS AND EXERCISES FOR THE CHAPTER

contained in the circumscribed circles of these four triangles (Theorem I-5.16). This defines the Simson lines of these triangles relative to Π, which in turn coincide with a line ε. This, because for each pair from these four triangles, the corresponding two Simson lines of Π, have two common points, therefore they are coincident. We consider the parabola having focus at Π

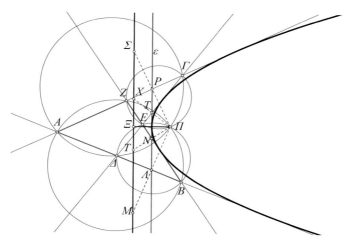

Fig. 6.68: Parabola of four lines in general position

and tangent at its vertex line ε. Using the Theorem 6.11, we prove that the original side-lines of the four triangles are tangents to the parabola.

Remark 6.11. Figure 6.69 shows the relations between the different lines we met in I-§ 5.8, as well as the parabola defined in the preceding exercise. Beginning with A', B', Γ' on the sides of triangle $AB\Gamma$, supposed to be collinear (they can be defined by a line which intersects the sides of the triangles), we construct the corresponding Miquel point M, contained in the circumcircle of the triangle (Corollary I-5.7). Point M is characterized by the fact that MA', MB', $M\Gamma'$ form the same angle ω with the respective sides of the triangle. That is why in the figure, the line of A', B', Γ' is called ω-Simson. Keeping M fixed and changing ω the corresponding ω-Simsons take different positions, which coincide with the tangents of a parabola κ (§ 6.5). The *Simson line* of M is taken for the special angle $\omega = 90°$ and is the tangent to the parabola κ at its *vertex*. The Steiner line of M (Theorem I-5.13) is the parallel to the Simson line, which passes through the orthocenter H of the triagle. This is the *directrix* of the parabola κ. Point M is the *focus* of the parabola κ. Special ω-Simson lines are the three sides of the triangle. Latter are themselves also tangents to the parabola and their points of contact with the parabola are found when two of A', B', Γ' coincide with the vertices of the triangle. Thus, for example the point of contact with $B\Gamma$ is the position taken by A' when Γ' coincides with B and B' coincides with Γ. An interesting exercise, accessible with our knowledge so far, is the determination of this contact point, using the aforementioned property.

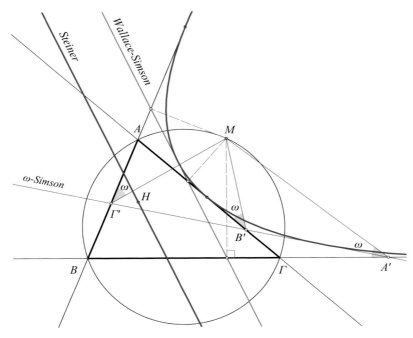

Fig. 6.69: Relations between lines of Simson, ω-Simson and Steiner

Exercise 6.43. Given is a parabola with focus at point Π and a line ε, which varies staying parallel to a fixed line ε_0 (See Figure 6.70). For each position of ε intersecting the parabola at points A and B we obtain an Archimedean triangle $AB\Gamma$ with circumcircle λ, as well as the triangle $A'B'\Gamma$ with circumcircle κ, where A', B' are the middles, respectively, of ΓA, ΓB. Show the following properties:

1. Triangle $A'B'\Gamma$ has its side $A'B'$ on a fixed line, which is the tangent ε_1 of the parabola, parallel to ε_0.
2. The middles M of the sides AB are contained in a line α parallel to the axis of the parabola and passing through the contact point Δ of ε_1 with the parabola.
3. The vertex Γ moves on the line α.
4. The center K of the circle κ moves on the orthogonal γ of ε_1 at its point Δ.
5. The circle κ is tangent to the respective λ, at point Γ.
6. The center O of λ moves on the line β orthogonal to α at its fixed point T.

6.8. COMMENTS AND EXERCISES FOR THE CHAPTER

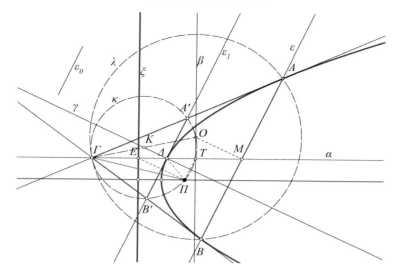

Fig. 6.70: Changing Archimedean triangles

Hint: (1), (2) and (3) follow directly from Theorem 6.12.

(4) is also a consequence of this proposition, since Δ is the middle of $A'B'$ and K lies on the medial line of this line segment.

(5) follows from the fact that the triangles ΓAB and $\Gamma A'B'$ are homothetic, relative to homothety center Γ.

(6) follows from the fact that T is the symmetric of Π relative to the line γ. This, because Π is on the symmedian $\Gamma \Pi$ from Γ of triangle $AB\Gamma$ (Theorem 6.14), consequently the arcs $\widehat{A'T}$, $\widehat{B'\Pi}$ on circle κ are equal and $T\Pi$ is parallel to $A'B'$. On the other hand, from the property of the parabola $|E\Delta| = |\Delta\Pi|$, follows that $E\Delta\Pi$ is isosceles, which implies that $E\Pi$ is orthogonal to $A'B'$, which is parallel to ΠT. It follows that $E\Pi T$ is a right triangle and Δ is the middle of ET, from which follows the claim.

Exercise 6.44. Let OA, OB be two tangents of a parabola, intersecting at O and A_1, B_1 be respective points on them, such that A_1B_1 is also a tangent to the parabola (See Figure 6.71). Let Π be the focus and ξ be the directrix of the parabola. Show that

1. For all possible positions of the tangent A_1B_1, triangles $A_1\Pi B_1$ are similar to $\Pi\Pi_1\Pi_2$, where Π_1, Π_2, are the projections of Π, respectively, on OA and OB.
2. If A_2, B_2 lying respectively on OA, OB, define a second tangent to the parabola, then the triangles ΠA_1A_2 and ΠB_1B_2 are similar.
3. If we pick on OA points $A_3, A_4, ...$, and on OA points $B_3, B_4, ...$, such that $|A_1A_2| = |A_2A_3| = |A_3A_4| = ...$ and $|B_1B_2| = |B_2B_3| = |B_3B_4| = ...$, then the lines A_3B_3, A_4B_4, ..., are also tangents to the parabola.

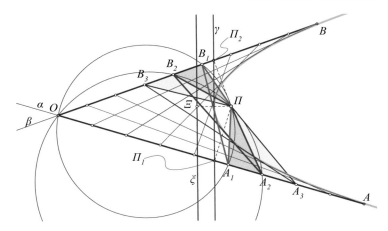

Fig. 6.71: Parabola as an envelope of lines

Exercise 6.45. We divide the sides of the triangle AOB into n equal segments with points $A_0 = O, A_1, A_2, ..., A_n = A$ and $B_0 = B, B_1, B_2, ..., B_n = O$, so that the lines A_iB_i for $i = 0, 1, ..., n$ are all tangents to a parabola κ. Determine the focus and the directrix of this parabola.

Exercise 6.46. Given is a line α, a point Π and a triangle $A_1\Pi B_1$ with its vertex A_1 on α and in such a way that its angles remain fixed (varying by similarity). Show that its vertex B_1 describes a line β. Also show that the lines A_1B_1 are tangents to a parabola, whose focus and directrix must be determined. How does this exercise completes Theorem I-2.28?

In the subsequent exercises we suppose that a parabola has been constructed if we determine its directrix and focus.

Exercise 6.47. Construct a parabola κ, for which are given three tangents and the position Π of its focus.

Exercise 6.48. Construct a parabola κ, for which are given three tangents and its directrix ξ.

Hint: Consider the triangle formed by the three tangents and determine at its circumcircle the point Π for which ξ is a Simson line (Exercise I-5.60).

Exercise 6.49. Find the geometric locus of the focals of all parabolas tangent to three given lines α, β and γ.

Hint: The circumcircle of the triangle formed by the three lines.

Exercise 6.50. Construct the parabola which is tangent to three given lines α, β and γ and has a given fourth line δ as its axis, assuming all four lines are in general position.

Hint: From one of the three vertices of the triangle $AB\Gamma$ formed by α, β, γ, say A, draw the parallel δ' to δ. The focus is the other intersection point of the circumcircle $(AB\Gamma)$ with the symmetric δ'' of δ' relative to the bisector of the angle $\widehat{BA\Gamma}$ (Theorem 6.14-1, -5).

Exercise 6.51. Construct the parabola, which is tangent to three given lines α, β and γ, given also the point of contact for one of them.

Hint: Using Theorem 6.15 extend the triangle formed by the three lines and define an Archimedean triangle of the parabola (Corollary 6.5).

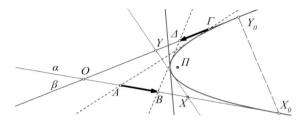

Fig. 6.72: Generalization of theorem of Thales

Exercise 6.52. Given are two non collinear and not intersecting line segments AB, $\Gamma\Delta$ and for variable λ we define the oriented line segments $AX = \lambda \cdot AB$ and $\Gamma Y = \lambda \cdot \Gamma\Delta$. Show that the lines XY are tangents to a parabola, whose focus and diretrix must be determined.

Hint: Determine the Archimedean triangle $X_0 O Y_0$, where O is the intersection point of the lines $\alpha = AB$ and $\beta = \Gamma\Delta$ (See Figure 6.72). Point X_0, for example, is determined from the λ for which $\lambda \cdot \Gamma\Delta = \Gamma O$, for which the corresponding XY coincides with α. Similarly we also find point Y_0. Lines $B\Delta$ and $A\Gamma$ are also tangents to the parabola.

Remark 6.12. The preceding exercise generalizes the theorem of Thales, where similarly produced lines XY are examined, but for segments with a common start point at O ($A \equiv \Gamma \equiv O$). Indeed, by keeping the lengths of the segments fixed, let us allow AB to slide on α and $\Gamma\Delta$ to slide on β, so that the initial points A and Γ of the two line segments tend to coincide with the intersection point O of the two lines. Then the focus of the parabola tends to O and the parabola tends to coincide with a half line ε through O, to which all lines XY are parallel.

Exercise 6.53. Show that two parabolas Σ, Σ' are always similar, in other words there exists a similarity Φ of the plane, so that $\Phi(\Sigma) = \Sigma'$.

Exercise 6.54. Show that two ellipses Σ, Σ' are similar, if and only if they have the same eccentricity. Show the corresponding property also for hyperbolas.

Exercise 6.55. Show that the geometric locus of the centers K of circles κ, which pass through a fixed point Π and intersect a fixed line ε under a fixed angle ω (that is the tangent to κ at the intersection point and ε form an angle of measure ω) is a conic with one focus at Π and corresponding directrix ε (See Figure 6.73).

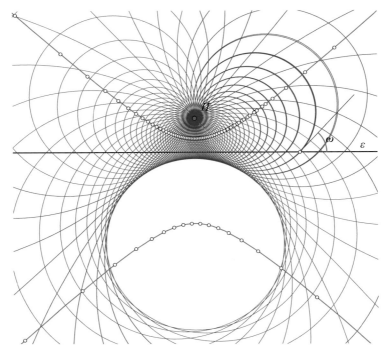

Fig. 6.73: Conic as locus of centers of circles

Exercise 6.56. Show that the white disk which shows up in the preceding figure is the interior of the circle κ_0 to which are tangent all the circles of the preceding locus. Determine its center and its position as a function of Π, ε and ω.

Exercise 6.57. Show that the conic of Exercise 6.55 is a hyperbola. Determine its center and the asymptotes, as well as the position of its other focus. Finally determine also its eccentricity as a function of Π, ε and ω.

Exercise 6.58. Show that the points P of a conic section satisfy an equation $ax + by + c = 0$, where x is the distance of P from fixed point Π and y is the distance of P from a fixed line ξ. Conversely, if the distances x, y, of P from Π and ξ, satisfy an equation like the preceding one, then they belong to a conic.

6.8. COMMENTS AND EXERCISES FOR THE CHAPTER

Determine the elements of the conic (focals, directrices and eccentricity) as functions of a, b and c.

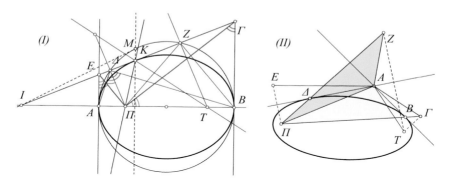

Fig. 6.74: Right angle $\widehat{E\Pi\Gamma}$ — Equal angles $\widehat{\Pi A \Delta} = \widehat{TAB}$

Exercise 6.59. Suppose that the tangent at the point K of an ellipse, with focals Π, T, intersects its auxiliary circle at points Δ, Z and the tangents at the endpoints of its major diameter AB at points Γ, E (See Figure 6.74-I). Show that:

1. The angles $\widehat{B\Pi\Gamma} = \widehat{AE\Pi}$ and $\widehat{E\Pi\Gamma}$ is a right angle.
2. Lines ΠE, $\Pi\Gamma$ are bisectors of the supplementary angles $\widehat{A\Pi K}$ and $\widehat{K\Pi B}$.
3. If a right triangle $E\Pi\Gamma$ rotates about the fixed right angle vertex Π and its other vertices move on two fixed parallels AE, $B\Gamma$, then the line $E\Gamma$ is tangent to an ellipse, whose one focus is Π.

Hint: For (1) observe first that the quadrilaterals $A\Pi\Delta E$, $\Pi B\Gamma\Delta$, $B\Gamma ZT$ are inscriptible to circle (Theorem 6.18). It follows that $\widehat{\Pi\Gamma B} = \widehat{\Pi\Delta B}$ and consequently $\widehat{B\Pi\Gamma} = \widehat{\Pi\Delta A}$, since the two angles are supplementary to these equal angles, because of the right triangles $\Pi B\Gamma$ and $A\Delta B$. Because of the inscriptible $\Pi\Delta EA$, we also have $\widehat{A\Delta\Pi} = \widehat{AE\Pi}$. The congruence of the two angles in the right triangles ΠAE and $\Pi B\Gamma$ implies the claim.

For (2) observe that $\Pi(IKE\Gamma)$ is a harmonic pencil, because the tangents at K and M pass through the same point I (Exercise 6.27).

(3) follows from the preceding properties.

Exercise 6.60. Show that for every point A external to the ellipse with focals at Π, T and tangents $A\Delta$, AB, for the respective angles holds $\widehat{\Pi A\Delta} = \widehat{TAB}$ (See Figure 6.74-II). Show the corresponding property for hyperbolas.

Hint: Consider the symmetric E, Z of Π, T relative to $A\Delta$ respectively and Γ of T relative to AB. Triangles ΠAZ and $\Pi A\Gamma$ are congruent, because they have respective sides equal ($|\Pi Z| = |\Pi\Gamma| = 2a$). It follows that $\widehat{\Pi AZ} = 2\widehat{\Pi A\Delta} + \widehat{EAZ}$ is equal to $\widehat{\Pi A\Gamma} = 2\widehat{TAB} + \widehat{\Pi AT}$.

Remark 6.13. If we interpret the point at infinity of the axis of the parabola, as a second focus of the parabola, then the preceding exercise generalizes property (5) of Theorem 6.14.

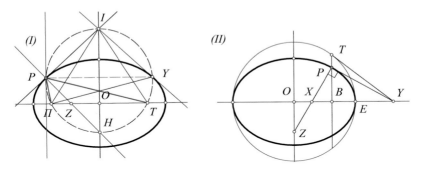

Fig. 6.75: Circle through the focals   Fixed ratio $\frac{PX}{PZ} = \frac{b^2}{a^2}$

Exercise 6.61. From a point I of the small axis draw the tangents to ellipse IP, IY. Show that the circle (IPY) passes through its focals.

Hint: If Π, T are the focals, $P\Pi TY$ is an isosceles trapezium (See Figure 6.75-I) and has circumcircle κ. The perpendicular PZ to the tangent is a bisector of the angle $\widehat{\Pi PT}$ (Theorem 6.18), therefore intersects the arc $\widehat{\Pi T}$ at its middle H and point I is a point of κ.

Exercise 6.62. The perpendicular PZ to the tangent to the ellipse at its point P intersects the major axis at X and the minor one at Z. Show that $\frac{PX}{PZ} = \frac{b^2}{a^2}$.

Hint: $\frac{PX}{PZ} = \frac{XB}{OB}$ and OB, XB are calculated as functions of the eccentric angle: $OB = a\cos(\phi)$ and $XB = \frac{PB^2}{BY} = \frac{b^2\cos(\phi)}{a}$ (See Figure 6.75-II).

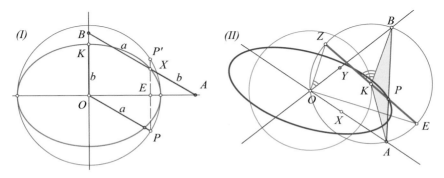

Fig. 6.76: Ellipse from moving segment , The analogue for non right angle

6.8. COMMENTS AND EXERCISES FOR THE CHAPTER

Exercise 6.63. A line segment of fixed length $d = a + b$, slides with its endpoints A, B respectively on two orthogonal axes. Show that one of its points X, which divides it into segments of lengths $|XB| = a$ and $|XA| = b$ describes an ellipse with axes a and b.

Hint: Let $a \geq b$ and draw from the intersection O of the orthogonal axes the parallel OP to BX (See Figure 6.76-I). Point P traces a fixed circle. It holds $\frac{PE}{EX} = \frac{a}{b}$, where E is the projection of X onto the axis where A moves. Apply Theorem 6.17 (3).

Exercise 6.64. Generalization of the preceding exercise for arbitrary angle $\widehat{XOY}$: A line segment of fixed length $d = |AB|$, slides with its endpoints A, B on two intersecting axes OX, OY. Show that one of its points P, which divides it into segments of fixed length $|PB| = a$ and $|PA| = b$ traces out an ellipse.

Hint: This exercise (as well as the next one), which is due to Lahire (1640-1718), is reduced to the preceding one (See Figure 6.76-II). Indeed, point O sees AB under a fixed angle $\widehat{XOY}$, therefore the circle $\kappa = (OAB)$ has a fixed radius. Consider the diameter of this circle EZ, which passes through point P. Its length is fixed and its endpoints move on two fixed orthogonal axes OE and OZ.

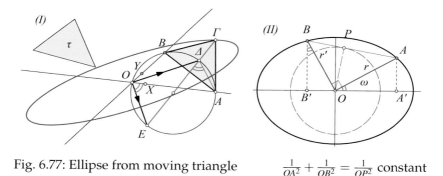

Fig. 6.77: Ellipse from moving triangle $\frac{1}{OA^2} + \frac{1}{OB^2} = \frac{1}{OP^2}$ constant

Exercise 6.65. The triangle $AB\Gamma$ is congruent to a fixed triangle τ and slides with its vertices A, B lying respectively on two intersecting axes OX, OY. Show that its third vertex Γ traces out an ellipse (See Figure 6.77-I).

Exercise 6.66. Show that the sum of the inverses of the squares of two orthogonal ellipse diameters is fixed.

Hint: It suffices to show that $\frac{1}{OA^2} + \frac{1}{OB^2} = \frac{1}{OP^2}$ is fixed, for the corresponding half diameters OA, OB (Exercise 3.32) (See Figure 6.77-II). If $r = |OA|$, $r' = |OB|$ and ω is the angle of OA with the main axis, then the projections of OA, OB on the major axis will be $x = OA' = r\cos(\omega)$, $x' = OB' = r'\sin(\omega)$ and on the minor axis $y = OA'' = r\sin(\omega)$, $y' = OB'' = r\cos(\omega)$. Calculate $\frac{1}{r^2} + \frac{1}{r'^2}$ taking into account the fact that (x,y) and (x',y') satisfy the equation $\frac{x^2}{a^2} + \frac{y^2}{b^2} = 1$.

Remark 6.14. In the preceding exercise we easily see that $\frac{1}{OA^2} + \frac{1}{OB^2} = \frac{1}{OP^2} = \frac{1}{a^2} + \frac{1}{b^2}$, where a and b are the axes of the ellipse and consequently chord AB is tangent to a fixed circle of radius equal to $\frac{ab}{c}$ where $c^2 = a^2 + b^2$.

Exercise 6.67. Prove the claims, related to the polars of conics, which were made in remark 6.6.

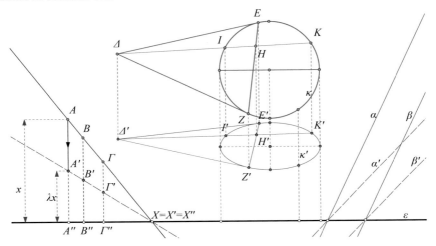

Fig. 6.78: Affinity relative to axis ε and ratio λ

Remark 6.15. Many properties of the ellipse can be proved by considering the *projection* of the circle in the sense found in Theorem 6.17. We have here a special case of an affinity, which we met in § 4.16. It concerns a mapping of the plane onto itself, defined by a line (axis) ε and a non zero number λ. This mapping fixes the points of ε and to every other point A corresponds point A' on the line AA'', such that the oriented segments are $A'A'' = \lambda \cdot AA''$, where A'' is the orthogonal projection of A onto ε (See Figure 6.78). It is easy to see that this mapping is a special case of an affinity and maps lines to lines, preserving the segment ratio $\frac{AB}{B\Gamma}$ of the same line: $\frac{AB}{B\Gamma} = \frac{A'B'}{B'\Gamma'}$. Parallel lines α, β are mapped respectively to parallels α', β'. Finally circles κ of diameter $2a$ are mapped into ellipses κ' with major axis a and minor axis $b = \lambda \dot{a}$, when $\lambda < 1$. In Theorem 6.17 the correspondence of the major circle to the ellipse, through the projection defined therein is a mapping, like the one we described here, with $\lambda = \frac{b}{a}$ and the circle κ having its center on ε. Analogous relations can be seen also in exercise 6.31.

A useful property of this mapping of the circle to the ellipse is that tangents to the circle κ (ΔE, ΔZ in figure 6.78) are mapped to corresponding tangents $(\Delta'E', \Delta'Z')$ of the ellipse κ'. Also the polar EZ of a point Δ relative to circle κ maps to the polar $E'Z'$ of the respective point Δ' relative to κ'.

Exercise 6.68. Let F be an affinity, in the preceding sense, with axis ξ and ratio λ. Let also κ be a parabola, whose directrix coincides with axis ξ (See Figure 6.79). Show that the projection $\kappa' = F(\kappa)$ of the parabola by F is also a parabola. Determine the focus and the directrix of κ'.

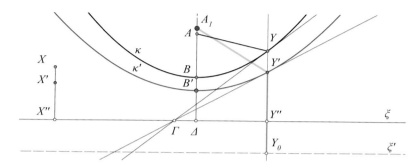

Fig. 6.79: Mapping of parabola through an affinity

Hint: Use the remark-1 of § 6.1 and Exercise 6.21 and suppose that the parabola κ is described by the equation $y = ax^2 + bx + c$. Take $x_0 = -\frac{b}{2a} = 0$, which simplifies the equation to $y = ax^2 + c$. The coincidence of the directrix of κ with the x-axis means $y_2 = c - \frac{1}{4a} = 0 \Leftrightarrow c = \frac{1}{4a}$. This way, the equation of the parabola takes the form $y = ax^2 + \frac{1}{4a}$. F maps a point of κ, described by (x,y), to the point $(x, y' = \lambda \cdot y)$, which implies that y' will satisfy equation $y' = (\lambda a)x^2 + \frac{\lambda}{4a}$.

Exercise 6.69. Express and prove the corresponding property for hyperbolas κ and their images $\kappa' = F(\kappa)$ by an affinity F, with axis ξ and ratio λ, assuming that one of the axes of the hyperbola coincides with axis ξ of F.

Exercise 6.70. Using an affinity F defined by a line ε and a ratio $\lambda > 0$, as in the preceding exercises, show that the areas $\{\varepsilon(p), \varepsilon(p')\}$ of a polygon p and its image $p' = F(p)$ under F are in the same ratio with that of the affinity: $\varepsilon(p')/\varepsilon(p) = \lambda$ (See Figure 6.80).

Hint: Show this, first for rectangles, parallelograms and triangles with one side parallel to ε. Then, for general triangles, and finally for polygons, by dividing them to triangles.

Remark 6.16. Last exercise makes plausible that the area of an ellipse is $(a \cdot b \cdot \pi)$, where $\{a, b\}$ are the axes of the ellipse. Although this is beyond the scope of these lessons, it is interesting to get an idea how this is done.

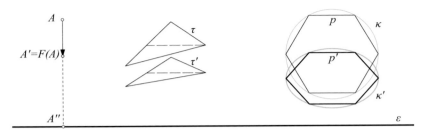

Fig. 6.80: Areas under the affinity are multiplied by λ

First, the area of the ellipse is considered, analogously to circles, as the limit of areas of inscribed to the ellipse polygons, which come "closer and closer" to the ellipse. Then, it is used the "continuity" of F, which is the property roughly expressed by the statement: "if points $\{X,Y\}$ are close to each other, then the corresponding image-points $\{F(X), F(Y)\}$ are also close to each other". This implies that, if the sequence of polygons $\{p_n\}$ inscribed in the circle $\kappa(O, a)$ comes closer and closer to the circle, as for example is the case with the sequence of regular polygons $\{\Pi_n\}$ of § 1.1, then the corresponding sequence of inscribed in the ellipse polygons $\{p'_n = F(p_n)\}$ comes also closer and closer to the ellipse $\kappa' = F(\kappa)$. Since, according to the preceding exercise, the areas of polygons are related by the ratio $\lambda = \frac{b}{a}$ of F : $\varepsilon(p'_n) = \lambda \varepsilon(p_n)$, and the limit of the sequence $\{\varepsilon(p_n)\}$ is the area of the circle $\pi \cdot a^2$, it follows that the area of the ellipse is the limit of the corresponding sequence of areas $\{\varepsilon(p'_n) = \lambda \varepsilon(p_n)\}$, which is $\lambda(\pi \cdot a^2) = \frac{b}{a}(\pi \cdot a^2) = \pi \cdot a \cdot b$.

Exercise 6.71. Given is an ellipse κ and a point P of its plane. Show that the ratio $\frac{|PA| \cdot |PB|}{|A'B'|^2}$, where A, B are the points of intersection with a line through P and $A'B'$ is the diameter of the ellipse parallel to AB, depends only on P and not on the direction of the line AB.

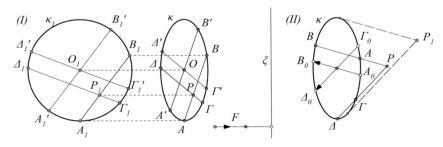

Fig. 6.81: Fixed point P \qquad\qquad Fixed directions

6.8. COMMENTS AND EXERCISES FOR THE CHAPTER

Hint: Consider the ellipse as the image $\kappa = F(\kappa_1)$ of a circle κ_1 by an affinity, as it is described in the preceding exercises (See Figure 6.81-I). Point P is the projection $P = F(P_1)$ of a point P_1 and the various chords AB, $\Gamma\Delta$, ... of the ellipse through P are projections of respective chords A_1B_1, $\Gamma_1\Delta_1$, ... through P_1. Similarly also the parallels of the chords, diameters of the ellipse $A'B'$, $\Gamma'\Delta'$, ... are projections of respective diameters of the circle $A'_1B'_1$, $\Gamma'_1\Delta'_1$, ... Because the affinity preserves the ratio of segments in parallel lines, we'll have

$$\frac{|PA|\cdot|PB|}{|A'B'|^2} = \frac{|PA|}{|A'B'|} \cdot \frac{|PB|}{|A'B'|} = \frac{|P_1A_1|}{|A'_1B'_1|} \cdot \frac{|P_1B_1|}{|A'_1B'_1|},$$

and similarly $\dfrac{|P\Gamma|\cdot|P\Delta|}{|\Gamma'\Delta'|^2} = \dfrac{|P_1\Gamma_1|}{|\Gamma'_1\Delta'_1|} \cdot \dfrac{|P_1\Delta_1|}{|\Gamma'_1\Delta'_1|}.$

The last products in the two equalities are expressed through the power of P_1 relative to the circle and are equal.

Exercise 6.72. Given is an ellipse κ and a point P of its plane. Show that the ratio $\frac{|PA|\cdot|PB|}{|P\Gamma|\cdot|P\Delta|}$, where AB and $\Gamma\Delta$ are chords through point P, which are parallel to two fixed diameters A_0B_0 and $\Gamma_0\Delta_0$ of the ellipse (See Figure 6.81-II), does not depend on the position of P.

Hint: Consider the ratio

$$\left(\frac{|PA|\cdot|PB|}{|P\Gamma|\cdot|P\Delta|}\right) \cdot \left(\frac{|\Gamma_0\Delta_0|^2}{|A_0B_0|^2}\right) = \left(\frac{|PA|\cdot|PB|}{|A_0B_0|^2}\right) / \left(\frac{|P\Gamma|\cdot|P\Delta|}{|\Gamma_0\Delta_0|^2}\right),$$

the value of which is 1, in view of the preceding exercise. This ratio does not change if we move P, yet keep fixed the directions of the diameters A_0B_0, $\Gamma_0\Delta_0$.

Remark 6.17. Note that the ratio of the preceding exercise, which remains independent of the position of P, gains a special meaning when P is in such a position (P_1) so that the secants PAB, $P\Gamma\Delta$ become tangents of the conic. Then points A, B coincide with a point of contact of a tangent parallel to A_0B_0 (See Figure 6.81-II) and a similar property holds for the other tangent. The ratio of the preceding exercise then is equal to the ratio of the squares of the tangents of the parallels to the fixed diameters A_0B_0 and $\Gamma_0\Delta_0$. The last two exercises and the subsequent one, which is a simple consequence, generalize for ellipses the properties of the power of a point relative to a circle (I-§ 4.1).

Exercise 6.73. Show that the ratio of the preceding exercise, which is equal to the ratio of the squares of the tangents of the parallels to the diameters A_0B_0, $\Gamma_0\Delta_0$, is equal also to the ratio $\frac{|A_0B_0|^2}{|\Gamma_0\Delta_0|^2}$ of the squares of these diameters.

Exercise 6.74. Point A of a diameter AOA' of an ellipse defines point A_1 of the auxiliary circle, through its intersection with the orthogonal AA_2 on the

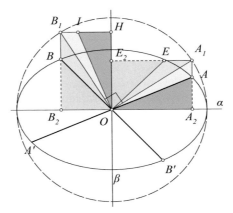

Fig. 6.82: Conjugate diameters

major axis α (See Figure 6.82). Subsequently A_1 is projected to point E_2 of the minor axis β. This forms a rectangle $OA_2A_1E_2$, where O is the center of the ellipse. This rectangle is divided into four triangles: OA_2A, OAA_1, OA_1E and OEE_2, where E is defined through the ratio $\frac{|A_1E_2|}{|EE_2|} = \frac{a}{b}$. We rotate the rectangle $OA_2A_1E_2$ by $\frac{\pi}{2}$ so that it takes the position OHB_1B_2. Show that after the rotation the segment E_2EA_1 passes to the segment B_2BB_1, where BOB' is the diameter conjugate to AOA'.

Hint: Application of Theorem 6.19 and of Theorem 6.17.

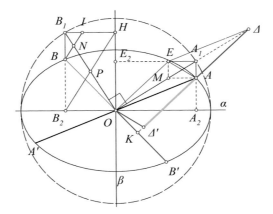

Fig. 6.83: Determination of axes

Exercise 6.75. Determine the direction and the length of the axes of an ellipse, for which are given, by position and length, two conjugate diameters AOA' and BOB'.

Hint: Using the preceding exercise and drawing the parallel $A\Delta$ and equal to OE, forms the parallelogram $OA\Delta E$ (See Figure 6.83), whose sides are equal to half the given diameters, and OE is orthogonal to OB. Therefore Δ can be constructed from the given data. Let Δ' be the symmetric of Δ relative to A. Then $\Delta'AEO$ is a parallelogram and $O\Delta'$ is parallel to EA. It follows that $\widehat{\Delta'OA_2} = \widehat{AEA_1}$ and $\widehat{A_2OA_1} = \widehat{EA_1O}$. Because, by construction, $\frac{|A_1E_2|}{|EE_2|} = \frac{|A_1A_2|}{|AA_2|} = \frac{a}{b}$, it follows that AA_1EM is a rectangle. Therefore $\widehat{AEA_1} = \widehat{EA_1M}$. This, because of the preceding equality between angles, implies that OA_2 is a bisector of the angle $\widehat{\Delta'OA_1}$, which can be constructed from the given data. This way the direction of the axes α and β is determined. Because A_1 is constructible, the length $a = |OA_1|$ determines the major axis and b is determined from a and the ratio $\frac{|AA_2|}{|A_1A_2|} = \frac{b}{a}$.

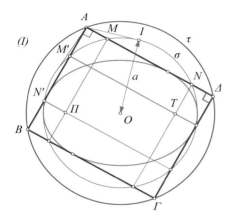
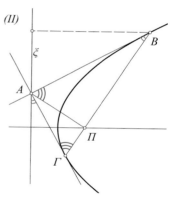

Fig. 6.84: Director circle of ellipse   Directrix of parabola

Exercise 6.76. For a given ellipse, show that the geometric locus of points A, from which tangents can be drawn which make at A a right angle, is a circle with center the center of the ellipse and radius $r = \sqrt{a^2+b^2}$, where a and b are the axes of the ellipse (See Figure 6.84-I).

Hint: According to Theorem 6.18(6) the projections M, N of the focals Π, T on the tangent $A\Delta$ are contained in the major circle $\sigma(O,a)$ of the ellipse. Therefore the power of the point A relative to this circle will be $|AI|^2 = |AM||AN| = |\Pi N'||TM'| = b^2$ (Theorem 6.18-7). Consequently $|AO|^2 = |AI|^2 + |OI|^2 = a^2 + b^2$.

The circle τ, defined in the preceding exercise is called **director circle** of the ellipse. Considering the symmetric Γ of A relative to O and drawing therefrom the tangents, we form a circumscribed rectangle of the ellipse. Circle τ is also the locus of the vertices of all circumscribed rectangles of the ellipse.

Exercise 6.77. Show that, for a given parabola, the geometric locus of points A from which tangents can be drawn, which intersect orthogonally at A is the directrix ξ of the parabola (See Figure 6.84-II).

Hint: Follows from proposition 6.8, in combination with properties (5,6) of the Archimedian triangle of Theorem 6.14.

Exercise 6.78. Construct 6 points of the hyperbola, which admits the triagle AOB as asymptotic with its side AB tangent to the hyperbola (See Figure 6.85-I).

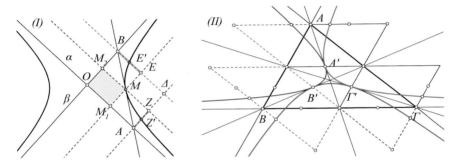

Fig. 6.85: Hyperbola with given asymptotic, Division into equiareal parts

Hint: One point of the hyperbola is the middle M of AB. Two additional points of the hyperbola are the middles E', Z' of the segments BE and AZ, where E, Z are the middles of the sides of the parallelogram $OA\Delta B$. The symmetric of M, E', Z' relative to O are also points of the hyperbola. How many additional points of the hyperbola can be constructed from these six points?

Exercise 6.79. Show that every line, which divides the triangle $AB\Gamma$ into two polygons of equal area, is tangent to a hyperbola, which admits an asymptotic triangle consisting of two sides and a median of the triangle, the median being tangent to the hyperbola. Every triangle defines this way three hyperbolas (See Figure 6.85-II).

Exercise 6.80. Let Γ be a point of the directrix ξ of the ellipse κ, relative to its focus A. Show that every line ε through Γ intersecting the ellipse (conic) at the points $\{\Delta, E\}$, forms the angle $\widehat{\Delta AE}$ at A which is bisected by the polar $p(\Gamma)$ of Γ relative to the ellipse (conic).

Exercise 6.81. Given two concentric circles $\{\kappa(O,b), \lambda(O,a)\}$ with $a > b$, show that the ellipse, which has these as respective minor and major circles, has its focals $\{A, B\}$ at the middles of the smaller sides of a rectangle which is inscribed in λ and has its bigger sides tangent to κ (See Figure 6.86-I). For which ellipses the circle κ is tangent to all the sides of the rectangle?

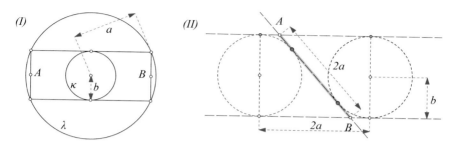

Fig. 6.86: Ellipse defined by a rectangle, Cylinder cut for a given ellipse

Exercise 6.82. Given two positive numbers $\{a > b\}$, construct a cylinder and two equal spheres inscribed in it, such that the intersection of the cylinder with a plane tangent to both spheres is an ellipse with axes $\{a, b\}$ (See Figure 6.86-II).

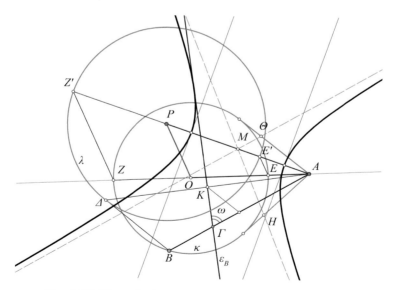

Fig. 6.87: Hyperbola generated from a variable angle

Exercise 6.83. Given are the circle κ, the constant $k > 0$, the angle ω and a point A outside the circle (See Figure 6.87). For every point B of the circle, consider the point Γ of the segment AB, for which $|A\Gamma|/|\Gamma B| = k$ and the line

ε_B passing through Γ and making the angle ω with AB. Show that all these lines ε_B are tangent to a hyperbola, of which we request the location of the focals and the asymptotes.

Hint: Consider the symmetric Δ of A relative to the line ε_B and show that the triangle $AB\Delta$ has constant angles, hence Δ is contained in a circle $\lambda(P)$ (Theorem I-3.21) and the lines ε_B coincide with the medial lines $K\Gamma$ of the segments $A\Delta$, hence they are tangent to a hyperbola (Exercise 6.32), which has its focals at $\{A,P\}$. The asymptotes of the hyperbola result from the positions which takes line ε_B, when B coincides with the contact points of the tangents to κ from A.

Many properties of the conics can be studied in a synthetic geometric way, as we did in the preceding exercises ([90], [91], [95], [101]). Many more, however, can be studied using the methods of analytic and projective Geometry ([94], [100]). Historically, the geometric method came first ([96]). The methods of analytic and projective geometry followed, which eventually led to, the today dominant, methods of Algebraic Geometry ([98]).

References

90. E. Askwith (1903) A course of pure Geometry. Cambridge University Press, Cambridge
91. W. Besant (1895) Conic Sections Treated Geometrically. George Bell and Sons, London
92. M. Bocher (1915) Plane Analytic Geometry. Henry Holt and Company, New York
93. O. Bottema, 1984. Archimedes Revisited, *Mathematics Magazine*, 57:224-225
94. M. Chasles (1865) Traite de Sections Coniques. Gauthier-Villars, Paris
95. E. Cockshott (1891) Geometrical Conics. MacMillan and Co., London
96. J. Coolidge (1968) A history of the Conic Sections and Quadric Surfaces. Dover, New York
97. T. Heath (1896) Apollonius of Perga, Treatise on conic sections. Cambridge University Press, Cambridge
98. D. Perrin (2008) Algebraic Geometry An Introduction. Springer, Heidelberg
99. J. Quinn, 1904. A linkage for describing the conic sections by continuous motion, *American Mathematical Monthly*, 11:12-13
100. G. Salmon (1917) A treatise on Conic Sections. Longmans, Green and Co., London
101. Ch. Taylor (1881) Geometry of Conics. Deighton Bell and Co, Cambridge

Chapter 7
Transformations in space

7.1 Isometries in space

> And the last, to perform enumerations everywhere so complete, and reviews so general, that I am certain I don't exclude anything.
>
> Descartes, *Discourse on the Method*, II

Transformations in space, analogously to those on the plane (§ 2.1), are processes through which to every point X of space corresponds another point Y of space, which we denote by $f(X)$. The terminology here, *Image, Prototype, Maps, Domain, Range, Composition of Transformations, ...,* etc. is the same with that of the aforementioned section and I do not repeat it.

Analogous is also the definition of the *isometry*. Formally, it is the same with that of the plane. The only thing that changes is the location of the points, which now are in space. We call then **isometry** of space, a transformation f of the space which preserves the distances between points, in other words, for every pair of points in space X, Y and their images $X' = f(X)$, $Y' = f(Y)$ holds

$$|X'Y'| = |XY|.$$

We can immediately see some consequences of the definition, completely analogous to similar properties of isometries of the plane, some of which I formulate as exercises.

Theorem 7.1. *An isometry maps a triangle $AB\Gamma$ to a congruent triangle $A'B'\Gamma'$, an angle to an equal angle, a line to a line and a plane to a plane.*

Proof. The first part is obvious, because by definition the isometry preserves the lengths of line segments, therefore the two triangles will have equal respective sides. The same way it is proved also that an angle $\widehat{XOY}$ maps to an equal angle $\widehat{X'O'Y'}$. For the second part we show first that the isometry maps lines to lines. This follows directly from the preceding part of the proof. In-

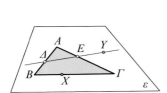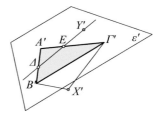

Fig. 7.1: Isometry preserves length, angles, lines, planes

deed, if X is a point of a line $B\Gamma$ and point X' is not on the corresponding line $B'\Gamma'$ (See Figure 7.1), then a genuine triangle $B'X'\Gamma'$ is created and for its lengths will hold

$$|B'X'|+|X'\Gamma'|=|BX|+|X\Gamma|=|B\Gamma|=|B'\Gamma'|,$$

which contradicts the triangle inequality. Thus, the isometry maps lines to lines. From this follows that it also maps a plane ε onto another plane ε'. Indeed if AB, $A\Gamma$ are two lines of plane ε and $A'B'$, $A'\Gamma'$ are their images, then for every other point Y of the plane ε we consider a line through Y which intersects AB, $A\Gamma$ respectively at points Δ, E. The images Δ', E' of these points through the isometry will be contained, according to the preceding part of the proof, respectively in lines $A'B'$, $A'\Gamma'$, therefore the line $\Delta'E'$ which contains Y' will also be contained in the plane of $A'B'$, $A'\Gamma'$.

Exercise 7.1. An isometry, which leaves two different points A, B fixed, leaves fixed also all the points of the line AB.

Exercise 7.2. An isometry, which leaves three non-collinear points fixed, leaves fixed also all the points of the plane, which passes through these three points.

Exercise 7.3. An isometry, which leaves four non-coplanar points fixed, leaves fixed also all the points of space, in other words it is the identity transformation e of space.

Exercise 7.4. Two isometries coincident at four non-coplanar points are coincident everywhere.

Exercise 7.5. An isometry f preserves parallelity of lines and planes. In other words, if $\{\varepsilon, \varepsilon'\}$ are respectively parallel lines or/and planes, then $\{f(\varepsilon), f(\varepsilon')\}$ are also parallel lines or/and planes.

Exercise 7.6. An isometry f preserves the angle between a line and a plane.

7.2. REFLECTIONS IN SPACE

Exercise 7.7. An isometry f maps a sphere Σ onto a sphere Σ' of equal radius. It also maps a circle κ onto a circle κ' of equal radius. If circle κ is contained in Σ, then κ' is also contained in Σ'.

Exercise 7.8. An isometry f maps a dihedral angle onto a congruent dihedral and a tetrahedron onto a congruent tetrahedron. More generally, it maps a polyhedron onto a congruent polyhedron.

As on the plane (I-§ 1.7), so in space we also have a notion of orientation, which can be defined through tetrahedra and depends on the order we use to place letters on the vertices. The tetrahedron $AB\Gamma\Delta$ is called **positively oriented**, if after placing a clock on the plane $AB\Gamma$ with its face towards Δ the direction $A \to B \to \Gamma$ is the same as that of its hands. If the direction $A \to B \to \Gamma$ is opposite to that of the hands of the clock, then the tetrahedron is called **negatively oriented**. We say that the transformation f **reverses the orientation** of a tetrahedron, when the tetrahedron $A'B'\Gamma'\Delta'$, which has vertices the images of the vertices of the tetrahedron $AB\Gamma\Delta$ through f has opposite orientation from that of $AB\Gamma\Delta$. If the two tetrahedrons $AB\Gamma\Delta$ and $A'B'\Gamma'\Delta'$ have the same orientation, then we say that f **preserves their orientation**. The transformations we consider in this chapter (are, as it is called, "continuous" and) have the property that, if they preserve the orientation of a tetrahedron, then they will preserve the orientation of every other too. Respectively, if they reverse the orientation of a tetrahedron, they will reverse the orientation of every other tetrahedron. It suffices then to examine what happens to the orientation of a single tetrahedron, in order to infer from it, the answer to whether the specific transformation preserves or reverses the orientation.

For the transformations which preserve the orientation of tetrahedra we also often say that they *preserve the orientation of space*, while for those which reverse the orientation of tetrahedra we say that they *reverse the orientation of space*.

7.2 Reflections in space

> And that can happen, you know, with the cleverest man, the psychologist, the literary man. The temperament reflects everything like a mirror! Gaze into it and admire what you see!
>
> F. Dostoevsky, *Crime and punishment*

A plane ε defines a transformation in space, called **reflection** or **mirroring** in ε, or relative to ε. Often ε is called the **axis** or **mirror** of the reflection. This is the analogue of reflections in the plane, relative to lines (§ 2.2). This reflection corresponds to every point X of the space: (i) X itself, if the point is

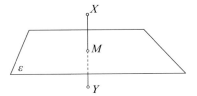

Fig. 7.2: Reflection in space

contained in the plane ε, (ii) point Y of space, such that ε will be the medial plane of XY, if X is not contained in ε (See Figure 7.2). Equivalently, if M is the projection of X onto ε, then Y will be in the extension of XM towards M and at equal distance with X from M.

Theorem 7.2. *A reflection in a plane ε is an isometry of the space, which reverses the orientation of space.*

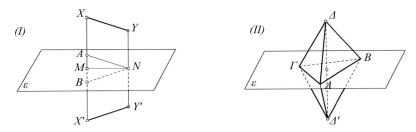

Fig. 7.3: Reflections in planes are isometries

Proof. Let us suppose that the points X, Y in space are not contained in plane ε, which defines the reflection, and X', Y' are their images (See Figure 7.3-I). Then, by definition, the line segments XX', YY' will be orthogonal to ε, therefore parallel and they will define the trapezium $XYY'X'$. We show that this trapezium is isosceles. For this, we draw parallels from the middle N of YY' respectively to YX and $Y'X'$, which intersect XX' respectively at points A and B. This creates the parallelograms $XANY$ and $X'BNY'$, for which hold $|AX| = |YN| = |NY'| = |BX'|$. However, if M is the middle of XX', then $|MX| = |MX'|$ and consequently that $|AM| = |MB|$. Also AB is, by its definition, orthogonal to MN, therefore the triangle ANB is isosceles and consequently we have $|XY| = |AN| = |BN| = |X'Y'|$. The proof for the case when one or both points X, Y are contained in ε or in the case where the isosceles degenerates to the line segment MN is even simpler.

That the reflection reverses the orientation, is seen easily by taking a tetrahedron $AB\Gamma\Delta$, of whose the basis $AB\Gamma$ is contained in the plane ε (See Figure 7.3-II). By the reflection in ε, this tetrahedron maps to the tetrahedron $AB\Gamma\Delta'$, where Δ' is the symmetric of Δ and we see immediately that these two tetrahedra have opposite orientations.

7.2. REFLECTIONS IN SPACE

Proposition 7.1. *For every reflection f holds $f \circ f = e$, in other words, the inverse of a reflection is the reflection itself.*

Proof. Obvious, and the same as that of Proposition 2.2.

Proposition 7.2. *If an isometry in space f, other than the identity, satisfies the relation $f \circ f = e$ and admits at least three non-collinear fixed points, then it coincides with the reflection relative to the plane defined by these three points.*

Exercise 7.9. Show that an isometry in space, which leaves fixed the points of a plane ε, and only these, coincides with the reflection relative to ε.

By analogy to the shapes of the plane, a shape in space Σ is called **symmetric relative to the plane** ε, when the reflection f relative to ε maps Σ to itself ($f(\Sigma) = \Sigma$). A plane ε, relative to which the shape Σ is symmetric, is called **plane of symmetry** of Σ. The most symmetric shape in space is the sphere.

Exercise 7.10. Show that every plane, that passes through the center of the sphere, is a plane of symmetry of the sphere. Show that shapes Σ, for which there exists a point O, such that every plane through O is a plane of symmetry of Σ, are unions of spheres centerred at O.

Exercise 7.11. How many and which are the planes of symmetry of a specific plane, or a dihedral angle?

Exercise 7.12. When does a trihedral angle has a plane of symmetry?

Exercise 7.13. Show that, if a trihedral angle has two planes of symmetry, then it also has a third and, consequently, has congruent faces and congruent dihedral angles.

Exercise 7.14. How many and which planes of symmetry has the cylinder and the cone?

Exercise 7.15. Show that, for two spheres of equal radii, there exists a reflection which maps the one onto the other.

As on the plane, so also in space a point O defines a **point symmetry** in O, or with respect to O, or relative to O. This transformation, to the point O corresponds O itself, and to every other point $X \neq O$ corresponds point Y, so that O will coincide with the middle of XY. We say that Y is the symmetric of X relative to O. The symmetry f relative to O is, like the reflection, its own inverse:

$$f \circ f = e.$$

A shape Σ is called **symmetric** in O, or relative to point O, when the symmetry f relative to O maps the shape to itself ($f(\Sigma) = \Sigma$). We also say then that O is a **center of symmetry** of the shape. The typical example is again the sphere.

Exercise 7.16. Show that in a line and in a plane any of their points is a center of symmetry for them.

Exercise 7.17. Which, among the Platonic solids, admits a center of symmetry?

Exercise 7.18. Show that the symmetry relative to a point is an isometry.

In space we also have a kind of symmetry which is defined through a line α (See Figure 7.4). We call **axial symmetry** relative to the line α in space the transformation which: (i) to every point X of the line α corresponds X itself and (ii) to every point X outside the line α corresponds point Y, so that α will be a medial line of XY. In other words point Y results by projecting point X onto α at M and extending XM towards M to the double of the distance $|XM|$. Every plane ε, orthogonal to the line α, maps through the axial symmetry

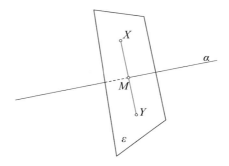

Fig. 7.4: Axial symmetry in space

relative to α to itself and consequently this defines a transformation of this plane to itself, which coincides with the symmetry relative to M, where M is the intersection point of the plane ε with the line α.

We call a shape Σ of the space **axial symmetric**, if there is a line α defining an axial symmetry f such that $f(\Sigma) = \Sigma$. The sphere is axial symmetric relative to every line through its centter. The cylinder and the right circular cone are axial symmetric relative to their axis.

Exercise 7.19. Show that a line, a plane and a sphere have infinitely many axes of symmetry.

Exercise 7.20. Show that a right conical surface and a right cylindrical surface are axial symmetric relative to their axis.

Exercise 7.21. Find the axes of symmetry of a rectangular parallelepiped.

Exercise 7.22. Which right prismatic surfaces admit an axis of symmetry?

Exercise 7.23. Which Platonic solids admit axes of symmetry and how many?

7.3. TRANSLATIONS IN SPACE

Exercise 7.24. Show that the axial symmetry relative to the line α is equal to the composition of two reflections relative to two orthogonal planes passing through α.

Hint: See also the proof of Theorem 2.6.

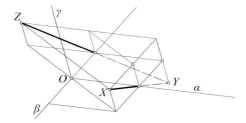

Fig. 7.5: Composition of axial symmetries of orthogonal axes

Exercise 7.25. Show that the composition of two axial symmetries relative to orthogonally intersecting axes $\{\alpha, \beta\}$ at O, is the axial symmetry relative to the axis γ, which is orthogonal to the plane of α, β at O (See Figure 7.5).

Exercise 7.26. Show that the composition of three reflections, relative to three planes, which pass through point O and intersect pairwise orthogonally, is the symmetry relative to O.

7.3 Translations in space

> At all times and in all places numerous men have existed who "did not believe in anything," precisely because "they did not question anything"; living, for them, meant simply abandoning oneself from one moment to the next, without any inner response or position in the face of any dilemma.
>
> *Ortega y Gasset, The origin of philosophy*

As in the plane (§ 2.3), so in space, an oriented line segment AB defines the transformation of **translation** by AB. This one corresponds, to every point X in space, a point Y, such that XY and AB are equal, parallel and equally oriented line segments. Equivalently: $ABYX$ is a parallelogram. Here too we consider the identity transformation as a **null translation**, i.e. a translation by a segment, whose endpoints coincide. All the properties of the translations of the plane, which are examined in the aforementioned section, can be carried over almost verbatim, along with their proofs, into corresponding properties of translations in space. I formulate some of them as exercises.

Exercise 7.27. Every translation is an isometry of space (Theorem 2.8).

Exercise 7.28. The composition of two or more translations is again a translation (Theorem 2.9, Corollary 2.3).

Exercise 7.29. The composition of translations along the oriented sides A_1A_2, A_2A_3, ..., $A_{k-1}A_k$, A_kA_1 of a (skew) closed polygon $A_1A_2...A_k$ is the identity transformation in space (Corollary 2.4).

Exercise 7.30. The composition of two reflections relative to two planes, whose distance is δ, is a translation along the line segment AB of length 2δ and direction orthogonal to the two planes (Theorem 2.10).

Exercise 7.31. The composition of two symmetries relative to axes, which are parallel and at distance δ, is a translation by a line segment AB of the plane of the parallels, orthogonal to the parallel axes and of length 2δ.

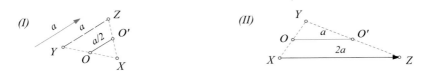

Fig. 7.6: Symmetry and translation   Symmetry and symmetry

Exercise 7.32. The composition of a symmetry relative to a point O and a translation α, is a symmetry to a point O' (Theorem 2.11, Fig. 7.6-I).

Exercise 7.33. The composition of two symmetries relative to two different points O, O' is a translation by the double of OO (Theorem 2.12, Fig. 7.6-II).

Exercise 7.34. The composition of ν symmetries relative to ν points A_1, A_2, ..., A_ν is, for even ν, a translation and for odd ν a symmetry (Corollary 2.5).

Exercise 7.35. Given ν different points $A_1, A_2, ..., A_\nu$, there exists exactly one (skew) polygon which has these points as middles of its successive sides if ν is odd. If ν is even, in general, there is no such polygon. If however there is one, there are infinitely many and in fact each point in space may be considered a vertex of such a polygon (Theorem 2.13).

Exercise 7.36. Given the tetrahedron $AB\Gamma\Delta$, determine the result of the composition of symmetries $f_\Delta \circ f_\Gamma \circ f_B \circ f_A$ on its vertices. How the order we take the vertices influences the result of the composition?

7.4 Rotations in space

> All that is distinctive of man, marking him off from the clay he walks upon or the potatoes he eats, occurs in his thought and emotions, in what we have agreed to call consciousness.
>
> *J. Dewey, The Role of Philosophy in the History of Civilization*

The main difference between rotations in space and those of the plane (§ 2.4) is that the former have an **axis of rotation** instead of a center of rotation. There exists, in other words, an entire line, which remains fixed during the rotation (its points are fixed points of the transformation). We also consider this line (axis) as being oriented, by choosing (arbitrarily) one of its directions as positive and its opposite as negative. The two notions of rotation, of plane and space, are however intimately related. One may say that the rotation in space is a kind of extension of the rotation on the plane. This is reflected in the property of a rotation in space, according to which, it defines a plane rotation on every plane orthogonal to its axis. We call then **rotation**

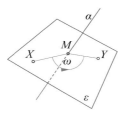

Fig. 7.7: Rotation in space

in space relative to the *oriented* line or **axis** α and oriented angle ω, the transformation which is defined as follows (See Figure 7.7): (i) The points of the axis remain fixed. (ii) For every point X outside the axis, we draw the plane ε through X, which is orthogonal to the axis and intersects it at point M. We choose its side (ε_+), which sees towards the positive direction of the axis. On this plane we define as *positive* the orientation of rotation which is opposite to the direction of the clock, the face of which sees towards the positive direction of the axis. To X the transformation corresponds Y, so that for the oriented angle holds $(\widehat{XMY}) = \omega$ and also $|MY| = |MX|$. In other words, after we determine the orientation in ε, we rotate X by applying to it the plane rotation of ε, with center M and angle ω.

Remark 7.1. In the case where the rotation angle is $|\omega| = \pi$ the rotation relative to axis α coincides with the symmetry relative to axis α. In every other case, of a rotation f by an angle which is not an integral multiple of π, changing, if necessary, the direction of the axis, we can define a rotation g, which gives the same result as f, for every point in space, and its angle is $0 < (\omega) < \pi$.

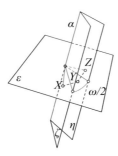

Fig. 7.8: Rotation as composition of two reflections

As in the case of the plane, so in space, the rotation is a composition of two reflections, this time relative to two planes ζ, η. Next proposition is proved exactly the same way as the corresponding one for planes (Proposition 2.6).

Proposition 7.3. *The composition of two reflections $f = h \circ g$, whose planes intersect along the line α making a dihedral angle of oriented measure ω with $|\omega| \leq \frac{\pi}{2}$, is a rotation with axis α and angle of rotation 2ω (See Figure 7.8).*

Theorem 7.3. *Every rotation in space is an isometry in space.*

Proof. A simple proof follows from the preceding proposition and the fact that reflections relative to planes are isometries in space.

Proposition 7.4. *The composition of two rotations with the same axis η and angles α and β is a rotation with the same axis and angle of rotation $\alpha + \beta$.*

Proof. The same as that of Proposition 2.5.

Proposition 7.5. *Every rotation f in space, relative to the axis α, defines on every plane ε, orthogonal to the line α a transformation f_ε of this plane. The transformation f_ε, of plane ε on itself, coincides with the rotation of the plane ε with center the point of intersection of ε with α and angle the rotation angle of f.*

Proof. Indeed, from the definition follows that f maps every plane ε orthogonal to the axis α to itself, leaving fixed the intersection point O of ε and α. It also defines an orientation of the plane considering as positive rotation the direction which is opposite to that of the clock, which has been placed on the plane with its face towards the positive direction of the axis α. f then defines a transformation f_ε of this plane. This transformation, according to the definition of rotation f, defines on ε exactly the same correspondence which is effected also by the rotation of ε relative to O and by the angle of f.

Proposition 7.6. *Every rotation f_ε, defined initially only for points of the plane ε, can be extended to a rotation f of the space, relative to the axis α coinciding with the line which is orthogonal to ε at the center O of the rotation f_ε and angle that of f_ε.*

7.4. ROTATIONS IN SPACE

Proof. This proposition, which is the converse of the preceding one, supposes that the plane ε is oriented. In other words one of the sides of the plane, ε_+ has been chosen and on it has been defined the positive rotation angle. The orientation of ε defines an orientation on the line α considering as positive the one towards the chosen side ε_+. Under these assumptions, f is defined as the rotation in space about axis α with angle that of f_ε.

Proposition 7.7. *The composition of two rotations relative to the axes α, β, which pass through point O, is a rotation relative to an axis γ which also passes through point O (See Figure 7.9).*

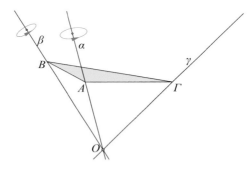

Fig. 7.9: Composition of rotations with intersecting axes

Proof. Let us represent the first rotation f as a composition of two reflections f_ζ, f_η relative to planes ζ, η which intersect along α. We can choose these planes so that the second η coincides with the plane of lines α, β. Analogously, we can represent the second rotation g as a composition of the reflections f_η, f_θ on the plane η and on one more plane θ which passes through β. Then the composition is:

$$g \circ f = (f_\theta \circ f_\eta) \circ (f_\eta \circ f_\zeta) = f_\theta \circ (f_\eta \circ f_\eta) \circ f_\zeta = f_\theta \circ f_\zeta.$$

Planes θ, η, passing by definition respectively through α, β, will intersect at O, therefore they will intersect along a line γ also passing through point O. The preceding equality shows that the composition $g \circ f$ is a rotation relative to the axis γ (Proposition 7.3).

Proposition 7.8. *The composition of two symmetries relative to axes α and β, which intersect at the point O making an angle ω with $|\omega| \leq \frac{\pi}{2}$, is a rotation f relative to the axis γ, which is orthogonal to the plane of α, β and passes through point O. The angle of rotation of f is 2ω (See Figure 7.10).*

Proof. The proof follows by applying the preceding proposition, since each axial symmetry is a rotation by π relative to its axis. The only additional

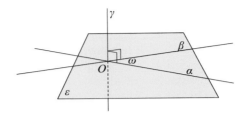

Fig. 7.10: Composition of two axial symmetries

characteristic in this case is that the rotation angle is determined directly (while in the preceding proposition it is not so obvious how the rotation angle depends on α and β, which however can be determined by resolving a spherical triangle).

Exercise 7.37. Can a translation in space equal a rotation in space?

Exercise 7.38. Which are all the isometries in space which map a line segment AB onto itself?

Exercise 7.39. An isometry in space, which fixes the points of a line α and only these, is a rotation with axis α.

Hint: If the isometry f maps a point X to X', draw the orthogonal XM towards α which intersects it at M. Point M is a fixed point of f, therefore MX' will be also orthogonal to α and the medial plane ε of XX' will contain α. Let f_ε be the reflection relative to ε. Then $f_\varepsilon \circ f$ fixes all the points of α as well as X. Consequently, it coincides with a reflection or the identity. It cannot, however, coincide with the identity, because of the hypothesis. If it did, then $f_\varepsilon \circ f = e \Rightarrow f = f_\varepsilon$, in other words, the initial f would be also a reflection and it would then fix also points lying outside the line α.

Exercise 7.40. An isometry in space, different from the identity, which fixes three non collinear points, fixes also every other point of the plane of the three points and coincides with the reflection relative to this plane.

Exercise 7.41. The composition of three reflections, relative to planes which intersect along the line α, is a reflection relative to a plane, which also contains α.

Hint: Consider the two of the three reflections and their composition, which is a rotation with axis α. These two planes can be rotated without altering their in-between angle, so that one of them coincides with the third of the given planes.

Exercise 7.42. Given two different but equal line segments AB and $A'B'$ in space, show that there always exists a rotation in space which maps A to A' and B to B', except in the special case where $AA'B'B$ is a non-rectangular parallelogram or AB, $A'B'$ are collinear and equally oriented.

Hint: The special case occurs when the medial planes ε, ζ of AA' and BB' do not intersect. When these planes coincide then the lines AA' and BB' are parallel and their endpoints are vertices of a trapezium. Then the symmetry relative to the axis α, which coincides with the line of the middles of the parallel sides of the trapezium (which is a rotation in space) does the job.

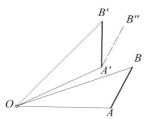

Fig. 7.11: Rotation which maps AB to $A'B'$

When ε, ζ intersect along a line α, then the rotation g with axis α which maps A to A', maps point B to point B'' (See Figure 7.11). There exists then an additional rotation g with axis OA' such that the rotation $g \circ f$ (Proposition 7.7) performs the desired action.

Exercise 7.43. Given two equal line segments AB and $A'B'$ in space, show that there always exists a rotation or a translation in space, which maps the one to the other.

Hint: A small difference from the preceding exercise: now the change of direction of the line segments is allowed.

7.5 Congruence or isometry in space

> But in fact the preference they have shown (the great novelists) for writing in images rather than in reasoned arguments is revelatory of a certain thought that is common to them all, convinced of the uselessness of any principle of explanation and sure of the educative message of perceptible appearance.
>
> *Albert Camus, Myth of Sisyphus p.110*

Isometries in space, lie at the root of the general definition of the notion of **congruence** or **isometry** of *shapes* in Euclidean space, which is the following:

Two shapes Σ and Σ' in space are congruent or isometric, if and only if there exists an isometry or congruence f in space, which maps the one onto the other ($f(\Sigma) = \Sigma'$).

Formally, the definition of the isometry in space is the same with the one of isometry or congruence between shapes of the plane. The only thing that changes is the domain where it is defined, which now is the space. In some treatises (for example [42]) the case of isometries, which can be represented as compositions of an odd number of reflections is put aside, like the symmetry relative to a point O. There, only shapes which can be mapped one to the other using isometries, expressible as compositions of an even number of reflections, are considered congruent. This corresponds to shapes which can be placed one on top of the other so that they coincide (which is Euclid's view). This, however, creates various problems, like for example, with shapes which are symmetric relative to a point, as is the sphere, the cube, the regular octahedron, etc. In this case it is somewhat unnatural to not consider the symmetry as already belonging to the officially accepted congruences. The analogue on the plane would be to separate the reflections and to not include them in the "official" congruences. We have here followed a different path, accepting as congruent, shapes which are mapped one to the other using any isometry, that is, a mapping which preserves distances between points, whether it preserves the orientation or reverses it. This is also the way congruence is treated in other geometries, different from the Euclidean. Our way therefore is in tune with the generally accepted view for congruence in other Geometries as well.

As is the case in the plane, so also in space, all the isometries are compositions of reflections. While on the plane every isometry can be written as a composition of at most three reflections, in space we'll see (Theorem 7.8), that every isometry can be written as a composition of at most four reflections. Knowing the compositions of two reflections, which are nothing but rotations and translations, it remains to see what kind of isometries are the compositions of three and four reflections. This leads to the so-called **rotoreflections**, **glide reflections** and **helicoidal** or **skrew displacements** ([42, II, p.91], [34, p.99]). We call **rotoreflection** the composition $f_\varepsilon \circ f$ of one rotation

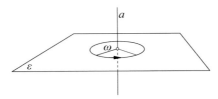

Fig. 7.12: Rotoreflection

f and one reflection f_ε on a plane ε orthogonal to the axis α of the rotation (See Figure 7.12). It is easy to see, that for such a composition the order is immaterial: $f_\varepsilon \circ f = f \circ f_\varepsilon$. Also, in the case where the rotation is performed by a flattened angle (of measure π) the composition coincides with a point symmetry. Therefore, rotoreflections include point symmetries as special cases.

7.5. CONGRUENCE OR ISOMETRY IN SPACE

The characteristic of a rotoreflection is that it fixes exactly one point. Next propositions prove this property.

Proposition 7.9. *The composition $f_\zeta \circ f_\varepsilon$ (or $f_\varepsilon \circ f_\zeta$) of a reflection relative to a plane ε and an axial symmetry relative to an axis ζ, not contained in the plane, is a rotoreflection.*

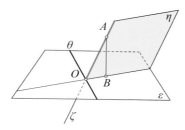

Fig. 7.13: Composition of reflection and axial symmetry

Proof. Consider the plane η, which passes through ζ and is orthogonal to ε (See Figure 7.13). Let κ be the intersection line of the planes ε and η and θ be the line of ε orthogonal to κ, and consequently to η, which passes through the intersection point O of ζ and ε. The composition $g = f_\zeta \circ f_\varepsilon$ maps each point X of θ to its symmetric X' relative to the intersection point O of ζ with ε. Consequently the composition $h = f_\eta \circ g$ fixes the points of the line θ. We also easily see that h fixes no other points except those of θ. Therefore h is a rotation (Exercise 7.39) and consequently $g = f_\eta \circ h$ satisfies the definition of a rotoreflection. The other ordering $f_\varepsilon \circ f_\zeta$ defines the inverse transformation of the preceding one.

Proposition 7.10. *A composition $f = f_\eta \circ f_\zeta \circ f_\varepsilon$ of three reflections relative to three planes ε, ζ, η, which pass through the same point O is a rotoreflection.*

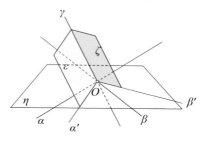

Fig. 7.14: Composition of three reflections

Proof. Suppose that the three planes intersect initially along the lines: α the planes $\{\varepsilon, \eta\}$, and β the planes $\{\zeta, \eta\}$, and γ the planes $\{\varepsilon, \zeta\}$ (See Figure

7.14). The transformation defined by the composition $f_\zeta \circ f_\varepsilon$ is a rotation relative to the axis γ and angle double that made by planes ε and ζ. Therefore this transformation does not change if we rotate the planes about the axis γ without changing their angle. We rotate them this way, so that in their new position, plane ζ will be orthogonal to η. In this case the transformation

$$f = f_\eta \circ f_\zeta \circ f_\varepsilon = (f_\eta \circ f_\zeta) \circ f_\varepsilon = f_{\beta'} \circ f_\varepsilon,$$

where $f_{\beta'}$ is the axial symmetry relative to the line β', which is the intersection of the plane ζ in its new position with the plane η. According to the preceding proposition, however, latter composition is a rotoreflection.

Proposition 7.11. *An isometry f in space, which fixes exactly one point O, is a rotoreflection.*

Proof. Indeed, let X be a point in space and $X' = f(X) \neq X$ (See Figure 7.15). Let also η be the medial plane of XX'. This plane contains O and defines a reflection f_η, such that the composition $f_\eta \circ f$ fixes points O and X, therefore all points of OX. This composition then coincides with, either a rotation

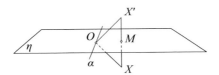

Fig. 7.15: Invariant line

about OX (Exercise 7.39), or with a reflection relative to a plane which contains OX (Exercise 7.40), or with the identity. The last two cases however cannot happen. If the last thing happened, it would be $f_\eta \circ f = e$, and then we would also have $f = f_\eta$, therefore f would fix more than one point. If it were $f_\eta \circ f = f_\theta$, where θ is a plane containing OX, then $f = f_\eta \circ f_\theta$ would be a rotation (Proposition 7.3) and would fix more than one point. Consequently the composition $f_\eta \circ f$ coincides with a rotation g about the axis OX. We have then $f = f_\eta \circ g$. The rotation g is written as a composition of two reflections relative to two planes intersecting by OX and the conclusion follows by applying the preceding proposition.

We call a *helicoidal* or *skrew displacement* the composition $g \circ f$ of a rotation f and a translation g parallel to the axis α of the rotation. It is easy to see, that for such a composition the order is immaterial: $g \circ f = f \circ g$. Also the helicoidal displacements include, as special cases, not only rotations (when the translation is *null*) but also translations (when the rotation is the identity). Next proposition gives a characterization of an isometry of this kind.

Theorem 7.4. *The composition of two axial symmetries relative to two lines in space is a helicoidal displacement (See Figure 7.16).*

7.5. CONGRUENCE OR ISOMETRY IN SPACE

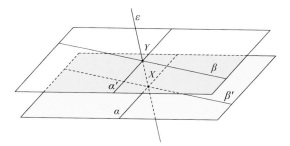

Fig. 7.16: Composition of axial symmetries

Proof. If the axes α, β of the two axial symmetries f_β, f_α intersect, then the composition $f = f_\beta \circ f_\alpha$ is a rotation (Proposition 7.8). If the lines α and β are skew, then the common orthogonal $\varepsilon = XY$ of the two skew lines is defined, and the two axial symmetries can be expressed as the compositions

$$f_\alpha = f_{\alpha\alpha'} \circ f_{\alpha\beta'}, \quad f_\beta = f_{\beta\alpha'} \circ f_{\beta\beta'}.$$

I denote with $\alpha\alpha'$, $\alpha\beta'$, etc., the planes which contain the respective lines. α' is the parallel to α from Y and β' is the parallel to β from X and $(\alpha\alpha', \alpha\beta')$, $(\beta\alpha', \beta\beta')$ are pairs of mutually orthogonal planes. Then

$$f = f_\beta \circ f_\alpha = (f_{\beta\alpha'} \circ f_{\beta\beta'}) \circ (f_{\alpha\alpha'} \circ f_{\alpha\beta'}) = f_{\beta\alpha'} \circ (f_{\beta\beta'} \circ f_{\alpha\alpha'} \circ f_{\alpha\beta'}).$$

The last three factors represent however a rotoreflection, in which the rotation and the reflection commute. This way we finally have

$$f = f_{\beta\alpha'} \circ (f_{\beta\beta'} \circ f_{\alpha\alpha'} \circ f_{\alpha\beta'}) = f_{\beta\alpha'} \circ (f_{\alpha\beta'} \circ f_{\beta\beta'} \circ f_{\alpha\alpha'})$$
$$= (f_{\beta\alpha'} \circ f_{\alpha\beta'}) \circ (f_{\beta\beta'} \circ f_{\alpha\alpha'}),$$

which is exactly the requested result. This, because the composition in the last parenthesis represents a rotation with axis ε and in the next to last a translation parallel to ε.

Exercise 7.44. Show that a reflection cannot be coincident with a helicoidal displacement.

Remark 7.2. The proof of the theorem shows, that the rotation angle of the helicoidal displacement is the angle between α, β' (or between their parallels α', β) and the axis of rotation coincides with the common orthogonal ε of the skew lines. Obviously if we rotate both skew lines about ε or we transpose them along ε, without varying their angle, we'll get the same helicoidal displacement. This observation leads to the proof of the next two propositions.

Corollary 7.1. *Every helicoidal displacement can be written as the composition of two axial symmetries.*

Theorem 7.5. *The composition $g \circ f$ of two helicoidal displacements f and g is again a helicoidal displacement.*

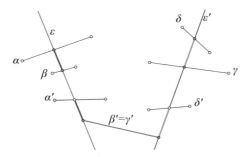

Fig. 7.17: Composition of helical transpositions

Proof. Let us suppose that the first helicoidal displacement is written as the composition of the axial symmetries $f = f_\beta \circ f_\alpha$ relative to the skew lines α, β, which have common orthogonal line ε, and the second $g = f_\delta \circ f_\gamma$ with skew lines γ, δ, whose common orthogonal is ε' (See Figure 7.17). According to the preceding remark, we can move and rotate the two systems (α, β) and (γ, δ), each on its common orthogonal, so that they arrive at positions respectively (α', β'), (γ', δ'), for which β', γ' coincide with the common orthogonal of ε and ε'. Then the composition of the two helicoidal displacements will be expressed through

$$g \circ f = (f_{\delta'} \circ f_{\gamma'}) \circ (f_{\beta'} \circ f_{\alpha'}) = f_{\delta'} \circ (f_{\gamma'} \circ f_{\beta'}) \circ f_{\alpha'} = f_{\delta'} \circ f_{\alpha'},$$

which is a helicoidal displacement.

Corollary 7.2. *The composition of arbitrary many helicoidal displacements is a helicoidal displacement.*

We call **glide reflection** the composition $g \circ f_\varepsilon$ of a reflection f_ε, relative to a plane ε, and a translation g by an oriented line segment AB, which is parallel to the plane ε. The definition includes reflections as special cases, in which the translation is null. Next theorem shows how this transformation can be written as composition of simpler ones.

Theorem 7.6. *The composition of three reflections $f_\gamma \circ f_\beta \circ f_\alpha$ respectively relative to three planes α, β, γ, which do not have, all three, a common point, is a glide reflection.*

Proof. If the three planes are pairwise parallel, then the composition of the two first is a translation and is written also as a composition of two other

7.5. CONGRUENCE OR ISOMETRY IN SPACE

planes parallel to the first α', β', provided their distance is the same as that between α, β. We can then write $f_\beta \circ f_\alpha$ as a composition $f_{\beta'} \circ f_{\alpha'}$ using as β' the plane γ. Then the initial composition gives the reflection relative to α':

$$f_\gamma \circ f_\beta \circ f_\alpha = f_\gamma \circ f_{\beta'} \circ f_{\alpha'} = (f_\gamma \circ f_{\beta'}) \circ f_{\alpha'} = f_{\alpha'}.$$

If the three planes are not pairwise parallel, then they pairwise intersect along three parallel lines $\Gamma\Gamma'$ planes α, β, AA' planes β, γ and BB' planes γ, α (See Figure 7.18). Rotating the dihedral angle of β, γ about its edge

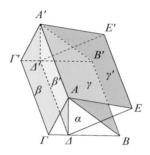

Fig. 7.18: Glide reflection

AA' without altering its angle, we place it at position β', γ' for which β' is orthogonal to the plane α. Then the composition

$$(f_\gamma \circ f_\beta) \circ f_\alpha = (f_{\gamma'} \circ f_{\beta'}) \circ f_\alpha = f_{\gamma'} \circ (f_{\beta'} \circ f_\alpha) = f_{\gamma'} \circ f_{AA'},$$

where $f_{AA'} = f_{\beta'} \circ f_\alpha$ is a composition of two reflections relative to orthog-

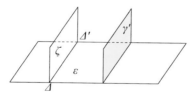

Fig. 7.19: Glide reflection in normal form

onal planes, therefore an axial symmetry relative to the line AA' of the intersection of the two planes. It is easy, however, to see that the composition $f_{\gamma'} \circ f_{AA'}$ is a glide reflection. Indeed we can write the axial symmetry $f_{AA'}$ as a composition of reflections $f_{AA'} = f_\zeta \circ f_\varepsilon$ on two planes ε, ζ, orthogonal and respectively parallel to γ' and passing through AA' (See Figure 7.19). Then the composition

$$f_{\gamma'} \circ f_{AA'} = f_{\gamma'} \circ (f_\zeta \circ f_\varepsilon) = (f_{\gamma'} \circ f_\zeta) \circ f_\varepsilon = h \circ f_\varepsilon,$$

where $h = f_\gamma \circ f_\zeta$, as a composition of two reflections in planes parallel and orthogonal to ε, is a translation along a line segment parallel to ε.

Theorem 7.7. *An isometry in space, which leaves no point fixed is either a glide reflection or a helicoidal displacement. In both cases the translation which participates in the isometry is not null.*

Proof. Suppose that the isometry f in space leaves no point fixed. Let us consider then an arbitrary point X and its image $X' = f(X) \neq X$. Let also f_ε be the reflection relative to the medial plane of XX'. Then obviously, the isometry $g = f_\varepsilon \circ f$ fixes point X. There exist then the following possibilities for g: (i) g fixes this and only this point, (ii) g fixes the points of a line through X and only these, (iii) g fixes the points of a plane through X and only these, and the trivial case (iv) g fixes every point. The last case cannot happen because then $g = e \Rightarrow f_\varepsilon \circ f = e \Rightarrow f = f_\varepsilon$, and f would fix all the points of plane ε, contrary to the assumption.

If (iii) happens, then g will coincide with a reflection f_η relative to a plane η (Exercise 7.9). Then $g = f_\varepsilon \circ f = f_\eta \Rightarrow f = f_\varepsilon \circ f_\eta$. Consequently f as a composition of reflections, is a rotation if planes ε, η intersect or a translation, if they are parallel. From the two, only the latter can happen, because otherwise f would fix the points of its axis of rotation. Therefore, in case (iii) f coincides with a translation (helicoidal with null rotation) and the theorem holds.

If (ii) happens, then g will coincide with a rotation, which is written as the composition $g = f_\eta \circ f_\zeta$ of two reflections relative to planes ζ and η. Then $f = f_\varepsilon \circ g = f_\varepsilon \circ f_\eta \circ f_\zeta$ will be a composition of three reflections. These three planes cannot have a common point, because then f would be a rotoreflection (Proposition 7.10) and would fix a point, contrary to the hypothesis. Consequently, the planes will not have common points and f will be a glide reflection (Theorem 7.6).

Finally if (i) happens, then g will coincide with a rotoreflection (Proposition 7.11), which is written as a composition of three reflections (Proposition 7.10). Then $f = f_\varepsilon \circ g$ will be the composition of four reflections and equivalently, two rotations, which are also helicoidal displacements. According to Theorem 7.5 the composition of the two latter ones will also be a helicoidal displacement and the theorem will hold.

The fact, as it was proved, that in the glide reflection or in the helicoidal displacement f a null translation cannot participate, follows from the assumption of the non-existence of a fixed point of f.

Theorem 7.8. *An isometry in space is either a rotoreflection, or a glide reflection or a helicoidal displacement. In all cases the isometry is written as a composition of at most four reflections.*

Proof. The proof is contained in the proof of the preceding theorem. If the isometry f leaves no point in space fixed, then, according to the preceding

7.5. CONGRUENCE OR ISOMETRY IN SPACE

proposition, it is either a glide reflection or a helicoidal displacement. If it fixes exactly one point, then it is a rotoreflection (Proposition 7.11). If it fixes exactly one line it is a rotation, that is a special case of a helicoidal displacement. Finally if it fixes exactly a plane it is a reflection, that is a special case of a rotoreflection.

Corollary 7.3. *In an isometry, which is written in different ways as a composition of reflections, the number of reflections is always even or always odd.*

Proof. Indeed, if we have two expressions of the isometry f with reflections:

$$f = f_k \circ f_{k-1} \circ \ldots \circ f_1 = g_m \circ g_{m-1} \circ \ldots \circ g_1,$$

then

$$e = f \circ f^{-1} = (f_k \circ f_{k-1} \circ \ldots \circ f_1) \circ (g_1 \circ g_2 \circ \ldots \circ g_m).$$

We can simplify the composition of reflections in the right hand side by quadruples. Each quadruple of reflections defines a helicoidal displacement and the composition of two helicoidal displacements, that is eight reflections, is replaced with a helicoidal displacement, in other words four reflections. Reducing then the quadruples, we'll arrive at a composition of 4 or 5 or 6 or 7 reflections. We must therefore show that the identity transformation cannot be written as a composition of 5 or 7 reflections. If we have then

$$e = h_1 \circ \ldots \circ h_k,$$

and $k = 7$ we compose both parts with an arbitrary reflection. If $k = 5$ we compose both parts with the last reflection. In the first case, on the right side results a composition of 8 reflections which is reduced to a helicoidal displacement. In the second case we have on the right side again a helicoidal displacement. It follows then that we have an equality of the form

$$h = h',$$

where the left is a reflection and the right is a helicoidal displacement. This however is impossible (Exercise 7.44).

Remark 7.3. The last corollary is connected to the phenomenon of preservation or reversal of orientation in space through an isometry (§ 7.1). It is easy to see that a reflection reverses the orientation. For this it suffices to consider a tetrahedron $AB\Gamma\Delta$ with the triangle $AB\Gamma$ contained in the plane which defines the reflection. Then the image of $A'B'\Gamma'\Delta'$ through the reflection fixes $AB\Gamma$, sends Δ to Δ' on the other side of the plane, so that $AB\Gamma\Delta$ and $A'B'\Gamma'\Delta'$ have opposite orientation (Theorem 7.2).

An isometry (congruence), which is written as a composition of an even number of reflections, preserves orientation and is called a **direct isometry**, while an isometry, which is written as a composition of an odd number of reflections, reverses the orientation and is called an **indirect isometry**.

7.6 Homotheties in space

> There is nothing in the world except curved empty space. Geometry bent one way here describes gravitation. Rippled another way somewhere else it manifests all the qualities of an electromagnetic wave. Excited at still another place, the magic material that is space shows itself as a particle. There is nothing that is foreign and "physical" immersed in space. Everything that is, is constructed out of geometry.
>
> J. Wheeler, Battelle Rencontres p.273

Formally the definition is the same with that of plane homotheties (§ 2.6): Given a number $\kappa \neq 0$ and point O in space, we call a **homothety** of center O and **ratio** κ the transformation which corresponds: (i) to point O the point O itself, (ii) to every other point $X \neq O$ in space, the point X' on the line OX, such that for the signed ratio holds

$$\frac{OX'}{OX} = \kappa.$$

A direct consequence of the definition is, that for every point O, the homothety with center O and ratio $\kappa = 1$ is the identity transformation. Often, when the ratio is $\kappa < 0$ we say that the transformation is an **antihomothety**. Its characteristic is that point O is between X and X'.

A homothety ($\kappa > 0$) preserves the orientation in space, while an antihomothety ($\kappa < 0$) reverses it. Latter follows from the fact that the antihomothety is a composition of a homothety, with the same center and opposite ratio, and a point symmetry relative to the center of the antihomothety. The properties, which I formulate as exercises, are similar to those of the plane and with similar proofs.

Exercise 7.45. The inverse transformation of a homothety is a homothety with the same center and ratio $\kappa' = \frac{1}{\kappa}$. The composition of two homotheties, with the same center, is a homothety with the same center and ratio the product of the ratios of the two homotheties.

Exercise 7.46. A homothety maps a line ε onto one parallel to it ε', and a plane α to a plane α' parallel to it.

Exercise 7.47. A homothety maps an angle to an equal angle and a dihedral angle to a congruent dihedral angle.

Exercise 7.48. For two spheres of different radii ρ, ρ' there exists one, if they are concentric, and two homotheties if they are not concentric, which map one to the other.

Exercise 7.49. The composition of two homotheties f and g, with different centers respectively O and P, and ratios κ and λ with $\kappa \cdot \lambda \neq 1$, is a homothety with center T on the line OP and ratio equal to $\kappa \cdot \lambda$ (Theorem 2.19).

7.7. SIMILARITIES IN SPACE

Exercise 7.50. The composition of two homotheties f and g with different centers respectively O and P, and ratios κ and λ with $\kappa \cdot \lambda = 1$, is a translation along a segment parallel to OP (Theorem 2.20).

Exercise 7.51. The composition $g \circ f$ of a homothety f and a translation g is a homothety (Theorem 2.21).

Exercise 7.52. Show that two triangles $AB\Gamma$ and $A'B'\Gamma'$, which have respective sides parallel are either contained in the same plane or in two parallel planes and there exists a homothety or translation which maps the one to the other.

Exercise 7.53. Show that for two tetrahedra $AB\Gamma\Delta$ and $A'B'\Gamma'\Delta'$, which have respective faces parallel, there exists a homothety or translation which maps one to the other.

7.7 Similarities in space

> Half the wrong conclusions at which mankind arrive are reached by the abuse of metaphors, and by mistaking general resemblance or imaginary similarity for real identity.
>
> H. J. T. Palmerston, Letter to H. Bulwer

Here also the definition is formally the same with that of the plane (§ 2.7): **Similarity** is called a transformation f in space, which multiplies the distances between points with a constant $\kappa > 0$, which is called **ratio** of the similarity. By definition then, for every pair of points X, Y the similarity corresponds points $X' = f(X), Y' = f(Y)$ which satisfy

$$|X'Y'| = \kappa \cdot |XY|.$$

This general definition includes isometries, for which we have $\kappa = 1$ and homotheties. Similarities not coincident with isometries, that is, similarities for which the ratio $\kappa \neq 1$, are called **proper** similarities. Obviously the composition of two similarities is again a similarity with ratio equal to the product of the ratios of the two similarities. In the case where the two ratios satisfy $\kappa \cdot \kappa' = 1$, the composition of the corresponding similarities is an isometry.

A similarity may preserve or reverse the orientation in space (§ 7.1). If the similarity f of ratio κ preserves the orientation, then by composing it with a homothety g of ratio $\frac{1}{\kappa}$, we get an isometry $g \circ f$ which also preserves the orientation. If the similarity reverses the orientation in space then by composing it with an antihomothety of ratio $-\frac{1}{\kappa}$ (which also reverses orientation) we get an isometry $g \circ f$ which preserves the orientation in space. This property is used in the proof of the following theorem.

Theorem 7.9. *Every proper similarity in space has a fixed point.*

Proof. ([34, p.104]) Let X be an arbitrary point in space and $Y = f(X)$ its image under the similarity f of ratio $\kappa \neq 1$. If $Y = X$, then this point is the fixed of f. If $Y \neq X$, then on the line XY we consider the unique point Z such that for the signed ratio holds $\frac{ZX}{ZY} = \kappa' = \pm\frac{1}{\kappa}$. We choose the positive sign if the similarity f preserves the orientation and the negative if the similarity reverses the orientation. We also consider the homothety g with center Z and ratio κ'. Then for the composition $h = g \circ f$ holds $h(X) = g(f(X)) = g(Y) = X$. This way then h, which is an isometry and preserves the orientation in space, admits X as a fixed point. According to Theorem 7.8 we conclude that h is a rotation, therefore there exists a line α, the axis of rotation, which remains fixed by h.

We consider the plane ε passing through Z and orthogonal to α. The isometry h maps it to itself, because it maps it to a plane also orthogonal to the line $h(\alpha) = \alpha$ and its intersection point with α remains fixed. The homothety g also maps the plane ε to itself. The latter because it maps it to a plane $g(\varepsilon)$ orthogonal to the line $g(\alpha)$, which however, because of the homothety, is parallel to α. Therefore $g(\varepsilon)$ is also orthogonal to α and also passes through Z, since the last one is a fixed point of g. Consequently $g(\varepsilon) = \varepsilon$.

This implies that the plane ε is invariant also relative to $h = g \circ f$ and relative to g, therefore also relative to the inverse of g and consequently relative to $f = g^{-1} \circ h$. The similarity f then maps the plane ε to itself and defines on it a similarity f_ε. According to Corollary 2.8 this similarity of the plane ε will admit a fixed point O, which will also be a fixed point of f. The uniqueness of the fixed point follows again from the fact that $\kappa > 1$. If there was also a second fixed point P then we would have $|OP| = |O'P'| = \kappa|OP| \Rightarrow \kappa = 1$, which is contradictory.

Corollary 7.4. *Every proper similarity in space is a composition of an isometry, which fixes a point O, and a homothety centerred at the point O.*

Proof. Indeed, from the preceding theorem we know that the similarity f admits a fixed point O. It follows that the composition $h = g \circ f$ with the homothety g, of center O and ratio $\frac{1}{\kappa}$, is an isometry and holds $f = g^{-1} \circ h$.

Exercise 7.54. Show that two similarities which coincide at four non coplanar points coincide everywhere.

Hint: Show first that the two similarities f and g have the same ratio κ. Then that $g \circ f^{-1}$ is an isometry which fixes four points.

Exercise 7.55. Show that an isometry f_ε of a plane ε in space to itself extends to an isometry f of the entire space to itself.

7.7. SIMILARITIES IN SPACE

Hint: The exercise supposes that f_ε is a transformation of the plane ε to itself, that is, it is defined only for points on the plane and for these points holds $|X'Y'| = |XY|$, where $X' = f_\varepsilon(X)$, $Y' = f_\varepsilon(Y)$. The exercise asks to define an isometry f, this time defined for every point in space, and which also maps ε to itself ($f(\varepsilon) = \varepsilon$) and for every point X of ε holds $f(X) = f_\varepsilon(X)$. Because every isometry on the plane is a composition of three reflections (Theorem 2.5), it suffices to show that a reflection of the plane ε is extended to an isometry in space. The latter however is obvious. If the reflection at ε has axis the line α it suffices for us to take the plane orthogonal to ε which also contains α and to consider the reflection f relative to this plane.

Exercise 7.56. Show that a similarity f_ε, of the plane ε in space to itself, extends to a similarity f of the entire space to itself.

Hint: If the similarity coincides with an isometry apply the preceding exercise. If f_ε is a proper similarity with ratio $\kappa \neq 1$, then it admits exactly one fixed point O on the plane ε (Corollary 2.8). Suppose f_O is the homothety in space with center O and ratio $\frac{1}{\kappa}$. Then the composition $g = f_O \circ f_\varepsilon$ is an isometry of ε. According to the preceding exercise g extends to an isometry g' in space. Then $h = f_O^{-1} \circ g'$ is a similarity extending f_ε.

Exercise 7.57. Show that for two different similarities f, g, which coincide at three non collinear points there exists a reflection h such that $g = h \circ f$.

Hint: Show first that the two similarities f and g have the same ratio κ. Next show that $g \circ f^{-1}$ is an isometry which fixes three points.

Exercise 7.58. Show that there exist exactly two similarities which map a triangle onto a similar triangle in space. One of them preserves the orientation in space, while the other reverses it.

Hint: If the two similar triangles $AB\Gamma$ and $A'B'\Gamma'$ belong to different planes ε and ε', intersecting along the line ζ, then with a rotation f of ε about ζ we map $AB\Gamma$ onto a congruent to it triangle $A''B''\Gamma''$ of plane ε'. There exists therefore a similarity g' of the plane ε' which maps $A''B''\Gamma''$ onto $A'B'\Gamma'$. Suppose g is the extension of g' in space (Exercise 7.56). Then the composition $h = g' \circ f$ is a similarity in space which maps $AB\Gamma$ onto $A'B'\Gamma'$. The rest follows from Exercise 7.57.

Exercise 7.59. Show that there exists exactly one similarity which maps a tetrahedron to a similar tetrahedron.

Hint: Suppose that the similar tetrahedra are $AB\Gamma\Delta$ and $A'B'\Gamma'\Delta'$. Apply Exercise 7.58 to the similar triangles $AB\Gamma$ and $A'B'\Gamma'$ and determine the two similarities f, f' which map $AB\Gamma$ onto $A'B'\Gamma'$. Then either $f(\Delta)$ or $f'(\Delta)$ will coincide with Δ' and will determine the unique similarity which is required by the exercise.

As it happens with isometries and congruence on the plane (§ 2.5) or/and in space (§ 7.5), so also with similarity transformations, they are at the root of a more general definition of similarity for shapes in space: Two shapes Σ, Σ' in space are called **similar**, when there exists a similarity f which maps one onto the other ($f(\Sigma) = \Sigma'$).

7.8 Archimedean solids

> ...these are not only the ones by divine Plato five shapes, that is the tetrahedron and hexahedron, octahedron and dodecahedron, with fifth the icosahedron, but also those found by Archimedes, thirteen in number, containing equilateral and equiangular but not similar polygons.
>
> *Pappus, Collection E'*

The **Archimedean solids** (or bodies) are *semi-regular* convex polyhedra whose faces, as also in Platonic solids (§ 4.6), are regular polygons. While in the Platonic solids all the faces are pairwise congruent, in the Archimedean solids we have faces of different shapes, which however are still constrained to be regular polygons. The exact definition of the Archimedean solid requires the use of the notion of isometry (congruence), and in fact an isometry which fixes the polyhedron. These are isometries in space which map the solid to itself. This defines then, to begin with, the **semi-regular** solids:

These are convex polyhedra, whose isometries leaving them invariant, act transitively on their vertices, and whose faces are regular polygons of at least two different kinds.

The phrase *act transitively on their vertices* means that, for any two vertices A, B of the polyhedron, there exists an isometry f of the space, which maps (replaces) the polyhedron onto itself and maps vertex A to B. A consequence of this property is that the polyhedral angles of the solid are pairwise congruent and the various polygons around each vertex are placed the same way. This leads also to the concise notation of the corresponding solid using a sequence, like for example, $(3,6,6)$, which means that each vertex of the solid is surrounded by a 3-angular and two 6-gonal faces.

Figure 7.20 shows one of the simplest Archimedean solids, the **truncated tetrahedron**, with corresponding symbol $(3,6,6)$. This results from the regular tetrahedron (Platonic solid), by cutting off its vertices, so that this makes four triangular and four hexagonal faces with all the edges equal. The line α_ε which is orthogonal to a face ε, at its center, defines the axis of some rotations in space, which map any vertex A of the polygonal face ε to another vertex B of the same face, mapping simultaneously the polyhedron onto itself. Composing rotations relative to similar orthogonal axes, with respect to the other faces, as well as reflections relative to the solid's planes of symmetry (if any), we get isometries in space, which fix the solid and map any vertex to any other.

7.8. ARCHIMEDEAN SOLIDS

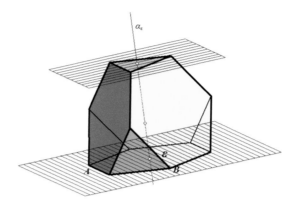

Fig. 7.20: Truncated tetrahedron $(3,6,6)$

As we'll see in the sequel, the semi-regular solids consist of three categories. The first consists of the **regular prisms** and the second contains the **antiprisms**. These categories contain both infinite many members. The third category of semi-regular solids includes only 13 solids, which are mentioned as *Archimedean*, in honor of Archimedes, who invented them, as mentioned by Pappus ([107, p.194], [104, p.405]). Figure 7.21 shows polyhedra

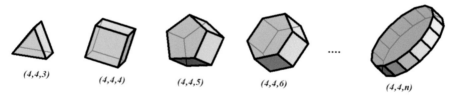

Fig. 7.21: Regular prisms $(4,4,n)$

which belong to the category of *regular prisms*. They are characterized by the fact that their lateral faces are squares and they have respectively symbols $(4,4,3)$, $(4,4,4)$ (cube), $(4,4,5)$, ... ,$(4,4,n)$, ...

Figure 7.22 shows polyhedra which belong to the category of *antiprisms*. They are characterized by two parallel faces which are regular n-gons and $2n$ in number lateral triangular faces. The corresponding symbols are $(3,3,3,3)$ (regular octahedron), $(3,3,3,4)$, $(3,3,3,5)$, ...

In the following, after a short discussion of three restrictions which hold for the faces around each vertex of a semi-regular polyhedron, we'll proceed to their classification, following Kepler [105], [103]. We have already met the reason of the first of these restrictions. It is expressed with Theorem 4.10, according to which the sum of the faces around the vertex of a convex polyhedral angle is less than 2π. Given that the angles of the regular polygons begin with $60°$ (equilateral triangle) and advance to $90°$, $108°$, $120°$, ... with sum $60 + 90 + 108 + 120 = 378 > 360$, which becomes even greater for polygons with more sides, it is impossible to see more than three kinds of regular polygons around a vertex. The first restriction then may be expressed in the form of next proposition.

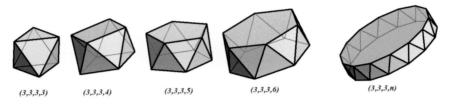

Fig. 7.22: Antiprisms $(3,3,3,n)$

Proposition 7.12. *Around each vertex of a semi-regular polyhedron there can be faces of at most 3 different kinds of regular polygons.*

The other two restrictions correspond to the two configurations of polygons around one vertex, which are summarized in the following figure.

Fig. 7.23: Impossible case at a vertex of a semi-regular polyhedron

Proposition 7.13. *Around a vertex of a semi-regular polyedron the configuration of {triangle, α-gon, β-gon, γ-gon} is impossible, when $\alpha \neq \gamma$, except the case $3 = \alpha = \beta \neq \gamma$.*

Proof. For the proof it suffices to try to place polygons around the triangle. Inevitably, the situation of figure 7.23 will result, where the angle in the position of the question mark cannot be determined, given that the polygons are regular therefore have equal angles and around each vertex the same combination of polygons must show up, having also the same order. The case $(3,3,3,n)$ is, however, possible and characterizes the antiprisms with base a regular n-gon.

Proposition 7.14. *Around a vertex of a semi-regular polyhedron, the configuration of {α-gon, with α odd integer, β-gon and γ-gon with $\gamma \neq \beta$} is impossible.*

7.8. ARCHIMEDEAN SOLIDS

Proof. For the proof is suffices, again, to try to place regular polygons according to this ordering. Inevitably a situation like that of figure 7.24 ($\alpha = 5$) will occur, where, in the position of the question mark, any regular polygon placed there leads to an incompatibility.

Fig. 7.24: Ordering inconsistency at a vertex of a semi-regular polyhedron

Theorem 7.10. *(Archimedean solids) Besides the regular prisms and antiprisms, there exist 13 more different semi-regular solids (See Figures 7.25, 7.26, 7.27).*

Proof. We examine the different possibilities, taking into consideration the preceding restrictions. As we'll see immediately, next possibilities exhaust all the cases of polyhedral angles which show up in semi-regular polyhedra.

1. **Two kinds of polygons on each vertex**

    1.1. Triangles + squares
      i. Triangles + 1 square only
      ii. Triangle + 2 squares
      iii. Triangles + $n > 2$ squares
    1.2. Triangles + pentagons
      i. Triangles + 1 pentagon only
      ii. Triangles + 2 pentagons
    1.3. Triangles + hexagons
      i. Triangles + 1 hexagon only
      ii. Triangles + 2 hexagons

    1.4. Triangles + n-gons ($n \geq 7$)
      i. Triangles + 1 n-gon only
      ii. Triangles + 2 n-gons
    1.5. Squares + n-gons ($n \geq 5$)
      i. Squares + 1 n-gon only
      ii. Squares + 2 n-gons
    1.6. Pentagons + n-gons ($n \geq 6$)
      i. Pentagons + 1 n-gon only
      ii. Pentagons + 2 n-gons

2. **Three kinds of polygons on each vertex**

    2.1. Triangles + squares + n-gons
    2.2. m-gons + n-gons + p-gons, without triangles

- (1.a.i) The square has angles of $90°$ and consequently it can coexist with, at most 4 triangles. Otherwise, the total sum of the faces around the vertex will exceed $360°$. The possible cases are: $(3,3,3,3,4)$, $(3,3,3,4)$ and $(3,3,4)$. From these the last one is impossible (restriction 7.14). The other two are possible and are realized in the so-called **truncated cube** and the **square antiprism** respectively.

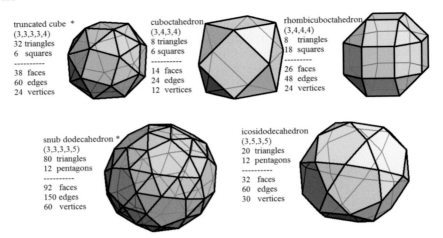

Fig. 7.25: Archimedean solids-I

- (1.a.ii) The two squares around a vertex have sum of angles 180° consequently they can coexist with at most 2 triangles. This results in cases $(3,3,4,4)$, $(3,4,3,4)$ and $(3,4,4)$. The case $(3,4,3,4)$ is rejected because of restriction 7.13. The other two are possible and realize in the so-called **cuboctahedron** and the regular triangular prism respectively.
- (1.a.iii) Three squares around a vertex have sum of angles 270° consequently they can coexist with only one triangle. The only possibility then in this category is $(3,4,4,4)$, which is realized in the so-called **rhombicuboctahedron**.
- (1.b.i) The pentagon with angle 108° can coexist with 4, at most triangles. Otherwise, the total sum of the faces around the vertex will exceed 360°. Then result the possibilities $(3,3,3,3,5)$, $(3,3,3,5)$, $(3,3,5)$, of which the last one is impossible according to restriction 7.14. The other two possibilities are realized in the so-called **snub dodecahedron** and the pentagonal antiprism.
- (1.b.ii) Two pentagons around a vertex have sum of angles 216° and can coexist with 2, at most, equilateral triangles. There result the possibilities $(3,3,5,5)$, $(3,5,3,5)$, $(3,5,5)$, of which the last one is impossible according to restriction 7.14. The first one is also rejected because of restriction 7.13. The only possibility in this category is, therefore, $(3,5,3,5)$, which is realized in the so-called **icosidodecahedron**.
- (1.c.i) The hexagon has angles of 120° and can coexist with 3, at most, equilateral triangles around a vertex. There result the two possibilities $(3,3,3,6)$ and $(3,3,6)$, of which the last one is impossible according to restriction 7.14. The only possibility in this category then is $(3,3,3,6)$, which is realized in the hexagonal antiprism.

7.8. ARCHIMEDEAN SOLIDS

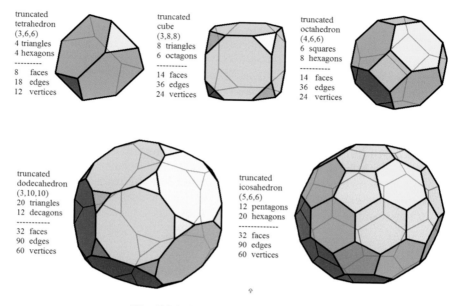

Fig. 7.26: Archimedean solids - II

- (1.c.ii) Two hexagons, with sum of angles 240° can coexist around a vertex with only one equilateral triangle. The only possibility in this case then, is, (3,6,6), which is realized in the **truncated tetrahedron**.
- (1.d.i) The case of one exactly n-gon and triangles, with $n \geq 7$ and respective angles greater than 128° allows the coexistence with 3, at most equilateral triangles around a vertex. The possibilities, consequently, in this category are $(3,3,3,n)$, $(3,3,n)$, of which the second is impossible because of restriction 7.14 and the first one is realized in the anriprism with base the regular n-gon.
- (1.d.ii) The case of exactly two n-gons and triangles, with $n \geq 7$ and respective angles greater than 128° allows the coexistence with one, at most, equilateral triangle around a vertex. The possibilities, then, in this category correspond to the symbol $(3, n, n)$, which, for odd n are excluded, because of restriction 7.14. Also, even numbers greater than 10 are excluded, since then the angle sum around a vertex exceeds 360°. Then, in this category, remain the possibilities $(3,8,8)$ and $(3,10,10)$, which are realized respectively in the **truncated cube** and the **truncated dodecahedron**.
- (1.e.i) The case of squares and exactly one n-gon around a vertex, with $n \geq 5$ is $(4,4,n)$, which corresponds to the regular prisms with base a regular n-gon.

- (1.e.ii) The case of squares and two exactly n-gons around a vertex, with $n \geq 5$ allows only one square and only solids with symbols of type $(4,n,n)$. These for $n \geq 8$ are rejected because they give a sum of angles around the vertex greater than $360°$. Also the type $(3,7,7)$ is rejected because of restriction 7.14. The only case in this category then is $(4,6,6)$, which is realized in the **truncated octahedron**.
- (1.f.i) The case of pentagons and exactly one n-gon, with $n \geq 6$ around a vertex, is rejected already from the respective type $(5,5,n)$ because such polygons give a sum of angles around the vertex greater than $360°$.
- (1.f.ii) For the same reason the case of pentagons and exactly two n-gons, with $n \geq 6$, is possible only for the solid with corresponding symbol $(5,6,6)$, which is realized in the **truncated icosahedron**. The preceding categories exhaust all the cases for vertices of semi-regular polyhedra where exactly two kinds of polygons show up.

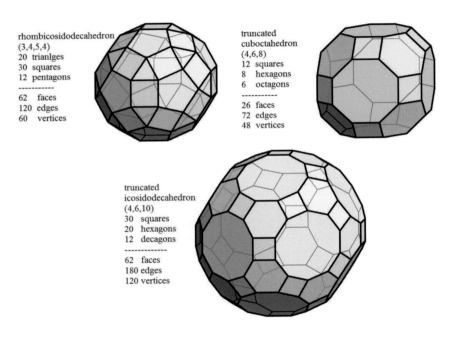

rhombicosidodecahedron
(3,4,5,4)
20 trianlges
30 squares
12 pentagons
----
62 faces
120 edges
60 vertices

truncated
cuboctahedron
(4,6,8)
12 squares
8 hexagons
6 octagons
----
26 faces
72 edges
48 vertices

truncated
icosidodecahedron
(4,6,10)
30 squares
20 hexagons
12 decagons
----
62 faces
180 edges
120 vertices

Fig. 7.27: Archimedean solids - III

- (2.a) This category includes polyhedra around whose vertices show up three kinds of polygons: triangles, squares and n-gons with $n \geq 5$. We again consider the cases:

7.8. ARCHIMEDEAN SOLIDS

1. Only a square ⇒ and at most 2 triangles. The symbols $(3,3,4,n)$, $(3,4,n)$ result, which are rejected, because of the restrictions 7.13 and 7.14 respectively.
2. Exactly 2 squares ⇒ and at most 1 triangle. The symbols $(3,4,4,n)$, $(3,4,5,4)$ result, of which the first is rejected, because of the restriction 7.13. The symbol $(3,4,5,4)$ is realized in the so-called **rhombicosidodecahedron**.

- (2.b) This last category includes polyhedra around the vertices of which show up three kinds of polygons, but not triangles. We see immediately that the symbol $(4,4,5,6)$ defines angles around a vertex, whose sum is greater than $360°$. It follows that the polygons around one vertex must all be different and according to the restriction 7.14, none of them contains an odd number. There result the possibilities $(4,6,8)$ and $(4,6,10)$, which are realized, respectively, in the **truncated cuboctahedron** and **truncated icosidodecahedron**.

The preceding analysis exhausts all possibilities of convex polyhedral angles, which may show up in semi-regular polyhedra, and shows that no more than the already mentioned exist.

Remark 7.4. The theorem deals with the determination of all possible semi-regular polyhedra. It does not address, however, the problem of existence. To take care of this matter, one must, like Euclid does for the Platonic solids, construct each solid, relying on the length of its edge or the radius of its circumscribed sphere. For most constructions, and specifically these which are characterized by the adjective **truncated**, but also for the cuboctahedron and icosidodechedron, the possibility of reduction to the construction of a Platonic body exists. For example, the truncated terahedron results by cutting off in a simple way the vertices of a regular tetrahedron. Also the truncated cuboctahedron results by cutting off vertices of the cuboctahedron, which results similarly from the cube.

Remark 7.5. Two of the Archimedean solids, the **truncated cube** and the **snub dodecahedron** show up in two, as they are called, **enantiomorphic** forms, which are nothing but their reflected images relative to a plane. According to our discussion for congruence in space (§ 7.5), the two enantiomorphic forms of each of these solids are isometric, but they cannot be placed one on top of the other, so that they coincide.

Remark 7.6. The Archimedean solids have come again to the foreground with the discovery in Chemistry (1996) of the **fullerens**, which are carbon molecules, whose atoms are on the vertices of polyhedra (See Figure 7.28). The most well known is C_{60}, which corresponds to the truncated icosahedron ([101]).

Fig. 7.28: Fullerene C_{60}, carbon atoms at vertices of a truncated icosahedron

7.9 Epilogue

> Books have the same enemies as men: fire, humidity, nonsense, time, and their own content.
>
> *Paul Valery, Thoughts and Aphorisms*

Looking back, at the material we discussed, in its entirety, we realize that most of the times we were concerned, not so much with properties which remain invariant relative to isometries, but rather with properties which remain invariant relative to similarities. Even measures of magnitudes which are preserved by isometries, like lengths, areas and volumes, change through similarities in a simple way, by multiplying the initial magnitudes respectively with κ, κ^2, κ^3, where κ is the ratio of similarity. This way, one could express the general rule, that Euclidean Geometry has as its object the set of properties of shapes in space, which remain invariant relative to similarities in space ([58, II, p.4]). There exist many interesting articles and books on the evolution of the idea of "Geometry" and the meaning of the word *Geometry*. I will mention one of the most important ones, that of S.S. Chern ([102]), wherein are contained references to relevant articles, as well as the books of Berger [100] and of Scriba and Schreiber [106].

Regarding the infrastructures, on which the discussed material relies, these are to be found in many places. To begin with, the basic structure in the entire system of Euclidean geometry is the one of the set of real numbers. Many of the weaknesses of Euclid's Elements stem from the fact, that the precise description of the structure of real numbers has been achieved after more than two milenia have been elapsed. The euclidean lines are, in essence, copies of the set of real numbers. A euclidean line is a "model" of the set of real numbers, and inversely, the set $\mathbb{R}$ of real numbers is an abstract mathematical model of the euclidean line. On this bidirectional correspondence relies the idea of the "analytic geometry" of Descartes, which, taking two orthogonal lines on the plane, represents it with the product $\mathbb{R} \times \mathbb{R}$. The structure of "vector space" emerges then at once. A structure we have been tacitly using all the time, working with directed segments AB, broken lines, parallelograms, polygons, middles of sides, centroids, which are notions related to "sums" of such directed segments = "vectors".

Another domain sustained by a multitude of infrastructures, is the one of relations and equations, which result in almost every theorem and exercise of the euclidean geometry. What are these equations? Which of them can be solved? Why there are solvable and non-solvable equations? How can be proved the unsolvability? Questions of this kind are related also to the "constructibility" problem of a segment of a given length with the exclusive use of the rule and compass. Whether, for example, a segment of length $\sqrt{\pi}$ or the side of the regular $n-$gon can be constructed. These questions lead to algebra, the "theory of fields" and the "theory of groups", which in turn lead to a multitude of other algebraic structures.

We encountered the notion of "limit" when trying to measure areas and volumes of polygons and polyhedra and lengths of simple curves, such as the circle. The "length", the "area", the "volume", in combination with the "limit", lead to the "theory of measure". Underneath the notion of "limit" lives the whole "infinitesimal calculus".

Finally another domain, which leads to substantial structures, is the one of the "symmetries of a shape". The underlying structure here is that of the "group". Many of the properties of the transformations, we have studied, in "translations", "rotations", "similarities", can be expressed in a unified way in the language of the "theory of groups". For example, the restriction of the different kinds of the platonic or/and archimedean solids to a small number, results from the fact that the group of isometries of the sphere has few concrete finite "subgroups". Thus, in this book, we make a good start. In many cases, underneath a simple theorem or exercise or question, we can find, if we dig a bit deeper, an important theory, a new world.

References

100. M. Berger (2010) Geometry Revealed. Springer, Heidelberg
101. F. Cataldo, A. Graovac, O. Ori (2011) The Mathematics and topology of fullerenes. Springer, Berlin
102. S. Chern, 1990. What Is Geometry?, *American Math. Monthly*, 97:679-686
103. P. Cromwell (1997) Polyhedra. Cambridge University Press, Cambridge
104. E. Dijksterhuis (1987) Archimedes. Princeton University Press, Princeton
105. J. Keppler (1619) Harmonices Mundi Libri V, Linz
106. Ch. Scriba, P. Schreiber (2005) 5000 Jahre Geometrie. Springer, Heidelberg
107. I. Thomas (1957) Greek Mathematics (selections) 2 vols. Harvard University Press, Cambridge

Index

π, 12
π approximation, 14, 45
π, value, 14

Acute triangle, 182
Adjacent angles, 253
Affinity, 391
Altitude, 156, 179, 245, 247, 258, 342
Altitude of cone, 296, 297
Altitude of prism, 193, 262, 282, 284
Altitude of pyramid, 177, 287, 289
Altitude of the cone, 200
Altitude of truncated cone, 206, 297
Analysis, 180
Angle measure, 18
Angle of line and plane, 150
Angle of parallelism, 105
Angle of similarity, 77
Angle of skew lines, 140
Angle of spherical lune, 222
Angle of spherical polygon, 227, 272
Angle oriented, 62
Angles of great circles, 221
Anthyphairesis, 33, 35
Anti-isometry, 54
Antihomothety, 73, 420, 422
Antiparallels, 121
Antiprism, 252, 425, 427, 428
Antisimilarity, 77, 81
Apex, 156, 177, 184, 232, 272, 317
Apollonian circle, 79, 81
Apollonius, 250, 365
Arbelos, 37
Arc length, 14, 17, 207
Archimedean axiom, 274, 280
Archimedean solid, 192, 252, 424, 427, 433
Archimedean tiling, 106, 109, 114

Archimedean triangle, 346, 349, 350, 354, 382, 396
Archimedes, 5, 14, 424
Arcs equal, 15
Area, 269, 391
Arithmetic mean, 294
Asymptote, 339, 368, 373, 374, 377, 398
Asymptotic triangle, 102, 374, 377, 396
Auxiliary circles, 369
Axes of ellipse, 388
Axial symmetry, 404, 405
Axiom of completeness, 3
Axioms, 3, 93, 139
Axioms, space, 127
Axis of antisimilarity, 78
Axis of cone, 200, 297, 317
Axis of cylindrical surface, 198, 315
Axis of parabola, 353
Axis of reflection, 401
Axis of rotation, 299, 306, 408
Axis of symmetry, 318, 404
Axonometric projection = Parallel projection, 234

Base of cone, 200
Base of pyramid, 177
Bases of a prism, 193
Bases of truncated cone, 206
Billiard, 116
Binet, 43
Bisecting plane, 306
Bisecting planes, 164
Bisector, 120, 158, 164

Cairo tiling, 114
Cavalieri, 306
Center of conic, 339

Index

Center of symmetry, 403, 404
center-line, 91, 250, 361
Central projection, 241
Centroid, 37
Centroid of tetrahedron, 187
Chord of conic, 337
Chord of sphere, 209
Circle area, 18, 22
Circle chain, 117
Circle of inversion, 90
Circle perimeter, 11
Circular directrix, 369
Circular section, 23
Circular sector, 22, 302
Circular sector area, 22
Circular segment, 303
Circumscribed conical surface, 205
Circumscribed parallelepiped, 183
Circumscribed polyhedron, 220
Circumscribed sphere, 220
Common orthogonal, 149
Commuting transformations, 58
Complete quadrilateral, 344
Composition of transformations, 47
Cone, 200
Cone angle, 201
Cone apex, 200, 201
Cone area, 261
Cone circular, 225
Cone frustum, 206
Cone generator, 202
Cone oblique circular, 253
Cone opening, 253
Cone opening = Cone angle, 201
Cone right circular, 200
Cone truncated, 206
Cone vertex = Cone apex, 200
Cones congruent, 200
Congruence, 49, 69, 411
Congruent prisms, 193
Congruent spherical lunes, 222
Congruent triangles, 70, 83, 235, 254
Congruent truncated cones, 206
Conic exterior, 337
Conic interior, 337
Conic section, 321
Conic section, proper, 318
Conic sections, 318
Conic singular, 318
Conic surface, 201
Conics with center, 318
Conjugate chords of conic, 340
Conjugate diameters, 340
Conjugate directions, 340, 362, 365

Contact point, 211
Continued fraction, 33, 45
Convex hull, 30
Convex polyheder, 424
Convex polyhedron, 188
Convex prism, 193
Convex pyramid, 177
Convex spherical polygons, 272
Cosine spherical 1st formula, 227
Cosine spherical 2nd formula, 230
Cross ratio, 90, 159, 343
Cube, 189, 193, 252, 308
Cubeoctahedron, 192, 428
Cylinder, 198, 267
Cylinder altitude, 198, 294
Cylinder area, 261, 262
Cylinder base, 198
Cylinder oblique circular, 199
Cylinder radius, 198
Cylinder unfolding, 199
Cylinder volume, 294
Cylinders congruent, 198
Cylindrical surface, 198, 200, 323

Dandelin, 320, 333, 397
Dandelin spheres, 321, 324, 329, 331, 333
Degenerate conic sections, 318
Desargues, 138, 244
Diagonal of parallelepiped, 193
Diametral, 210
Diametrically opposite, 276
Diametrically opposite = Diametral, 210
Dihedral, 161, 166, 245
Dihedral acute, 162
Dihedral angle , 161
Dihedral complementary, 162
Dihedral congruent, 162
Dihedral face, 161
Dihedral measure, 161
Dihedral obtuse, 162
Dihedral right, 162
Dihedral supplementary, 162
Dihedrals, 192
Dihedrals of polyhedron, 188
Dihedrals of the trihedral, 167
Direct isometry, 54, 419
Direct similarity, 77
Director circle, 395
Directrices of ellipse, 330
Directrices of hyperbola, 330
Directrix, 318, 325, 327, 329, 331
Distance from plane, 144, 145
Distance line, 103–105
Distance of two planes, 145

Index

Distance of parallel from plane, 145
Distance of skew lines, 145
Dodecahedron, 191, 233, 250
Domain, 47
Dual platonic solid, 192

Eccentric angle, 357, 363, 388
Eccentricity, 330, 335, 341, 354, 357, 366, 385
Edge of dihedral, 161
Edges of polyhedral angle, 177
Edges of polyhedron, 188
Edges of pyramid, 177
Edges of trihedral, 167
Ellipse, 317, 320, 322, 327, 354, 358, 395
Ellipse, area, 391
Ellipse, auxiliary circles, 358, 361, 387
Ellipse, axes, 355, 361
Ellipse, center, 355
Ellipse, diameter, 355
Ellipse, half parameter, 357
Ellipse, major axis, 355
Ellipse, major circle, 357
Ellipse, minor axis, 355
Ellipse, minor circle, 357
Ellipse, secondary circle, 357
Enantiomorphic, 112, 431
Envelope, 346
Equality, 28
Equilateral, 186, 227, 232, 309, 311
Escribed polyhedron, 220
Euler, 278
Euler characteristic, 278
Exterior points, 337

Face, 193
Face of polyhedron, 188
Faces of trihedral, 167
Fibonacci, 42
Fixed point of transformation, 51
Focal distance, 354, 366
Focal parameter, 342, 344, 354, 357, 366
Focal point, 318
Focals, 318
Focals = Focal points, 318
Focus, 318
Ford, 39
Ford circles, 39

Galileo, 21
Gauss, 207
Gelder, 45
Generator, 192
Generators of cone, 200

Generators of surface, 198
Genus of polyhedron, 278
Geometric locus, 143, 149, 154, 157, 164, 250, 317, 325, 331, 343, 352, 361, 374, 384, 386, 395
Glide reflection, 70, 412, 416, 417
Golden section, 42, 101, 251
Golden section ratio, 36
Great circle, 210

H-circle, 96, 97
H-isometry, 99
H-line, 93
H-line segment, 95
H-medial line, 99
H-middle, 97
H-plane, 93, 99
H-point, 93
H-point at infinity, 93
H-reflection, 93, 99, 104
H-regular polygon, 101
H-triangle, 100
H-trigonometry, 106
Half spaces, 127
Half turn, 63
Harmonic quadrilateral, 92
Height of spherical zone, 271
Helicoidal displacement, 414, 416
Helicoidal displacements, 412
Hexagon symmetric, 108
Hippocrates, 26
Hjelmslev, 124
Homothetic polygons, 153
Homothety, 73, 117, 420, 422
Homothety center, 73, 420
Homothety ratio, 73, 420
Horizon, 93, 103
Horocycle, 103
Horocycle, limit point, 104
Hyperbola, 318, 324, 329, 339, 366, 372, 385
Hyperbola, axes, 367
Hyperbola, center, 367
Hyperbola, conjugate axis, 369
Hyperbola, conjugate hyperbola, 376
Hyperbola, diameters, 367
Hyperbola, focal points, 367
Hyperbola, major circle, 369
Hyperbola, primary circle = major circle, 369
Hyperbola, rectangular, 320, 377
Hyperbola, transverse axis, 369, 373
Hyperbola, vertices, 369
Hyperparallels, 94, 95
Hyperparallels, common orthogonal, 101

Icosaehedron, 233
Icosahedron, 191, 252
Icosidodecahedron, 192, 428
Identity transformation, 48
Image, 47
Indirect isometry, 419
Inscribed conical surface, 205
Inscriptible in sphere, 220
Interior of trihedral, 167
Interior points, 337
Intersection of planes, 134
Inverse transformation, 48
Inversion, 89, 95
Inversion center, 90
Involution, 54
Isometries, 424
Isometries of the plane = Concgruences of the plane, 49
Isometry, 50, 64, 399, 400, 403, 406, 408, 418, 423
Isoperimetric inequality, 27
Isosceles, 20, 156, 300, 346
Isosceles trapezium, 297, 309, 388

Kepler, 306

Lateral area, 261–264, 267, 271, 314
Lateral edges, 177, 193
Lateral faces, 177
Lateral surface, 201, 261, 293
Latus rectum of conic, 342
Legendre, 291
Leibnitz, 5
Lhuilier, 196
Limit, 303
Limiting parallels, 94
Lindemann, 13
Line at infinity, 330
Line parallel to plane, 135
Line, in space, 127
Lobatsevsky, 92
Logarithmic spiral, 89
Lunule, 23, 26

Maps = transforms, 47
Maximal area, 20, 29, 39, 118, 208, 256, 267, 282
Maximal lateral area, 267
Measure in radians, 15
Medial line, 99, 143, 345, 361, 372
Medial plane, 143, 149, 160, 187, 212
Medians of tetrahedron, 187
Menelaus theorem, 74
Middle parallel, 155

Middle section, 206
Middle sphere, 221
Middle term, of fractions, 40
Minimal angle, 151
Minimal area, 286, 364
Minimal distance, 124, 150, 159, 282
Minor circle = Small cirlce, 210
Miquel point, 381
Mirror, 401
Mirror = Reflection axis, 51
Mirroring = Reflection, 50, 401
Monge's point, 246, 247
Monohedral tiling, 107

Nappe of conic surface, 201
Negatively oriented, 63, 401
Newton, 5
Non coplanar lines, 137
Non coplanar lines = Skew lines, 129, 139
Null translation, 55, 405

Oblique circular cone, 253
Opposite, 193
Order of tiling, 109
Orientation, 54, 62, 67, 79, 83, 170, 401, 407, 419, 422
Orientation preserving transformation, 54
Orientation reversing transformation, 54
Oriented angle, 62, 407
Orthocenter, 174, 182, 245, 247, 348, 378
Orthocentric tetrahedron, 247
Orthodiagonal, 37
Orthogonal axonometric projection, 235
Orthogonal cut, 194, 195, 197, 294, 295
Orthogonal from point, 98
Orthogonal lines, 157
Orthogonal pencil, 97, 104
Orthogonal planes, 163, 405
Orthogonal section, 193
Orthogonal sides, 256
Orthogonal to plane, 141, 145, 151
Orthogonals in space, 136
Ovoid, 24

Parabola, 318, 320, 326, 331, 344, 354, 380, 384
Parabola, axis, 345, 347, 380, 384
Parabola, diameter, 345
Parabola, directrix, 348, 384
Parabola, parameter, 344
Parabola, vertex, 344
Parallel chords, 337
Parallel circle = Small circle, 210
Parallel planes, 129, 136, 152

Index   439

Parallel projection, 234
Parallelepiped, 193, 248, 267, 404
Parallels in space, 129
Penrose tiling, 114
Pentagon, 114
Perimeter, 3, 11, 13
Perspective horizon, 241
Perspective projection, 242–244
Perspective projection = Central
   projection, 241
Plane, 127
Plane of symmetry, 403
Platonic body = Platonic solid, 189
Platonic solid, 220, 232, 248, 311, 404, 424,
   433
Platonic solids, 186
Pohlke, 236
Pohlke's theorem, 236
Poincare, 92
Point of contact, 213
Point symmetry, 52, 62, 403
Polar relative to conic, 343, 362
Pole of circle, 211
Pole relative to conic, 343
Polygon regular, 11, 12, 16, 28
Polygon regular, perimeter, 11
Polyhedral angle, 177
Polyhedral angles of polyhedron, 188
Polyhedron, 188, 220, 279, 293
Polyhedron area, 261
Polyhedron circumscribed, 305
Positively oriented, 63, 401
Power of point, 249
Preimage, 47
Preserves the orientation, 401
Primary, 357
Prism, 193, 282
Prism volume, 282
Prismatic surface, 192, 195, 196, 286
Prismatoid, 310
Projection of line, 165
Projection of point, 144
Projection of shape, 144
Projection onto a plane, 138
Proper conic sections, 318
Proper similarity, 77, 421
Pursuit problems, 89
Pyramid, 177, 179, 220, 287
Pyramid frustum, 178
Pyramid volume, 287, 289

Quadrant, 161
Quadrilateral inscriptible, 91

Radians, 14, 15
Radical axis, 250
Radical plane, 249, 250
Range, 47
Ratio, 77, 421
Real numbers $\mathbb{R}$, 3
Rectangular hyperbola, 377, 378
Rectangular parallelepiped, 193, 220, 280
Reflection, 50, 60, 72, 73, 401, 414, 418
Reflection axis, 51, 83
Reflective similarity = Antisimilarity, 77
Regular dodecahedron, 189
Regular icosahedron, 189
Regular octahedron, 189
Regular pentagon, 251
Regular polygon, 107
Regular polygon area, 18
Regular polyhedral angles, 185
Regular polyhedron, 189
Regular prism, 193
Regular pyramid, 184
Regular tetrahedron, 189
Regular tiling, 107
Reverses orientation, 401
Rhombicosidodecahedron, 431
Rhombicuboctahedron, 428
Rhombus, 379
Right circular cylindrical, 198
Right prism, 193
Right triangle, 387
Rotation, 63, 64, 68, 79, 122
Rotation in space, 407
Rotational similarity, 77
Rotoreflection, 412, 418

Salinon, 25
Scale, 77
Semi latus rectum, 342
Semiregular tiling, 108
Sequence, 3
Sequence bounded, 3
Sequence Fibonacci, 42
Sequence increasing, 3, 6, 11, 17, 44
Sequence limit, 14, 44
Sequence member, 4
Sequence of polygons, 7
Sequence term, 7
Sequence, Haros-Farey, 40
Sequence, limit of, 3–5, 7, 11
Sequence, upper bound, 4
Signed measure of angle, 62
Similarity, 77, 85, 421, 422, 424
Similarity center, 122
Similarity direct, 77

Similarity proper, 77, 421
Similarity reflective = Antisimilarity, 77
Similarity rotational, 77
Simson line, 381
Sine spherical formula, 231
Singular conic sections, 318
Skew lines, 129, 139, 145, 149, 157, 159, 415
Skew lines distance, 145
Skew orthogonal, 145, 156, 174
Skew orthogonals, 140, 155
Skew quadrilateral, 141, 160
Small circle, 210
Snub dodecahedron, 428
Solid angle, 177
Space, 127
Sphere, 191, 209, 213, 215, 219, 220, 224, 232, 248, 254, 267, 271, 273, 276, 293, 298, 309, 311, 321, 333, 397, 401, 403, 412, 420
Sphere area, 269
Sphere center, 209
Sphere diameter, 209
Sphere exterior, 209
Sphere interior, 209
Sphere radius, 209
Sphere volume, 298, 302
Spheres congruent, 210
Spherical area axioms, 272
Spherical cap, 271
Spherical cosine 1st formula, 227
Spherical cosine 2nd formula, 230
Spherical excess, 276
Spherical lune, 221, 222, 276
Spherical polygon, 224, 225, 276
Spherical polygon vertex, 225
Spherical polygon, angle, 224
Spherical polygons, 272
Spherical polyhedra, 220
Spherical ring, 303
Spherical sector, 302
Spherical sine formula, 231
Spherical triangle, 223, 227, 233, 275, 276
Spherical triangle, apex, 224
Spherical triangle, right, 230
Spherical triangle, side, 224
Spherical triangle, side length, 224
Spherical triangle, vertex, 224
Spherical triangles, congruent, 224
Spherical zone, 270, 271, 304
Spherical zone/cap area, 271
Squaring the circle, impossible, 13
Steiner line, 348
Stereometry, 127
Supplementary triangle, 232

Symmedian, 348
Symmetric, 53, 403
Symmetries, 52
Symmetry, 404, 406, 416

Tangent cone, 214
Tangent plane, 202, 211
Tangent to conic, 336, 379
Tangent to sphere, 213
Tangent to the conic, 335
Tetraheder, 401, 406, 423
Tetrahedron, 180, 219, 245, 306
tetrahedron, 244, 254
Thales in space, 152
Three orthogonals theorem, 148
Tile of type p_n, 124
Tiling, 106, 276
Tiling pentagonal, 108
Transcendental, 13
Transformation, 47, 399
Translation, 55, 62, 75, 405, 410, 414
Trapezium, 179, 194
Triangle construction, 86
Triangle inequalities, 31
Triangle inequality, 31, 359, 371
Triangles homothetic, 74
Trihedral, 167, 225, 403
Trihedral angle, 205
Trihedral angle = Trihedral, 167
Trihedral supplementary, 228, 229, 232
Trihedrals of the pyramid, 177
Trihedrals vertical, 171
Triply orthogonal tetrahedron, 247
Truncated cone, 264
Truncated cone volume, 297
Truncated cube, 429
Truncated cuboctahedron, 431
Truncated dodecahedron, 429
Truncated icosahedron, 430
Truncated icosidodecahedron, 431
Truncated octahedron, 430
Truncated prism, 194, 245
Truncated pyramid, 178
Truncated pyramid volume, 292
Truncated tetrahedron, 424, 429

Ultraparallels = Hyperparallels, 94
Unfolding, 267
Unfolding of cone, 207
Unfolding of plane onto other, 196
Unfolding of pyramid, 182
Unfolding of tetrahedron, 182
Unfolding of trihedral, 169

Index

Vanishing line, 241
Vertical trihedrals, 171
Vertices of polyhedron, 188
Volume, 279

Volume axioms, 279
Volume of cone, 296
Volume of oblique prism, 283
Volume of spherical zone, 304